FILTER AIDS AND MATERIALS

FILTER AIDS AND MATERIALS
Technology and Applications

Henry T. Driscoll

NOYES DATA CORPORATION
Park Ridge, New Jersey, U.S.A.
1977

Library of Congress Catalog Card Number: 77-71926
ISBN: 0-8155-0658-9
Printed in the United States

Published in the United States of America by
Noyes Data Corporation
Noyes Building, Park Ridge, New Jersey 07656

FOREWORD

The detailed, descriptive information in this book is based on U.S. patents issued since 1970 that deal with filter aids and materials.

This book serves a double purpose in that it supplies detailed technical information and can be used as a guide to the U.S. patent literature in this field. By indicating all the information that is significant, and eliminating legal jargon and juristic phraseology, this book presents an advanced, technically oriented review of filter aids and materials.

The U.S. patent literature is the largest and most comprehensive collection of technical information in the world. There is more practical, commercial, timely process information assembled here than is available from any other source. The technical information obtained from a patent is extremely reliable and comprehensive; sufficient information must be included to avoid rejection for "insufficient disclosure." These patents include practically all of those issued on the subject in the United States during the period under review; there has been no bias in the selection of patents for inclusion.

The patent literature covers a substantial amount of information not available in the journal literature. The patent literature is a prime source of basic commercially useful information. This information is overlooked by those who rely primarily on the periodical journal literature. It is realized that there is a lag between a patent application on a new process development and the granting of a patent, but it is felt that this may roughly parallel or even anticipate the lag in putting that development into commercial practice.

Many of these patents are being utilized commercially. Whether used or not, they offer opportunities for technological transfer. Also, a major purpose of this book is to describe the number of technical possibilities available, which may open up profitable areas of research and development. The information contained in this book will allow you to establish a sound background before launching into research in this field.

Advanced composition and production methods developed by Noyes Data are employed to bring our new durably bound books to you in a minimum of time. Special techniques are used to close the gap between "manuscript" and "completed book." Industrial technology is progressing so rapidly that time-honored, conventional typesetting, binding and shipping methods are no longer suitable. We have bypassed the delays in the conventional book publishing cycle and provide the user with an effective and convenient means of reviewing up-to-date information in depth.

The Table of Contents is organized in such a way as to serve as a subject index. Other indexes by company, inventor and patent number help in providing easy access to the information contained in this book.

15 Reasons Why the U.S. Patent Office Literature Is Important to You –

1. The U.S. patent literature is the largest and most comprehensive collection of technical information in the world. There is more practical commercial process information assembled here than is available from any other source.

2. The technical information obtained from the patent literature is extremely comprehensive; sufficient information must be included to avoid rejection for "insufficient disclosure."

3. The patent literature is a prime source of basic commercially utilizable information. This information is overlooked by those who rely primarily on the periodical journal literature.

4. An important feature of the patent literature is that it can serve to avoid duplication of research and development.

5. Patents, unlike periodical literature, are bound by definition to contain new information, data and ideas.

6. It can serve as a source of new ideas in a different but related field, and may be outside the patent protection offered the original invention.

7. Since claims are narrowly defined, much valuable information is included that may be outside the legal protection afforded by the claims.

8. Patents discuss the difficulties associated with previous research, development or production techniques, and offer a specific method of overcoming problems. This gives clues to current process information that has not been published in periodicals or books.

9. Can aid in process design by providing a selection of alternate techniques. A powerful research and engineering tool.

10. Obtain licenses – many U.S. chemical patents have not been developed commercially.

11. Patents provide an excellent starting point for the next investigator.

12. Frequently, innovations derived from research are first disclosed in the patent literature, prior to coverage in the periodical literature.

13. Patents offer a most valuable method of keeping abreast of latest technologies, serving an individual's own "current awareness" program.

14. Copies of U.S. patents are easily obtained from the U.S. Patent Office at 50¢ a copy.

15. It is a creative source of ideas for those with imagination.

CONTENTS AND SUBJECT INDEX

INTRODUCTION

Filter media can generally be classified as being one of two types, depth filters and surface filters. A depth filter removes suspended material from the fluid passed through the filter by collecting it not only on the surface of the element but also within the pores. A depth filter has a considerable thickness, and has a plurality of pores of distinct length. The longer the pores, the higher the dirt capacity of the filter, because there is more room for dirt along the pores. Most depth filters are made of masses of fibers, or other particulate material, held together by mechanical means or by bonding. One or several layers of such materials can be employed, and these layers can vary in porosity. In most cases, however, the greater percentage of contaminants unable to pass through the filter is trapped at the surface of the filter.

A surface filter removes suspended material from the fluid passed through the filter by collecting such material on its surface, and the material thus removed forms a filter cake or bed upon the filter. This material naturally obstructs the openings in the surface of the filter, because the fluid must flow through this material, which thus effectively reduces the diameter of the filter openings to the size of the pores in the filter cake. This reduction in effective diameter of pore openings in the filter increases the pressure differential required to maintain flow through the filter.

Sand, other inorganic particulate materials and anthracite coal fines are most commonly used as the granular filter medium. Where backwashing is critical, the coal particles are particularly valuable due to ease of fluidization and washing. In recent years, plastic particles, particularly polyvinyl chloride, have been used in deep bed filtration units.

A number of common fibrous materials have also been employed through the years, such as cotton, glass wool and fluffed wood pulp.

In the continuous types of filters one generally uses an initiating medium which may be either woven-mesh metallic or nonmetallic screens or cloths. Many metals and special steel alloys are used, with the choice largely being dictated by the corrosivity of the materials being filtered. Commercially, metallic filter cloths are available in variety of weaves such as plain, twill, corduroy-twilled and stranded.

While cotton and wool are most often used, many synthetic fibers are finding wide use in continuous filters. Thus polyamide, nylon and acrylic fibers are routinely used commercially for highly demanding environments employing continuous filtration.

One of the more stable and inert types of filters is woven from glass yarns to produce cloths of the desired texture and weave. Glass fiber may also be used in combination with asbestos and cellulosics to provide durable filter assemblies.

This book provides an in-depth look at the many filtration processes which have been described in the patent literature in recent years. Many of the processes are general in nature and can be applied to standard industrial filtration requirements. However, an effort has been made to include some of the less commonly considered filtration needs of the industry, which are, in many cases, highly demanding in terms of filter type and construction. Thus there are chapters devoted to processes developed specifically for molten aluminum, bacterial and medical applications, gel filtration chromatography and polymer filtration.

For a detailed discussion of the use of membranes in industrial separation the reader is referred to another Noyes Data book:

P.R. Keller, *Membrane Technology and Industrial Separation Techniques, 1976*

PARTICULATE FILTER MEDIA

CONVENTIONAL FILTER MEDIA

Sand Filter with Thin Precoat Layer

D.R. MacPherson; U.S. Patent 3,680,699; August 1, 1972; assigned to Johns-Manville Corporation describe an improved process for water clarification. A precoat layer containing less than 0.25 pound of porous granular material per square foot is deposited on the top surface of a sand filter. The porous granular material in the precoat layer is selected to consist essentially of particles larger than the effective size of the sand in the sand filter. The water to be clarified is passed through the precoat layer of porous granular material, and subsequently through the sand filter.

When the water to be clarified contains a substantial amount of colloidal solids, a body feed of porous coarse granular material is added to the water to be clarified prior to its passage through the precoat layer. The body feed material, when utilized, also consists essentially of particles larger than the effective size of the sand in the sand filter layer.

Preferably, the sand filter is backwashed when the pressure differential across the filter becomes excessive and the sand filter is then again coated with a precoat layer of less than 0.25 pound of filter aid per square foot of filter area before resuming filtration operations.

The filter bed comprises a lower layer of sand and a precoat layer comprising less than 0.25 pound of coarse porous granular material per square foot of filter bed. The precoat layer is positioned on top of the layer of sand, and the particles of the precoat layer are larger than the effective size of the sand in the layer of sand.

The following examples illustrate the process. All percentages are weight percentages unless otherwise noted. All screen sizes referred to are U.S. Standard Series, unless otherwise noted.

Examples 1 through 3: Turbid water containing ferric hydroxide is filtered through a pressure sand filter using the techniques and apparatus illustrated in Figure 1.1. At the start of the filter cycle, a precoat hopper **10** is filled with a predetermined quantity of filter aid and the filter aid is fed into a flooded funnel **12**. A portion of the influent water is sent through a pipe **14** to a booster pump **16** which boosts the water pressure up to about 90 psi. The water from the pump (about 30 to 50 gallons per minute) is sent through a suitable eductor **18** and then into a pressure sand filter generally **20**. The velocity of the water from pump **16** through the eductor creates a vacuum at the top opening of the eductor which causes the water and filter aid material from the flooded funnel to be drawn into a header pipe **22**.

The header pipe is used to distribute the coarse filter aid throughout the surface area of the sand filter bed, because the velocity of flow in the main header **24** is usually too low to keep the filter aid in suspension. Holes of 3/8 inch diameter are provided on the top of header **22** about 10 inches apart throughout the length of the header which extends the length of filter **20**, and a 3/8 inch hole is provided at the end of the header. This arrangement insures uniform distribution of the coarse filter aid on the surface of the sand.

FIGURE 1.1: WATER CLARIFICATION PROCESS

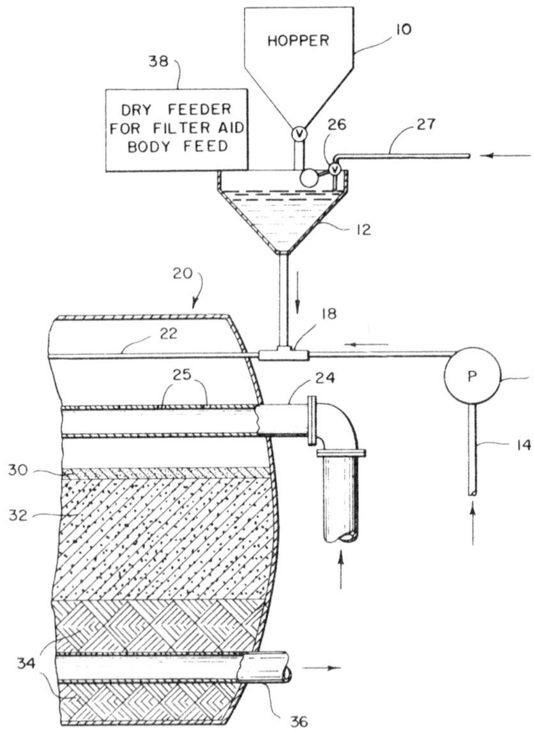

Source: U.S. Patent 3,680,699

During filtration, the raw influent passes through header **24** and out through holes **25** (about one inch in diameter) in the top of the header. Water is forced under pressure of about 60 psi through precoat layer **30** and sand filter layer **32** to remove solids from the influent water. The sand in layer **32** is considerably finer than AAWU standard fine sand because of repeated backwashing over a number of years. The clarified water passes through a gravel bed **34** and is removed from filter **20** through effluent header **36**.

The filter aid for the body feed is provided by a dry feeder **38** which supplies filter aid at a predetermined rate to flooded funnel **12**. Conventional fittings are provided to backwash the pressure sand filter **20**.

In the installation shown, it is desired to lower the iron content of the effluent from filter **20** to no more than 0.3 ppm from the approximately 3.1 to 3.5 ppm level at which it enters header **24**. The following table compares the cycle lengths and total throughput that are achieved using an ordinary sand filter with the results obtained when the same sand filter is coated with a porous granular material in accordance with the process.

Example	Filter Rate (gpm)	. . . Total Iron (ppm). . . . Influent Average	Effluent End of Test	Cycle Length (hr)	Throughput (gal)
1	244	3.5	0.3	11.25	165,000
2	305	3.1	0.3	15.65	296,000
3	361	3.4	0.3	3	65,000

In the procedure of Examples 1 and 2, the sand filter is coated with 0.17 lb of flux calcined diatomaceous earth per 1,000 gallons of water, about half the filter aid is supplied as a precoat and about half as a body feed. In Example 1, the filter aid has a particle size of from 20 to 60 mesh, as defined in ASTM E11, and in Example 2, the filter aid varies in size from 30 to 60 mesh.

The results of Examples 1 and 2 in which the process produces 300 to 500% increases in cycle length and thoughput compared to the prior art sand filter of Example 3 are typical of the improved results which are obtained by the process.

Examples 4 through 7: Two pressure sand filtration columns (4¾ inches inside diameter, clear heavy-wall Lucite) are set up so that both can be run in parallel. Twenty-six inches of filter sand are placed in each column. This sand contains about 60% by weight of 20 to 30 mesh particles and about 40% by weight of 30 to 50 mesh particles.

A high pressure, six-stage Moyno pump system dispenses a body feed of flux-calcined diatomaceous earth with a portion of the influent into the sand bed. The flow rates through the sand filters are measured with Hagan ring-balance meters. The pressure differentials across each of the sand filters are indicated with a conventional 24 hour recording Foxboro pressure meter placed across the influent and effluent lines for each filter.

In all tests, a filter aid is used in conjunction with one of the sand filters, and the other plain sand filter is run in parallel under identical conditions. Periodically, both sand filters are used for blank runs to determine that each filter is behaving in a similar manner. The filter aid utilized in Examples 4 through 7 is a flux-calcined diatomaceous earth (Celite) having particles in the size ranges listed in the following table. The table below gives the gain in water throughput with granular precoat for parallel pressure sand filters using coarse, flux calcined diatomaceous earth as precoat and body feed at constant filter rate of 12 gsfm.

Gain in Water Throughput

	. . . . Raw Raritan River Water			Alum Floculated Water
Example	**4**	**5**	**6**	**7**
Column designation				
A	Sand	Sand	Celite	Celite
B	Celite	Celite	Sand	Sand
Celite filter aid used	30/45	30/45	30/45	30/45
Influent				
Pressure psi	40	40	80	80
Turbidity ppm*	5-7	3-7	2-5	0.5-2
Length running hour				
To breakthrough sand**	6.5	6.5	6.0	3.6
Total	23.2	16.5	16.0	9.6
Diatomaceous earth used				
Precoat psf	0.54	0.54	0.18	0.18
Body feed psf	2.08	1.24	0.84	0.18
Total per 1,000 gal	0.157	0.149	0.091	0.052
Throughput gsf				
Sand up to break-through	4,650	4,650	4,310	2,590
Diatomaceous earth	16,700	11,900	11,500	6,910***

*Measured with Hellige turbidimeter. Water quality of effluent as measured by turbidity through column including filter aid was 50 to 100% better than through sand.

**Pressure differentials at breakthrough varied from 0 to 3.0 psi.

***This test could have been continued much longer than 9.6 hours.

The results of this table show the significant increase in throughput that can be achieved by the process while actually producing an increase in water quality.

Granulated Walnut Shells

G. Hirs; U.S. Patent 3,780,861; December 25, 1973; assigned to Hydromation Filter Company describes a method for filtering contaminants from suspension in liquids by utilizing, as a filter medium, granulated shells of black walnut (*Juglans nigra* L.).

The shells of black walnuts are light in weight, with a specific gravity of 1.3 to 1.4. The shells also are relatively strong, having a modulus of elasticity of 170,000 psi, and the shells are relatively nonabrasive when compared to sand or anthracite. This last property is somewhat surprising in view of the common use of walnut shells as a blasting grit for metal finishing. However, black walnut shells are utilized as soft grit which does not cause pitting of the metal part to be cleaned.

To illustrate the utilization of granular black walnut shells as a filter medium, the following table is presented. In obtaining the figures in this table, liquid containing a standard test dirt consisting of Fe_3O_4 dirt in the indicated amount was passed through 12 inches of 20 to 50 mesh granular black walnut shells. The dirt was added at a rate of 250 mg/l. The filter flow rate was 10 gallons per minute per square foot of cross sectional area of the shells. The following results were obtained:

Gallons of Filtered Liquid	Pounds of Dirt Added per Cubic Foot of Media	Filter Effluent Clarity (mg/l)
20	0.5	0.24
40	1.0	0.36
60	1.5	3.2
80	2.0	9.9
100	2.5	14.5

The above table illustrates that granular black walnut shells have a dirt holding capacity which is substantially equivalent to that of silica sand.

Other favorable characteristics of black walnut shells when utilized as a filter medium are that they are oil and water resistant, they have a hardness in Mohs of 3, and are for all practical purposes chemically inert. Also, granular black walnut shells are commercially available in all practical mesh sizes for use in deep bed filtration at a cost which is about one-sixth the cost of many inexpensive plastic materials.

Anthracite Coal and Walnut Shells

A process described by *G. Hirs; U.S. Patent 3,900,395; August 19, 1975; assigned to Hydromation Filter Company* involves the use of granular black walnut shells in combination with anthracite coal as a deep bed filter method and apparatus. The granules of black walnut shells are larger than the fines of anthracite coal in the filter and, having a specific gravity of approximately 1.4 (as compared with a specific gravity of from 1.5 to 1.7 for anthracite), intermix with the anthracite coal fines after initial backwashing of the filter bed.

The presence of the three-dimensionally angular and larger sized granular walnut shells on the surface of the flaky anthracite coal fines opens the surface of the anthracite coal bed for more effective penetration by dirt particles, resulting in less surface loading and longer filtration cycles. This particular mixture of granular walnut shells with anthracite coal results in a synergistic effect, with the combination producing more effective filtration than either walnut shells or anthracite coal alone.

In addition, the intermixed granular walnut shells break up the flat, impervious surface of the fine, flaky particles of anthracite so that they can also be effective in the filtering process. Without the intermixed walnut shells, the anthracite fines tend to lay flat much as the slats of a venetion blind. The walnut shells open these blinding fines to take full advantage of filtration surface phenomenon, which predominates in most deep bed filtration.

Black walnut shells have certain inherent features which make them a particularly adaptable and unique material for use in a filter bed in combination with anthracite coal. First, the material is a low-cost waste by-product, as are some of the anthracite coals used in the filtering field. Further, the material has an ideal specific gravity for use with anthracite coal since it will tend to float or stay on top of the larger anthracite granules and not intermix therewith. Further, it fractures into an ideal three dimensional angular shape for effective intermixing with the anthracite coal fines and is readily wettable in an aqueous solution. Walnut shells therefore have a definite advantage over plastic granules, since the latter are costly and hydrophobic.

Referring to Figure 1.2, reference numeral **10** refers to the filter vessel which houses the multimedia filtering arrangement **19** forming the process. Upper part **11** of the filter bed is comprised primarily of granulated black walnut shells having a specific gravity of approximately 1.4 and a size greater than that of the fines of anthracite coal.

Additionally, intermixed with the walnut shells are fines of anthracite coal, which, but for the presence of the shells, would accumulate at the top of the lower portion of the filter bed. Reference numeral **12** refers to the lower part of the filter bed, which is comprised primarily of anthracite coal in combination with relatively few, if any, granules of black walnuts. In the preferred case the filter media bed **19** is formed by a lower bed of anthracite coal having a depth of 42 inches and an upper bed of black walnut shells having a depth of six inches.

In forming the complete filter bed, the lower material of anthracite is placed within the filter vessel first, backwashed to remove most of the very fine material, and allowed to settle back in place before the shells are layered thereover. These two layers, preferably, then are intermixed at the interface by backwashing.

FIGURE 1.2: FILTRATION UNIT EMPLOYING WALNUT SHELLS AND ANTHRACITE COAL AS FILTER MEDIA

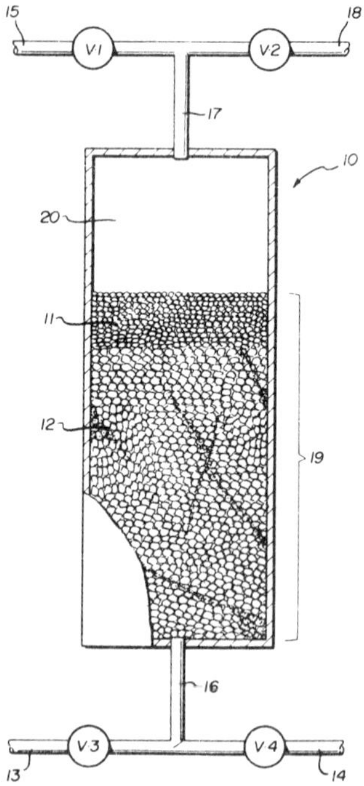

Source: U.S. Patent 3,900,395

Other standard features of the illustrated deep bed filtering system include inlet and outlet conduits **13, 14, 15, 16, 17** and **18** and regulator valves **V1, V2, V3,** and **V4.**

In operation, contaminated liquid enters through conduit **15**, open valve **V1** and into filter vessel **10** through conduit **17**. The contaminated liquid passes through the multimedia granular filter bed **19** and out through conduit **16**, open valve **V3**, and conduit **13**. During the filtration cycle backwash valves **V2** and **V4** are closed.

After a period of use, filter media bed **19** becomes loaded with entrapped particulate contaminants and must be backwashed. This is accomplished by closing valves **V1** and **V3**, opening valves **V4** and **V2** and flowing backwash water into and through the filter vessel through conduits **14** and **16**.

The filter bed **19** will expand into space **20** and will release entrapped contaminants which will flow with the backwash water out through conduits **17** and **18.** After the backwash cycle has been completed, valves **V2** and **V4** are closed, filter media bed **19** settles to the position shown in the drawing, valves **V1** and **V3** are opened, and the filtration cycle resumes.

The following examples demonstrate the effectiveness of the filter media bed of the process. For purposes of these examples number 1 and number 1½ anthracite refer to commercially available Anthrafilt with the following sieve analysis:

Anthracite

U.S. Standard Mesh Size	No. 1 (percent by weight)	No. 1½ (percent by weight)
Retained on No. 8	0	5.5
Retained on No. 16	37.2	62.5
Retained on No. 20	44.3	24.0
Retained on No. 30	13.7	5.5
Passing No. 30	4.8	2.5

The granular walnut shell designations referred to have the following sieve analysis:

Granular Walnut Shells

U.S. Standard Mesh Size	No. 20/30 (percent by weight)	No. 12/20 (percent by weight)
Retained on No. 14	0	2.0
Retained on No. 16	0	29.5
Retained on No. 20	14.4	66.5
Retained on No. 30	73.5	2.0
Passing No. 30	12.1	0
Retained on No. 40	10.6	
Passing No. 40	1.5	

In each example the dirt loading comprised of 100 ppm particulate contaminants consisting of Georgia kaolin hydrate of which the medium particle size was 0.55 micron. Polyelectrolyte filter aid (Nalco 8861) was added at the rate of 0.5 ppm. In all cases the total depth of the filter media bed was 48 inches. All of the tests were run at a flow rate of 10 gpm per square foot of surface inlet area and at substantially the same temperature to eliminate liquid viscosity as a material factor. Each test was terminated when a 6 foot pressure differential was reached across the filter bed or when the rate of dirt removal dropped below 80%, which ever first occurred.

Example 1: In this example the test was run with 48 inches of No. 1 anthracite. The terminal pressure differential across the filter bed of 6 feet was reached after 285 gallons of liquid were filtered per square foot of inlet surface area. Composite effluent clarity was excellent, being less than 5 Jackson Turbidity Units (JTU). On observation it was apparent that a surface loading condition predominated and caused the terminal pressure differential to be reached in a relatively short time.

Example 2: The filter bed for this test run was comprised of 42 inches of No. 1 anthracite topped with 6 inches of No. 20/30 granular black walnut shells. Terminal pressure differential was reached after 1,840 gallons of contaminated liquid were filtered per square foot of inlet surface area. The composite effluent clarity was below 5 JTU. This filter was more than 6 times as effective as the filter in Example 1, the primary difference in effectiveness being attributed to the presence of granular walnut shells to open up the surface of the anthracite filtering media.

Example 3: In this test run 42 inches of No. 1 anthracite coal was again used, but was topped by 6 inches of No. 12/20 granular black walnut shells. In this particular test 2,300 gallons were filtered per square foot of inlet surface area before the run was terminated on the basis of deterioration of effluent quality, rather than terminal pressure differential. The composite effluent clarity was less than 25 JTU.

In this example visual observation showed that the walnut shells intermixed with the anthracite fines and also with some of the larger granules of anthracite at lower levels of the filter media bed. This greater depth of intermixing produced a more predominant synergistic effect resulting in overall improved filter efficiency.

Example 4: In this test, 48 inches of No. 1½ anthracite were used without the aid of walnut shells. The run was terminated as a result of effluent clarity deterioration after 2,100 gallons of contaminated liquid were filtered per square foot of inlet surface area. The composite effluent clarity was approximately 21 JTU. The larger granule size of No. 1½ anthracite and the lesser amount of fines resulted in a much longer run than that experienced in Example 1.

Example 5: In this final test, 42 inches of No. 1½ anthracite were used in conjunction with 6 inches of No. 12/20 granular black walnut shells. The test was terminated on the basis of effluent deterioration after 4,700 gallons of contaminated liquid were filtered per square foot inlet surface area. Here again the dramatic effectiveness of using granular black walnut shells in combination with anthracite coal is well demonstrated.

PLASTIC FILTER MEDIA

Discrete Particles of Polyvinyl Chloride

F.T. Tulley and B.C. Harris, Jr.; U.S. Patent 3,544,457; December 1, 1970; assigned to Ethyl Corporation describe a process and apparatus for removing solids from a liquid stream. The liquid is passed through a filter containing discrete particles of a plastic material having irregularly, nonuniformly shaped surfaces. The particles of plastic material may be supported on gravel or rocks or by other media. It is preferred that the plastic particles have a size from about 16 mesh to about 25 mesh. The preferred material for the plastic particles is polyvinyl chloride particles which have been formed by the suspension process.

The process is characterized by the extremely high throughput of liquids through the filter made using discrete plastic particles versus that obtained using sand. Throughput for a conventional sand filled filter is about 2 to 3 gal/min/ft^2 of filter area. Throughput for the apparatus of the process using discrete, irregularly, nonuniformly shaped plastic particles from about 16 to 25 mesh is about 12 to 16 gal/min/ft^2. Additionally the process provides a filtration system wherein one can obtain an effluent with about 0.2 ppm turbidity by 6 cycles at a rate of about 15 to 16 gal/min/ft^2 as opposed to 230 cycles to obtain the same turbidity when using the same flow rate on a filter containing the conventional size sand particles.

Referring to Figure 1.3a it will be seen that the filter material is contained within vessel **10** which may be a large masonry structure or concrete structure if used for municipal or industrial large volume water treatment. The liquid may be passed through the filter under pressure if desired. The vessel is provided with a liquid inlet line **12** carrying the fluid to be treated. In the bottom of the vessel there is provided a plurality of gathering lines **14** for collection of the filtered liquid from the bottom of the vessel.

Each collection line is provided with a plurality of slots or openings **15** to admit the treated liquids to the interior of the line. A header **16** ties the plurality of lines together and conveys the liquid to the discharge line **17**. A backwash line **18** is connected into a pipe tee at the bottom of the vessel to provide backwash water for clarifying and removing the sludge from the top of the filter bed. A trough **11** having weir slots **13** carries away the backwash liquid and solids.

Immediately on the top of the collection line is a plurality of large rocks **19** and on top of that is placed a layer of smaller sized pebbles **20**. Immediately above the pebbles is a layer of discrete, irregularly, nonuniformly shaped plastic material **21**. This plastic filter media as shown in Figure 1.3a has a graded size, the largest particles being on the top and the smaller particles being on the bottom layers.

As shown in Figure 1.3b the vessel **10** contains rocks **19** and pebbles **20** and a layer of plastic filter media **22** wherein the particle size of the filter media is substantially the same. Sometimes it is desirable to provide a filter having all particles of the same size.

It is preferred that the plastic particles used to fill the filter have a particle size which will be fairly evenly distributed between about 10 and 60 mesh particle size. However, for some services the particle size should range from about 16 to 40 mesh. In particularly critical filtration areas it has been found that the particle size for most efficient filtration should range from about 16 to 25 mesh. In physical dimension the particles should have a maximum dimension of from 2 to 10 mm, preferably from 4 to 8 mm. The plastic filter media should be placed on a rock and pebble bed of about 20 inches when used for industrial

FIGURE 1.3: FLUID TREATMENT FILTER

a.

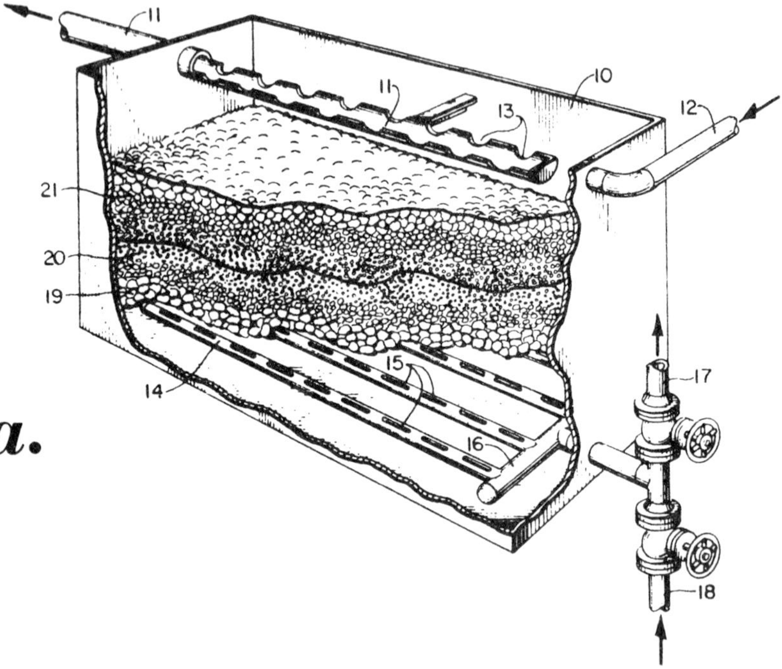

b.

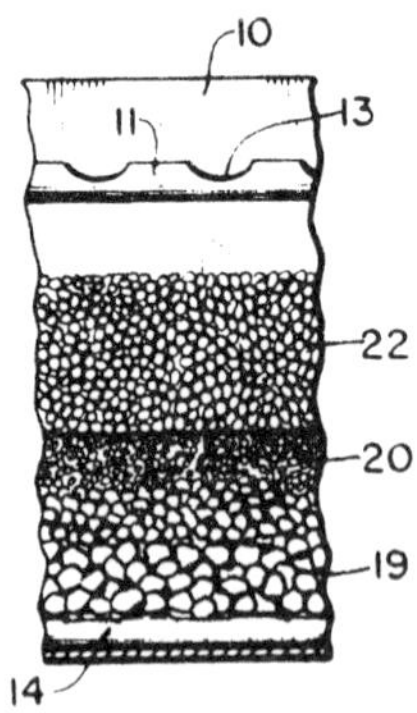

Source: U.S. Patent 3,544,457

and municipal water filtration. It has been found that about 20 inches of plastic material filter media on top of the rock and pebble bed is satisfactory. However, other depths of support and filter media will be found suitable depending upon the configuration of the filter vessel and the throughput of liquid over the filter. The pebble and rock support should range from a minimum size of about 1 mm up to about 2½ inches in diameter maximum. However, it is not essential that rocks and pebbles be utilized to support the plastic filter media since it can be supported on fine mesh wire screen or other suitable material which will retain the plastic filter media within the treatment vessel.

The plastic material chosen for the filter media may be any plastic material, synthetic or natural, which has the requisite physical properties for the service in which the filter is utilized. The plastic filter media may be selected from thermoplastics and thermosetting plastics.

One of the most preferred plastic materials for forming the plastic media used in the filter is polyvinyl chloride. The polyvinyl chloride may be prepared by either the emulsion, bulk or suspension process. It has been found that when polyvinyl chloride is prepared by the emulsion process described in U.S. Patent 2,843,576 that there is provided a granular porous, polyvinyl chloride in aggregate form which has a particle size eminently suitable for use in the present filter.

Under proper conditions of concentration of an emulsifying agent the procedure set forth in the above mentioned patent can yield a vinyl chloride having a particle size within the desired 16 to 40 mesh with a very high percentage being in the 16 to 25 mesh range. The bulk density of the polyvinyl chloride should be from about 20 to 60 lb/ft^3. These particles of polyvinyl chloride are characterized by an extremely high surface area per unit volume and an extremely irregular configuration for surface. The polymer has a popcorn appearance characterizing a high surface area and a low density because of its porosity. Another suitable commercially available polyvinyl chloride is Ethyl SR225 PVC resin made by a suspension process.

In a test unit constructed utilizing as the plastic filter media polyvinyl chloride particles prepared by a suspension process (Ethyl SR225 PVC resin) having a particle size of from about 16 to 40 mesh extremely high efficiencies were realized. In a test unit containing about one cubic foot of filter media using river water which had been treated by standard flocculating treatments and chlorinated prior to introduction into the filter, it was found that a rate of 15 to 16 gallons per minute per square foot could be passed through the filter and that the effluent water would have a turbidity of 0.2 ppm after only 6 cycles through the filter bed.

In a like size filter utilizing sand in place of the polyvinyl chloride particles it required 230 passes at the same flow rate through the sand before the turbidity reached the 0.2 ppm level. Thus it is seen from the foregoing that extremely high filtration rates can be obtained with very good turbidity levels using the process.

The filters may be backwashed as shown in Figures 1.3a and 1.3b after the initial filtration level drops off (e.g., 20% drop) using treated backwash water or other liquid to remove the sludge and dirt and accumulated solid matter from the top two or three inches of the plastic filter media.

The process is characterized in that it can maintain high flow rates for a longer period of time without necessity of backwashing. Additionally the random shaped plastic filter media provides a much less turbid effluent at a much higher flow rate than does the conventional sand filter bed.

Preparation of Porous Polyvinyl Chloride Particles

W.E. Burt; U.S. Patent 3,842,006; October 15, 1974; assigned to Ethyl Corporation describes a process for the manufacture of porous polyvinyl chloride particles suitable for efficient filtration of solids from liquid streams.

The process is characterized by the extremely high throughput of liquids through filters made from the polyvinyl chloride particles of the process. Throughput for a conventional rapid sand filter is from about 2 to 3 gal/min/ft^2 of filter area. Throughput for the apparatus of the process using discrete, porous polyvinyl chloride particles having a size of from about 6 to 60 mesh is about 18 gallons per minute per square foot. The filtration efficiency of a filter prepared from the polyvinyl chloride filter particles of the process is greater than 70%, whereas the typical swimming pool filter made of sand has a filtration efficiency of approximately 50%.

Referring to Figure 1.4, numeral **20** indicates an endless belt conveyor which is extended at one end over a driving roller **18** and at the other end over an idler roller **21**. The roller **18** can be driven from a motor **19** through a belt **30**.

A feed hopper **11** is positioned above the belt, preferably above the idler roller **21** and is provided with an outlet **31** which extends transversely to the belt. Provision can be made, if desired, to vibrate the hopper **11** according to conventional practice to insure the free, uniform flow of particles across the desired width of the belt. The surface of the belt on which the particles are distributed is a plain surface and preferably is formed of or treated with a material to which the particles do not adhere during the operation of the process.

Polyvinyl chloride particles **12** are distributed across the desired width of the belt and a loose layer or bed of individual particles is formed of desired thickness by stationary leveling knife **13** which extends transversely to the belt. The knife is preferably adjustable in a vertical plane to increase or decrease the distance between its lower edge and the belt. The knife permits reasonably accurate adjustment of the thickness of the bed of particles **12** to achieve uniform sintering.

The number **16** indicates the heating zone through which the bed or sheet of particles is advanced towards the discharge end of the conveyor. Provision is made, such as by induction heating or by radiant heating coils, to produce and maintain in the bed of particles a temperature sufficiently high to cause incipient or surface fusion of individual particles as the sheet is advanced through the zone.

Either or both the temperature maintained in the heating zone **16** and the speed of the belt can be adjusted to obtain the desired fusion of the surface of the particles without gross deformation of the shapes of the individual particles. The bed or layer of individual particles which enters the heating zone becomes, during its passage through this zone, a microporous sheet of individual particles firmly attached at a multiplicity of points on their surface to adjacent particles.

FIGURE 1.4: PREPARATION OF THERMOPLASTIC FILTERING MEDIA

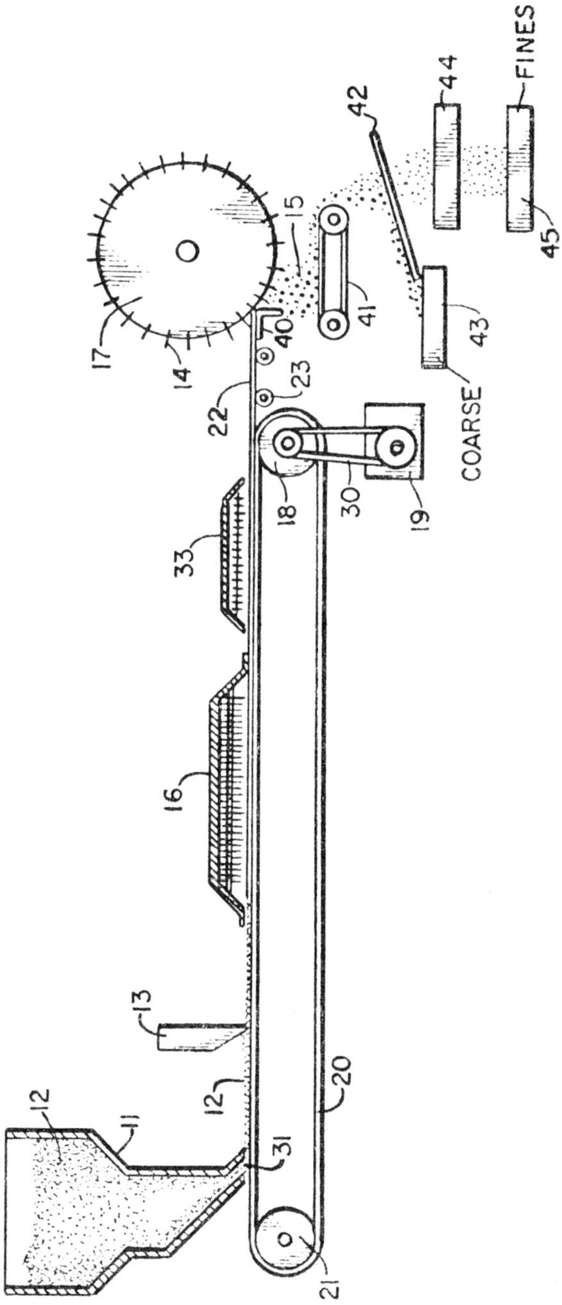

Source: U.S. Patent 3,842,006

On leaving the heating zone the sheet is cooled to about atmospheric temperature. The rate of cooling can be increased by passing it through a cooling zone **33** in which air circulation or in which refrigerating coils are provided. On cooling to atmospheric temperature, the sheet is strong and rigid.

The rigid sheet passes from endless belt conveyor **20** over rollers **23** to a grinder. The grinder consists of grinder cylinder **17** and grinder plate **40**. The teeth **14** of the grinder cylinder mesh with matching teeth in the grinder plate thereby comminuting the rigid sheet **22** into small particles **15**. However, if desired, the sintered sheet may be comminuted by other means such as a dicer or by passing the sheet into the nip of a pair of rotating hard rubber or metal rolls.

The particles fall from the grinder plate onto endless conveyor belt **41**, which conveys the particles to sizing screen **42**. The screen openings in the sizing screen are sufficiently small to exclude particles larger than the maximum particle size desired. A preferred screen opening is about 12 mesh.

The large particles which remain on the screen **42** are removed from the screen by any conventional means, such as inclining the screen and vibrating the screen. The large particles removed from the screen are placed in a container such as box **43**. The particles in the box are deposited on conveyor **22** at any point between cooling zone **33** and grinder cylinder **17** and recycled through the grinder and sizing screens.

The particles which pass through the screen fall onto screen-box **44** which has a sizing screen for a bottom. The openings in screen **44** are chosen sufficiently small to retain only the smallest desired particles. A preferred screen opening for screen **44** is about 50 mesh. Provision can be made, if desired, to vibrate screen **44** and screen **42** according to conventional practice to insure the free flow of particles through each screen. The particles passing through screen **44** fall into box **45**. The contents of box **45** is then transferred by any conventional means to hopper **11** for resintering. The particles remaining on the screen **44** have a maximum size smaller than the screen openings in screen **42** and a minimum size larger than the size of the screen openings in screen **44**. The particles remaining on screen **44** are removed for use as filtration media.

The particles prepared in accordance with the process are highly suitable for use as filtration media. A large variety of polyvinyl chloride resins are suitable for preparing filter media. Suitable polyvinyl chloride particles may be prepared by either the emulsion or suspension process.

Preferred polyvinyl chloride suspension resins are SM-175 and SM-200 (Ethyl Corporation). SM-175 is a polyvinyl chloride resin prepared by the suspension process and has a particle size distribution by weight of about 5% between 60 and 80 mesh, and about 35 to 40% between 80 and 100 mesh, about 40 to 45% between 100 and 140 mesh, about 10 to 15% between 140 and 200 mesh, and about 10% below 200 mesh in diameter (all mesh sizes are U.S. Standard Screen).

SM-200 is a polyvinyl chloride resin prepared by the suspension process and has a particle size distribution by weight of about 3 to 15% between 60 and 80 mesh, about 30 to 50% between 80 and 100 mesh, about 30 to 50% between 140 and 200 mesh, and about 0 to 3% smaller than 200 mesh.

Emulsion resins suitable for sintering and comminuting in accordance with the process preferably include EH-230 and EH-240 (Ethyl Corporation). EH-230 has a particle size distribution by weight of about 0 to 1% between 80 and 100 mesh, about 90% between about 100 to 200 mesh, and about 0 to 10% smaller than 200 mesh. EH-240 has a particle size distribution by weight of about 2% between 140 and 170 mesh, about 6% between 170 and 230 mesh, about 15% between 230 and 270 mesh, about 15% between 270 and 325 mesh, about 15% between 325 and 400 mesh, and about 47% smaller than 400 mesh.

Several resins were sintered and comminuted in accordance with the process, and filtration tests were run upon these particles. The following examples illustrate the process.

Example 1: Polyvinyl chloride filtration media particles were prepared by sintering EH-230 and grinding the sintered sheet into particles. The total time the polyvinyl chloride particles were within the heating zone was about 5 to 6 minutes and the total time within the cooling zone was 2 to 3 minutes. The heating zone was operated at a temperature of 210°C, and the cooling was accomplished by blowing air at 35°C.

Screen **42** had a mesh size of 12 mesh and screen **44** had a mesh size of 30 mesh. The particle size distribution of the filter media was about 19.8% between 12 and 16 mesh, 38% between 16 and 20 mesh, 24.5% between 20 and 35 mesh, 21.1% between 25 and 30 mesh with about 5.5% smaller than 30 mesh. Bulk density was 21 lb/ft^3. Porosite was 33%. The filter media was tested according to the National Sanitation Foundation testing procedure and gave a five turnover turbidity reduction (filtration efficiency) of 78%.

Example 2: Polyvinyl chloride filter media was prepared as in Example 1 with the exception that the particle size distribution of the filter media was about 23.8% between 12 and 16 mesh, about 44.6% between 16 and 20 mesh, about 18.1% between 20 and 25 mesh, about 9.7% between 25 and 30 mesh and approximately 3.4% smaller than 30 mesh. The polyvinyl chloride filter media was tested according to the National Sanitation Foundation procedure and gave a five turnover turbidity reduction (filtration efficiency) of 72%.

Examples 3 to 7: A typical swimming pool filter sand was tested according to the National Sanitation Foundation procedure. The sand filter had a filtration efficiency of about 50%. Examples 5 through 7 were prepared as in Example 1, but a different filtration testing procedure was used. Example 3 is sand tested according to the different testing procedure. The testing procedure used for Examples 4 through 6, referred to as the Mapico Black procedure, was as follows.

Apparatus and Procedure for Filtration Tests – The laboratory filtration unit included a rectangular tank for containing the turbid water. The filter media was contained in a hollow metal cylinder 48 inches in height and 2¼ inches in diameter. The cylindrical container was located in an upright position, with the top of the container having an opening therein for admission of fluids. The bottom of the container had a hole therein for passage of the effluent. The filter media was supported by a metal support located one inch from the bottom of the container. The support had holes therein smaller than the smallest sized filter particle for passage of the effluent.

The tank was filled with turbid fluid and the turbid fluid was pumped into the top of the filter media container. The turbid fluid would then flow through the filter and out of the hole in the bottom of the container. The filter effluent was pumped continuously through a Hach CR (continuous reading) Turbidimeter and returned to the rectangular tank. The water capacity of the tank system was 12.55 gallons. The turbidity of the filter effluent was read on the master indicator connected to the Hach Turbidimeter. The turbidity of the fluid was measured in Jackson Turbidity Units (JTU).

The tank was filled with clear tap water and the filter media added to the desired depth in the 2¼ inch diameter filter media container to form the filter bed. The pump supplying the feed vessel was started and the filter bed backwashed for 5 minutes. The filter bed was allowed to settle, classifying the particles, fines on top and coarse at the bottom of the filter bed.

Forward flow of clean tap water (6 gpm/ft^2) downward through the filter bed was started and allowed to run until the turbidimeter shows less than 1.0 JTU which is a baseline and indicates the unit was working properly. The pressure drop through the fresh filter bed was measured. The proper amount of Mapico Black, Fe_3O_4 having particle size range of 0.3 to 0.8 micron, suspended in 400 milliliters of tap water in a Waring blender was added to the reservoir water with stirring to start the test. The peak turbidity was determined and periodic readings were recorded over a 4 to 5 hour period. Terminal pressure drop through the filter bed was measured and the filter bed was backwashed for about 5 min to permit observation of how well the soil (Mapico Black) separates from the filter media and to determine filter bed expansion versus backwash flow rate.

Example No.	Filter Material	Particle Size (U.S. mesh)	Bed Height (inches)	Head Loss (inches) Initial	Final	H	Length of run (hr)	Turbidity of . . Effluent (JTU). . Initial	Peak	Final
4	Sand	28-35	19	22	24	2	7	1.1	19.5	5.0
5	EH-230	12-26	21	7	7.25	0.25	5	0.9	16.0	2.7
6	EH-230	26-40	20	26.5	42.75	16.25	5	0.4	5.0	2.2
7	SM-175	10-30	21	7.5	8.5	1	5	0.7	19.0	3.2

Operating Conditions for Filtration Tests — (1) The filter medium bed height was 19 to 21 inches for the evaluation of the different media and the effect of different particle size of the media on filtration effectiveness. (2) The forward flow rate (downward through the filter bed) was 6 gallons per minute (gpm) per square foot of filter bed area for both the 12 to 26 and 26 to 40 sieve number PVC particles. The backwash flow rates were 17 gpm/ft^2 for the 12 to 26 mesh PVC particles and 4 to 7.5 gpm/ft^2 for the 26 to 40 sieve number PVC particles. (3) The turbidity standard used to evaluate the filter media was 0.94 gram of Mapico Black (iron oxide, particle size 0.3 to 0.8 micron) slurried in 12.55 gallons of tap water to give a turbidity of 20 JTU.

It can be seen from the examples that filter media prepared from polyvinyl chloride in accordance with the process is very efficient in reducing the turbidity of fluids such as water. A typical swimming pool filter sand was tested (Example 3) according to the National Sanitation Foundation procedure and efficiencies of approximately 50% were obtained. Filtration media prepared in Examples 1 and 2 of the process were tested according to the same procedure and had filtration efficiencies greater than 70%, which is approximately 50% greater than typical sand filters.

The final turbidity of the effluent filtered in accordance with Examples 5, 6, and 7 (2.7, 2.2, and 3.2, respectively) was approximately one-half as large as the final turbidity of effluent filtered in the sand filter of Example 4 (5.0).

The filtered media prepared in accordance with the process can be used in virtually all types of filtering apparatus which employ solid particulate filter media such as sand. The filter media may be enclosed completely within the filter, such as in a cylindrical filter having filter media tightly contained therein, an entrance port at one end of the cylinder, and an exit port at the other end of the cylinder.

Example 8: A three-component inverted filter bed was prepared from (a) 210 grams of filter media prepared in accordance with Example 1, having a particle size of 12 to 16 U.S. Mesh, bulk density of 0.565, and a specific gravity of 1.4; (b) 197 grams of filter media prepared as in Example 1 with the exception that 20% barium sulfate was added to the EH-230 latex prior to spray drying the latex, the filter media having a particle size of 16 to 25 U.S. Mesh, bulk density of 0.505, and a specific gravity of 2.02; and (c) 202 grams of EH-230 prepared in accordance with Example 1 with the exception that 33% barium sulfate was added prior to spray drying EH-230 latex, the filter media having a particle size of 25 to 30 U.S. Mesh, bulk density 0.515, and a specific gravity of 2.43. All the particles were mixed together and then added to the filter media container. The bed height was 6 inches. The filter media container used was the same bed used in the Mapico Black Filtration Procedure.

On backwashing the filter bed at 20 gpm/ft^2, the coarse polyvinyl chloride filter media came to the top of the bed and gave a fair degree of separation from the two heavier fractions. Most of the finer, heavier particles of filter media migrated to the bottom and the intermediate filter particles remained in the middle.

Backflushing Technique Using Polyvinyl Chloride Bed

G. Hirs and R.H. Wykoff; U.S. Patent 3,557,955; January 26, 1971; assigned to Hydromation Engineering Company describe a method for filtering which provides the filtering of extremely small particles, greater dirt retention in the filter medium, higher flow rates through the filter medium, a smaller volume of filter medium, and improved backflushing or backwashing to obtain a cleaner medium. The medium is composed of finely divided particulate material which in the preferred case is polyvinyl chloride, polyethylene or wood sawdust, these materials being light and having great strength to provide stable, fine particles capable of being backflushed.

While the apparatus can assume many different configurations, liquid permeable retainers preferably are provided at both the inlet and outlet sides of the filter medium. Where the filter medium is located between retainers, it is important to reverse the liquid flow at least through the inlet retainer, during backflushing to remove any deposits from that retainer. It is possible to use only one liquid permeable retainer together with agitation and hydraulic flow during backflushing to provide a simplified filter which is effective in some applications. The process also features backflushing of a particulate filter medium which is very compact and contains relatively great dirt loads. Such backflushing utilizes both mechanical and hydraulic forces, and may be accomplished in several different ways. The media can be compacted, if desired, to reduce the size of the voids therein, thereby increasing the effectiveness of the filter medium in entrapping and holding extremely small particles of dirt. The filter medium is compacted prior to the filtration operation, such compaction being accomplished in any one of several ways, e.g., by the use of fluid pressure or a differential flow rate.

Surfactant-Treated Polyvinyl Chloride Particles

A.J. Haefner; U.S. Patent 3,568,846; March 9, 1971; assigned to Ethyl Corporation describes a filter media which is prepared by subjecting particles of a plastic material having discrete, irregular, nonuniform shaped surfaces to treatment with a quantity of a surfactant or wetting agent sufficient to permit the plastic particles to be readily wetted by water.

Among the wetting agents which are eminently suitable are soaps such as fatty acid salts of some of the alkali metals such as potassium or sodium with long chain fatty acids. Commercially available soaps made from mixed fatty acids containing about 80% tallow fatty acids and about 20% coconut oil fatty acids are quite suitable. Such soaps may contain sodium or potassium salts of lauric acid and stearic acid together with oleic and other 12 to 22 carbon atom fatty acids.

Example 1: When a quantity of polyvinyl chloride (Ethyl SR 225 PVC resins made by a suspension process) was placed in a container approximately 50% of the resin floated on top of the water either as individual particles or agglomerates. A solution of 0.01% sodium lauryl sulfate in water was prepared and the same quantity of the same lot of polyvinyl chloride resin was added to this solution with stirring. The resin particles were quickly wetted out and sank to the bottom of the container after stirring ceased.

After dewatering the treated resin was placed in a closed filter tank and upon introduction of turbid water immediately wetted out and settled into a uniform bed in the lower portion of the tank. On backwashing the individual particles were cleanly separated and no tendency to agglomerate or to float to the top of the water was evidenced.

Example 2: 100 grams of the same polyvinyl chloride resin used in Example 1 was placed in a large beaker. To that beaker was added 1 cc of a 2% solution of sodium lauryl sulfate by spraying over the resin as the resin was agitated. 200 grams of water was added to the beaker containing the treated resin particles with vigorous stirring. When the stirring was stopped all the resin particles quickly settled to the bottom of the beaker.

In addition to the anionic soaps and detergents the cationic soaps are also suitable wetting agents. Particularly suitable are the quaternary ammonium compounds in which the aliphatic radical can be straight chain or branched chain and containing from about 6 to 18 carbon atoms, e.g., myristamidopropyl dimethylbenzyl ammonium chloride.

One of the preferred detergent surfactant materials is sodium dioctylsulfosuccinate. The particle size of the plastic particles most suited for use as filter media should be between about 10 to 60 mesh.

Electrophotographic Imaging Systems

M. Sato; U.S. Patent 3,544,458; December 1, 1970; assigned to Xerox Corporation describes a process and apparatus for selectively separating particles from mixtures, suspensions and colloidal suspensions by contacting the suspensions with solid particles capable of generating an electrostatic field with the particles and thereby hold them. An application of the process is in liquid developers and other suspensions utilized in electrophotographic imaging systems.

Figure 1.5a is a representation of a system using a filter mechanism. A tank **10** is shown holding a suspension **11** having solid particles **12** suspended in a liquid carrier. The suspension is brought to a filter housing **13** by any suitable means such as pump **P** via a conduit. Alternatively, a gravity system could be used to bring the suspension to the filter housing. Within the housing are several layers of filter material **14** formed as solid particles. There is an egress port **15** through which the liquid carrier **16** passes into a tank **17** which contains the decontaminated liquid carrier. The suspension particles **12** are attracted to and held by the filter material particles **14** through the use of electrostatic forces.

The filtering system functions on a suspension having a carrier which could be a solid, liquid or gas and an electrically different particle suspended in the carrier such that the charge on the particle is different from that on the carrier. This charge difference may be either positive or negative with respect to the carrier and the carrier may be at a neutral potential having no charge thereon whatsoever. The filter material must acquire a charge opposite to that of the particle when the suspension moves past the filter material. The suspension is passed through a pile of filtering material by any means such as a pump or gravity operating on the suspension or by shaking or jogging the housing holding the filtering material or by any other suitable means to create a flow of the suspension through the filtering material.

Figure 1.5b is a blowup schematic of the suspension particles being electrostatically attracted to and held by filter material particles. The figure shows a blowup segment of filtering material **14** and solid suspended particles **12** held in a liquid carrier **16** moving past the filtering material. The charge shown on the filter material is negative and the attractive charge on the suspension material would be positive. This causes the suspension particles to be attracted to the filtering material in the particular liquid carrier. An example of such a charge relationship would include having a filtering material of polyvinyl chloride particles.

If the liquid carrier is mineral oil the surface of the polyvinyl chloride will acquire a negative charge when the mineral oil flows over it. If the solid matter suspended in the mineral oil suspension is glass, it will acquire a positive charge in the mineral oil and will be attracted to the polyvinyl chloride particles as the suspension flows past the particles. In this manner, the glass particles would be removed from the mineral oil allowing the mineral oil, designated **16**, to pass through the filter mechanism in a purified state having the glass particles removed therefrom.

It is known that when solid particles are present in a given nonconductive or insulative liquid, the solid particles are charged to either a positive or a negative polarity. For example, when such solid matter as polyvinyl chloride, polyethylene, polyvinylidene chloride, fluoric resin, or nitrocellulose is submerged in a

FIGURE 1.5: FILTRATION TECHNIQUE EMPLOYING AN ELECTROSTATIC FIELD

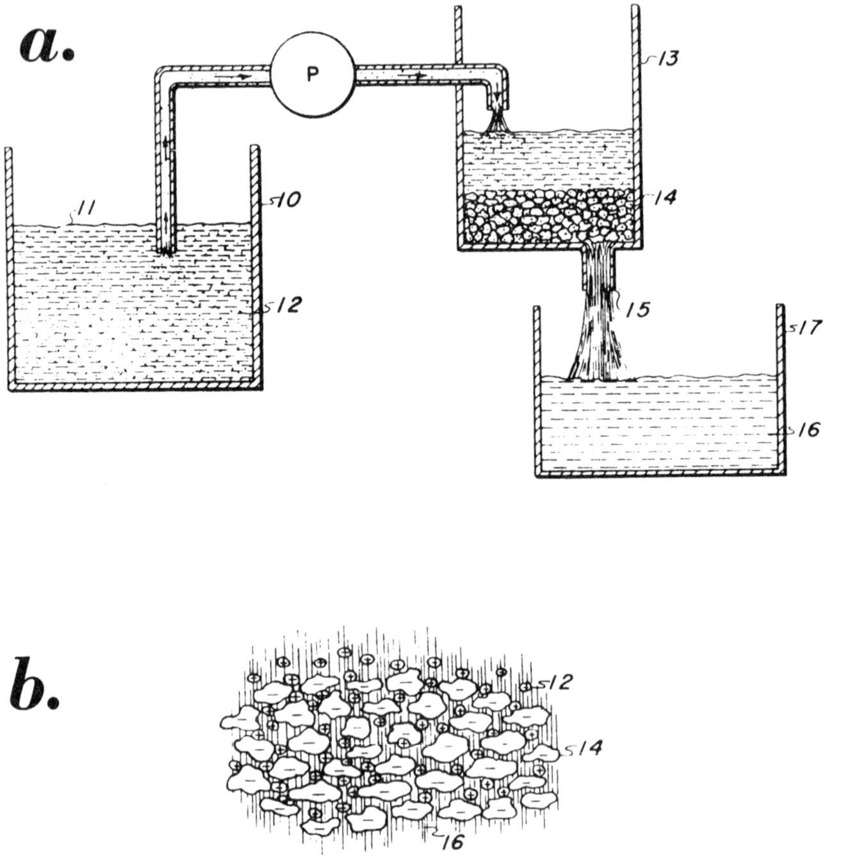

Source: U.S. Patent 3,544,458

mineral oil, the surface of such solid matter is charged negatively. Such solid matter as protein, polyamide resin, ethylcellulose, or glass is charged positively in a mineral oil. If a solid particle of polyvinyl chloride is submerged in a mineral oil wherein suspended particles are charged positively, then the suspended particles are caused to adhere to the surface of this solid.

The mechanism of this phenomenon is such that Coulomb's attraction generated between the negative charge on the surface of polyvinyl chloride and the positive charge of the suspended particles causes the latter to be attracted and deposited on the surface of polyvinyl chloride.

Unlike the principle of filtration by filter paper which serves to impede the passage of suspended particles mechanically, the action of this process relies on on electrostatic attraction of the suspended particles. Accordingly, solid particles are used for attracting suspended particles. The filtering material preferably has a much greater particle size than the suspended particles. For example, the filtering material may have a particle size of the order of a fraction of a millimeter to several millimeters, while suspended particles may be colloidal having a size of the order of from 0.1 to 1.0 μ. These sizes are such that if the suspended particles are not attracted electrostatically, they would pass freely through the spaces between the filtering material. The filtering material may comprise globular, cylindrical, flat, amorphous, fibrous, or textural shapes and qualities.

Since the principle of this process is in the electrostatic phenomenon, the charge relationship between the suspended particles and the filtering material is important. Accordingly, if the electric charge on the suspended particles has the like sign with that of the electric charge on the surface of the filtering material, then the surface of the filtering material does not attract suspended particles. In addition, dyes and resins which remain dissolved in an insulative liquid carrier cannot be removed although particles suspended therein can be filtered.

According to this process, fine particles which are suspended as colloid in such a highly insulative liquid as gasoline or kerosene can be filtered safely and efficiently. However the dye which is dissolved in gasoline for the purpose of coloration will not be removed.

When the surface of the filtering material becomes covered with suspended particles throughout, the electric charge on the surface of the filtering material is replaced by the electric charge of suspended particles and consequently the filtering material will no longer attract more suspended particles of the same kind or charge. To continue the operation of the filter with the same filtering material, the suspended particles can be washed off the filtering material to again expose the surface of the filtering material for further action.

Alternatively, a suspension containing suspended particles of opposite charge to the original suspended particles can be moved through the filter to allow these suspended particles to adhere to the former suspended particles of opposite charge which are already on the surface. By the latter means, the filtering material can be used repeatedly without being washed.

A particular use for this filtering system is in the field of electrophotography using a developing process called liquid development of electrophotographic images. This process utilizes a liquid developer which has colored, toner particles suspended in an insulative liquid. After an electrostatic latent image is formed on a sensitive member as, for example, described in U.S. Patent 2,297,691 or 2,357,809, a visible image can be formed by means of the liquid developer.

The sensitive member is cleaned with an insulative liquid called cleaning liquid to remove extraneous matter and toner particles not forming the image. Consequently, the cleaning liquid will become contaminated with charged particles picked up thereby. In order to prevent this, it is necessary to remove such charged particles from within the cleaning liquid. This process is suitable for this purpose. Also, this process is suitable for the filtration of the liquid developer which is composed of a solid charged toner particle in an insulative liquid carrier.

It has been found that good filtering results are achieved, especially with colloidal suspensions, if the filtering material is piled to a thickness corresponding to several hundred or even several thousand layers or more of the particles of the filtering material. For example, when glass globules 0.2 mm across are used as the filtering material, the layer thickness should exceed several centimeters. As the particle diameter is increased, so should the thickness of the filtering material.

In cases where the particle size of the filtering material is as large as several millimeters, it is desirable to incorporate therein another filtering material which has a particle diameter small enough to prevent colloidal particles from passing freely through the opening of the original filtering material.

Example 1: A piece of 200 mesh metal gauze was fastened to the lower end of a length of glass tubing about 30 cm in height and 10 cm in inside diameter. Glass beads having a particle diameter of about 0.2 to 0.3 mm were placed in the cylinder to a height of about 25 cm. Then, a green colloidal liquid having pigment particles of copper phthalocyanine suspended in kerosene was poured from above the receptacle. The liquid which came out of the metal gauze at the bottom of the receptacle was colorless and transparent and contained no pigment at all.

Example 2: Chips (irregular shape having 0.1 to 1.0 mm of diameter) cut with a saw from a rigid plate of polyvinyl chloride were cleaned sufficiently with methanol to remove machine oil. Thereafter, they were placed in a receptacle similar to that of Example 1. When a colloidal liquid having positively charged zinc oxide 0.05 to 0.5 mμ in particle diameter suspended in cyclohexane was poured from above, the zinc oxide particles adhered to the surface of polyvinyl chloride chips and colorless transparent cyclohexane containing no zinc oxide flowed out of the metal gauze at the bottom of the receptacle.

Example 3: The procedures of Example 2 were followed, except the colloidal liquid suspension used was composed of particles obtained by coating carbon black particles with polyvinyl chloride resin and were suspended in a mineral oil. The colloidal particles were charged negatively. When the colloidal liquid was poured from above the receptacle, the liquid flowing out of the metal gauze at the bottom of the receptacle was found to contain a large volume of colloidal particles. The result suggests that no filtration can be achieved in the case of such combination.

Example 4: Oil yellow, which is an oil-soluble dye, was dissolved in a mineral oil. When the resultant solution was passed through the filtering materials of Examples 1 and 2, the dye concentration in the solution remained unchanged. This suggests that no filtration is obtainable in such case as this.

Example 5: When the filtration was continued in the procedure of Example 2, the entire surface of the filtering material eventually became covered with colloidal particles and, as a consequence, no further filtration could be obtained. Therefore, a colloidal liquid containing particles bound to acquire negative charge in the liquid, for example, the colloidal particles used in Example 1, was poured into a receptacle. Consequently, copper phthalocyanine particles deposited themselves on the zinc oxide particles which had already covered the surface of the filtering material, with a result that this colloidal liquid was filtered completely. This indicates that when the entire surface of the filtering material is covered throughout with one kind of colloidal particles, it is possible to have another kind of colloidal particles deposited thereon, and that this coating can be repeated. According to this procedure, the step of washing off deposited particles can be omitted even when the filtering material becomes covered by colloidal particles throughout.

Example 6: In the same receptacle as used in Example 1, brass beads 0.2 to 0.4 mm in diameter were placed in the lower half portion of the receptacle as the filtering material bound to be charged negatively and glass beads 0.2 to 0.3 millimeter in diameter were placed in the upper half portion as the filtering material bound to be charged positively. By this combination, removal of colloidal particles could be accomplished by passing the liquid through the two layers of filtering material in the receptacle, no matter whether the particles present in the colloidal liquid are charged positively or negatively.

Triboelectric Filtration for Toner Particles

According to a process described by *M. Sato; U.S. Patent 3,909,383; Sept. 30, 1975* insulating liquids contaminated with suspended particles and dissolved material are regenerated by initially removing the suspended particles by triboelectric filtration and thereafter removing the dissolved material by adsorption of the dissolved material onto a suitable surface. The following examples illustrate the process. Parts and percentages are by weight unless otherwise indicated.

Example 1: An electrostatic latent image formed on an imaging surface of a zinc oxide binder plate is briefly immersed in kerosene and thereafter developed with a liquid developer containing about 300 parts cyclohexane, about 200 parts kerosene and about 2 parts of a commercial black offset ink. The commercial black offset ink comprises a dye, rosin-modified phenol-formaldehyde resin varnish and a pigment. Upon mixing of the foregoing ingredients, the resulting developer contains dissolved rosin-modified phenol-formaldehyde resin varnish as well as colloidal particles comprising the pigment particles, dye and rosin-modified phenol-formaldehyde resin varnish adsorbed on the surface of the pigment particles.

Development of the electrostatic latent image is effected by immersion of the binder plate in the liquid developer for about one minute. Upon removal of the binder plate, it is found that the developed plate is coated on the imaging surface as well as the rear surface with a uniform, thin layer of liquid developer. The deposited layer of liquid developer is then dried on the surface of the plate. A heavy deposit of toner particles is observed on the background areas of the surface as well as on the rear surface of the plate.

Example 2: The process described in control Example 1 is repeated except that prior to drying the developed surface of the binder plate, the plate is gently washed by immersing it in a bath of about 1 liter of isoparaffin. During the washing step, the liquid developer adhering to the background areas of the binder plate diffuses into the cleaning agent. The plate is thereafter removed from the cleaning bath and dried by exposure to a current of warm air. Upon repeating the development and washing steps with 10 additional binder plates, the cleaning liquid changes from a clear transparent liquid to a conspicuously colored liquid having markedly reduced cleaning power. Since the rosin-modified phenol-formaldehyde resin varnish is more soluble in the isoparaffin cleaning liquid than in the cyclohexane carrier liquid, rapid contamination of the cleaning liquid with the varnish is observed. The contaminated cleaning liquid is then subjected to an electrophoretic regeneration treatment. In conducting the electrophoretic regeneration treatment, an electrophoretic device is employed comprising a stack of 101 metallic plates having a thickness of about 1 millimeter and sides of 20 centimeters by about 30 centimeters, each metallic plate being separated from each adjacent metallic plate by a distance of about 2 millimeters.

Every other plate in the stack is connected electrically to one electrode of a source of DC potential and the remaining plates are connected to the other electrode of the DC potential source. The stack is inserted in a container having an inside dimension of about 20 x 30 x 30 centimeters. After the container is filled with the contaminated cleaning liquid, a DC potential of about 1,000 volts is applied to the alternately connected metal plates.

Complete removal of the toner particles from the cleaning liquid is effected in from about 3 to 10 seconds. Because the toner particles are positively charged in the above described developer, these particles are attracted and deposited on the plate connected to the cathode of the DC potential source.

The regenerated liquid is drained through a valve at the bottom of the container while the electrical potential remains applied to the plates. The resulting regenerated liquid is transparent and free of suspended toner particles. Unfortunately, the regenerated liquid is characterized by yellow tint due to the presence of dissolved varnish. This partially regenerated cleaning liquid is then employed to wash freshly developed imaging surfaces.

Although the resulting washed imaging surfaces have reduced background toner deposits, the deposited varnish coating on the imaging surface interferes with subsequently deposited toner particles when the imaging surface is imaged two additional times in a multicolor process. Particularly poor image quality is obtained when the deposited varnish coating is nonuniform.

Example 3: The process described in control Example 2 is repeated except that the partially regenerated cleaning liquid is subjected to a second regeneration treatment which removes the dissolved varnish. The second regeneration treatment is effected with the aid of a device comprising a metallic cylinder having an inside diameter of about 20 centimeters and a height of about 40 centimeters. A valve is provided at the bottom of the cylinder to permit removal of treated liquid.

The cylinder is filled to about 80% of its capacity with about 30 to 60 mesh particles of crystalline metal aluminosilicate (Linde molecular sieve 13 X). These particles contain tiny pores having a nominal pore diameter of about 10 A. The regenerated carrier liquid containing the dissolved varnish is poured into the top of the cylinder and removed through the valve at the bottom of the cylinder. The liquid removed from the bottom of the cylinder is substantially free of any dissolved varnish. Some entrained molecular sieve particles, however, are observed in the regenerated liquid initially removed from the cylinder. These entrained particles may easily be electrophoretically removed from the cleaning liquid by the technique described in Example 2.

Example 4: A cleaning liquid contaminated with toner and varnish in substantially the same manner as described in Example 2 is treated in a triboelectric filtering device. This device comprises a metallic cylinder having an inside diameter of about 20 centimeters and a height of about 100 centimeters. A 200 mesh metallic screen is positioned at approximately the middle of the cylinder. An outlet is provided at the bottom of the cylinder below the lower metallic screen. The lower half of the cylinder is filled with the same molecular sieve material described in Example 3 while the upper half of the cylinder is filled with brass shavings.

These brass shavings range in size from about 1/3 millimeter to 3 centimeters in length and from about 30 microns to about 1/3 millimeter in thickness. When the contaminated cleaning liquid is poured into the top of the cylinder, the brass shavings become negatively charged upon contact with the cleaning liquid. Thus, the positively charged toner particles are attracted to and deposit on the surface of the brass filings during passage of the liquid through the upper half of the cylinder. The dissolved varnish is removed by adsorption during passage of the cleaning liquid through the lower half of the cylinder. The regenerated liquid removed from the outlet at the bottom of the cylinder is colorless and transparent.

Example 5: A metallic cylinder having an inside diameter of about 20 centimeters and a height of about 120 centimeters is divided into three equal portions along its height by 200 mesh metallic screens positioned inside the cylinder. The uppermost portion is filled with polyvinyl chloride shavings, the middle portion with a layer of about 30 to 60 mesh particles of molecular sieve material 13 X, and the lowermost portion with a layer of a mixture of polyvinyl chloride particles and glass beads having a diameter of from about 0.3 to 0.4 millimeter. An outlet is provided at the bottom of the cylinder.

The polyvinyl chloride particles are granular and have an average diameter ranging from about 1/3 millimeter to about 3 millimeters. A contaminated cleaning liquid formed by a process substantially identical to the process described in Example 2 is introduced at the top of the cylinder. The suspended toner particles are triboelectrically removed from the contaminated cleaning liquid by the polyvinyl chloride particles. The dissolved varnish is removed from the cleaning liquid in the middle portion of the cylinder by adsorption onto the molecular sieve material.

As the treated clear and colorless cleaning liquid continues into the lowermost portion of the cylinder with entrained molecular sieve particles, the cleaning liquid causes the glass beads to be positively charged and the polyvinyl chloride particles to be negatively charged. Thus, any colloidal molecular sieve particles originating from the middle portion of the cylinder will be removed by either the charged glass beads or the charged polyvinyl chloride particles depending upon the polarity of charge carried by the molecular sieve particles.

Corrugated Cylinder with Randomly Packed Plastic Material

British Patent 1,082,661 describes a process and plant for biologically treating wastewater in an aerated filter in which the wastewater may be passed intermittently through a packing medium comprising pieces of an inert expanded polymeric material having a large number of cavities.

While filtration media comprising randomly packed pieces of inert expanded polymeric material having the shape of a small thin-walled cylinder will provide an adequately large surface area for most purposes, the wall thickness of such cylinders will have to be such that the cylinders are capable of withstanding the compressive forces existing in the filtration media. It will be realized that in the lower regions of the filtration media the compressive forces imposed on the pieces will be very high because, apart from having to support the pieces located above them, a considerable weight of aerobic microorganic growth will have to be supported, as high as 3.3 kg/m^2 of surface area available and the weight of the wastewater itself will be considerable.

J. Lunt; U.S. Patent 3,882,027; May 6, 1975; assigned to Imperial Chemical Industries Limited, England has found that by imparting a plurality of grooves or corrugations to the wall of such a cylinder, the rigidity of the cylinder is so greatly increased that cylinders having very thin walls can be used as a filtration medium for treating wastewater.

According to the process a packing medium for biologically treating wastewater in an aerated filter comprises cylindrical pieces of an inert organic polymeric material, the walls of the cylindrical pieces being corrugated. In preference the packing medium is a cylindrical piece in which the wall of the cylinder is corrugated and terminates in planes which are transverse to the main axis of the cylinder.

Alternatively, however, the wall of the cylindrical piece may terminate in planes which are oblique to the main axis of the cylinder. In this case it is preferred that the wall of a cylindrical piece terminates, at each end of a cylindrical piece, in planes which are included at an angle in the range 85° to 65°, and more preferably 85° to 75°, to the main axis of the cylindrical piece.

A corrugated cylinder having ends which are oblique to the main axis of the cylinder has been found to pack more randomly and wetted more readily after being randomly packed than a corrugated cylinder having ends which lie in planes at right angles to the main axis of the cylinder. Even so, such cylinders are generally more difficult and expensive to produce and so, in an art where cost is all important, those cylinders which terminate in planes which are transverse to the main axis of the cylinder are preferred.

Though the cylindrical pieces may be longitudinally corrugated, it is preferred that they are transversely corrugated. Cylinders with plain walls have a greater compressive strength in the direction along the main axis than across the main axis and the formation of transverse corrugations increases the compressive strength across the main axis albeit at the expense of strength along the main axis, the degree being dependent on depth and frequency of the corrugations. In a filter bed or tower, some compression along the main axis is acceptable whereas compression across the main axis may lead to blockage of the filter by accumulation of solids.

While the diameter and length of the cylindrical pieces may be of the same order of magnitude it is preferred that the length of a cylindrical piece is less than its diameter. This has the advantage that it reduces the likelihood of blocking occurring between adjacent pieces in the filter and, in addition, reduces the chance of pieces in the lower region of the biological filter adopting positions such that their main axes are horizontal which might interfere with flow of wastewater through the filter and lead to less complete utilization of available surface and impair satisfactory ventilation of the filter.

Suitable inert polymeric materials for use in the manufacture of these pieces include polyvinyl chloride, polystyrene, polyethylene, polypropylene and polyamide resins. Certain polyester resins, acrylic resins and cellulose derivatives which are sufficiently inert to the particular wastewater can also be used. These polymeric materials may be used in the cellular form to reduce cost without reducing rigidity.

The cylindrical pieces may be produced in any suitable manner. Conveniently the polymeric material is extruded through a suitable die to produce a continuous tube having an uncorrugated wall, the wall of the tube then being deformed by a suitable molding device to provide transverse or oblique corrugations in the wall. The transversely or obliquely corrugated tube is optionally perforated at several positions before being cut up into a plurality of short cylindrical pieces. Alternatively the cylindrical pieces may be produced by injection molding.

An advantage of this packing media over conventional random packing media is that it has a lower bulk density than conventional random media. Consequently the structure of a purification plant employing the new media, and hence its cost, can be reduced. The weight reduction which can be achieved by employing this packing medium also makes easier the construction of taller purification plants which can offer a substantial saving in ground area in congested industrial areas and other places in which land is at a premium. Also the packings will accept high hydraulic loads due to their higher voidage compared with stone, clinker, etc.

Of more significance is the high rigidity offered by the corrugated cylindrical packing media of the process. As an indication of the order of rigidity which can be achieved, a 51 mm diameter, 51 mm long, extruded plain cylinder of rigid polyvinyl chloride having a wall thickness of 0.4 mm required a load (at 20°C) of 0.2 Newton applied along the upper surface parallel to the main axis to give an instantaneous deflection of 1% across the axis of the cylinder. A transversely corrugated cylinder of rigid polyvinyl chloride of the same dimensions but with sine wave corrugations of depth 2.5 mm x pitch 4 mm required a load of 18 Newtons to give the same deflection.

As a further comparison a plain cylinder of rigid polyvinyl chloride of the same diameter and length but of 2 mm wall thickness required a load of 22 Newtons to give the same deflection. A plain wall cylinder of the same dimensions but fabricated from expanded polyvinyl chloride of 60% degree of expansion and of wall thickness 1 mm showed a deflection of 1 mm under a load of 2 Newtons. In each case, loads required to give the same deflection when applied in the same direction as the main axis were much higher.

Example: In this example, the filter medium used were randomly packed cylindrical pieces of polyvinyl chloride. The pieces had an outside diameter of 34 millimeters and a length of 38 millimeters. The wall thickness was 0.40 mm and the wall terminated in planes which were transverse to the main axis of the cylinders. The walls were transversely corrugated, the corrugations being individual and having a pitch of 3 mm, a depth of 2 mm and having a castellated form. The cylindrical pieces were randomly dumped into a square section tower which was 6 ft deep and had a 16 ft^2 cross section.

The filter so formed was used to treat domestic effluent which had been subjected to a primary biological treatment process followed by primary settlement.

In the process the effluent flowed over the surfaces of the cylindrical pieces and bacteria and other microorganisms grew as a slimy film on the wetted surfaces of the pieces by feeding on the organic impurity absorbed from the effluent. In time a balanced community of organisms developed on the wetted surfaces within the filter.

When a balanced state was reached, the filter was capable of removing 90% BOD (biochemical oxygen demand) from a domestic effluent feed having a mean BOD of 197 mg/l which was fed intermittently to the filter at a hydraulic load of 1.73 m^3/m^3/day.

A similar filter was assembled which in the balanced state was capable of removing 81% BOD from a domestic effluent feed having a mean BOD of 176 mg/l which was fed continuously to the filter at a hydraulic load of 3.5 m^3/m^3/day.

Insoluble Porous Polymeric-Phenolic Complexes

M.M. Fein, N.D. Field and E.P. Williams; U.S. Patent 3,954,682; May 4, 1976; assigned to GAF Corporation describe a process for the preparation of phenolic complexes of water-insoluble but water-swellable crosslinked N-vinyllactam or N-alkyl-N-vinylamide polymers in the form of porous beads or granules. The complexes are particularly suited for use in situations where phenolic materials in limited quantities are desired such as, for example, in canisters or columns which are used to treat liquids or water. The complexes are prepared by adsorbing a phenolic material onto the porous granular or porous bead-like crosslinked N-vinyllactam or N-alkyl-N-vinylamide polymers. The following examples illustrate the process.

Example 1: Into a two gallon, stainless steel stirred autoclave the following ingredients were charged: 3.3 grams azobisisobutyronitrile previously dissolved in 660 grams of vinylpyrrolidone; 33 grams Dow divinylbenzene mixture (50 to 60, assay 55%); 2,150 grams distilled water; 10 grams 10% by weight dibasic sodium phosphate buffer solution; 570 grams anhydrous Na_2SO_4 (reagent grade).

The autoclave was then purged of air by applying a vacuum to 25 mm. The vacuum was then released with nitrogen and the procedure repeated two more times. The materials were then heated with stirring (210 rpm, 4 blade turbine) to 65°C over a period of one hour and held at the temperature for a period of three hours at a pressure of 3 to 4 psig. A side flange was removed from the autoclave, and a dispersion of 1.5 grams azobisisobutyronitrile in 50 ml of CP ethanol were added through the port.

The flange was then replaced and the reaction continued for an additional hour at 65°C. The temperature was then raised to 85°C over a period of one-half hour and then held at that temperature for an additional 2 hour period (8 to 10 psig). The reaction product was then cooled and discharged into a Buchner funnel and washed with distilled water to remove any remaining sodium sulfate until a portion of the filtrate was tested with a few milliliters of saturated barium chloride solution and was found to be less turbid than a barium chloride control test with tap water.

After allowing the excess water to drain, the wet porous beads weighed 3,125 grams. A 100 gram portion of these porous beads were then dried in a high-vacuum oven at a temperature of between 65° and 70°C, and the dried porous beads again weighed, so as to determine a 95.1% yield.

Example 2: Into a 500 ml resin flask equipped with thermometer, stirrer, gas inlet, dropping funnel, and condenser connected to gas outlet were charged: 40.0 grams (0.36 mol) distilled vinylpyrrolidone, 0.12 gram azobisisobutyronitrile, 240.0 grams distilled water, 1.6 grams (0.010 mol) methylenebisacrylamide, 0.46 gram 10% by weight Na_2HPO_4 solution and 40.0 grams anhydrous sodium sulfate.

Air was removed by closing the gas inlet and applying enough vacuum at the gas outlet to cause violent ebullition for 30 seconds. The vacuum was then released by allowing nitrogen to enter the gas inlet. This procedure was repeated two more times and then a positive pressure of nitrogen maintained by connecting the gas outlet to a mineral oil blow-outlet having a depth of one inch of oil. The flask was heated with stirring in a constant temperature bath for 4 hours at a reaction temperature of 50° to 65°C. At the end of this time a solution of 0.04 gram azobisisobutyronitrile, 0.4 gram methylenebisacrylamide, 10.0 grams CP ethanol and 10.0 grams distilled water was charged to the dropping funnel and the air removed by violent ebullition and replaced with nitrogen. This procedure was repeated two more times by suitable adapters connected to the dropping funnel.

The above solution was then added to the contents of the flask and the heating continued for 2 additional hours at 62° to 65°C. At the end of this time the beads produced were filtered and washed with one gallon of distilled water and then dried in a vacuum oven at 40°C.

The dried white porous beads weighing 41.5 grams (98.8% yield), were substantially ash-free, whiteness being due to the light scattering caused by the pores within the beads. The porosity was confirmed by microscope examination.

Example 3: Following the procedure of Example 2, 40.0 grams (0.36 mol) distilled vinylpyrrolidone, 0.4 gram azobisisobutyronitrile, 200.0 grams distilled water, 1.2 gram ethylene glycol dimethacrylate, 0.5 gram 10% by weight Na_2HPO_4 solution and 33.0 grams anhydrous sodium sulfate were charged to the resin flask and heated with stirring at a reaction temperature of 50° to 69°C for a total of 5½ hours. The porous beads were filtered and washed with 1 gal distilled water and dried in a vacuum oven at 45° to 50°C. The dried white porous beads weighed 39.0 grams (94.6% yield).

Example 4: 2.96 grams of polymer porous granules prepared according to Example 1 were slurried in a 10% solution of n-heptyl-p-hydroxybenzoate in CP ethanol for 19 hours, filtered and dried for 7 days in a high vacuum at 45°C. It was found that 1.44 grams of n-heptyl-p-hydroxybenzoate had been adsorbed by the porous granules.

Gradient Filter Bed Using Plastic Encapsulated Particles

M. Lerner and G. Erlich; U.S. Patent 3,704,786; December 5, 1972 describes a particulate filter bed in which the particles define a generally uniform gradient of removal ratings of decreasing coarseness from top to bottom, and a method of producing the filter particles. The bed comprises a multiplicity of particles of uniformly decreasing size and increasing specific gravity from top to bottom.

The particles have cores of approximately equal size and specific gravity and outer coatings of varying thicknesses, but equal specific gravity. The specific gravity of the outer coating is substantially less than the specific gravity of the core, so that the overall specific gravity of the larger particles is less than that of the smaller particles. The method comprises encapsulating the higher specific gravity core material with lesser specific gravity material by spray coating.

Figure 1.6a is a cross-sectional view of a filter assembly containing the particulate filter media. The filter assembly shown in Figure 1.6a comprises a tank-like housing **1** having an inlet **2** and an outlet **3**, and a bed **4** of particulate filter media **5** disposed in the tank in the line of flow from the inlet to the outlet. A perforated plate **6** or other suitable means disposed in the tank beneath bed **4** prevents the particles **5** from being carried out by effluent filter flow.

The particles **5** are of randomly varying size ranging from about 0.005" to 0.125" in diameter and are positioned in the tank in generally decreasing order from top to bottom so that filter bed **4** comprises a uniform gradient of decreasingly sized filter flow passages ranging from coarser on top to finer on the bottom.

This is accomplished by employing particles the specific gravities of which are inversely proportional to their size, the smaller particles having greater specific gravities than the larger particles. The specific gravity differential between particles of different sizes effectuates through the force of gravity a generally uniform stacking effect of larger particles upon smaller particles.

There are, of course, in any random distribution of particle sizes groups of particles having the same size and these particles will tend to be arranged at the same vertical level. However, the number of equal size particles is preferably small, so that, the formation of definitive layers or strata particularly in the central portions of the bed **4** is kept to a minimum.

The gradient of decreasingly sized flow passages through which the flow proceeds within bed **4** greatly increases the contaminant removal efficiency over conventional filter beds having commingled particles of varying size or stratified layers of varied size particles. Since the uppermost particles are largest, there is little tendency for the top of the bed to become prematurely plugged with large contaminant particles. Likewise, small contaminant particles which would ordinarily pass through a filter bed are entrained in the lower portions of the bed.

The cleanability of the filter bed of the process through backwashing is also very good. When the bed is subjected to reverse backwash flow the flow passages through which the entrained contaminant particles must pass to be discharged increase in size as the contaminant particles rise in the bed, thus easily releasing them.

Figure 1.6b is a cross-sectional view of several coated particles and their relationship to one another. As shown in Figure 1.6b the particles **5** of the filter bed **4** comprise an inner core **10** of silica sand and an outer coating **12** of thermoplastic resin encapsulating the core. All of the cores **10** are of approximately equal size, while the outer coatings are of varying thicknesses. In this manner, since the specific gravity of the thermoplastic coating is approximately one-half that of the sand, the overall specific gravity of each particle **5** is inversely proportional to its size, the larger particles having a lesser specific gravity than the smaller particles.

FIGURE 1.6: PARTICULATE FILTER MEDIA

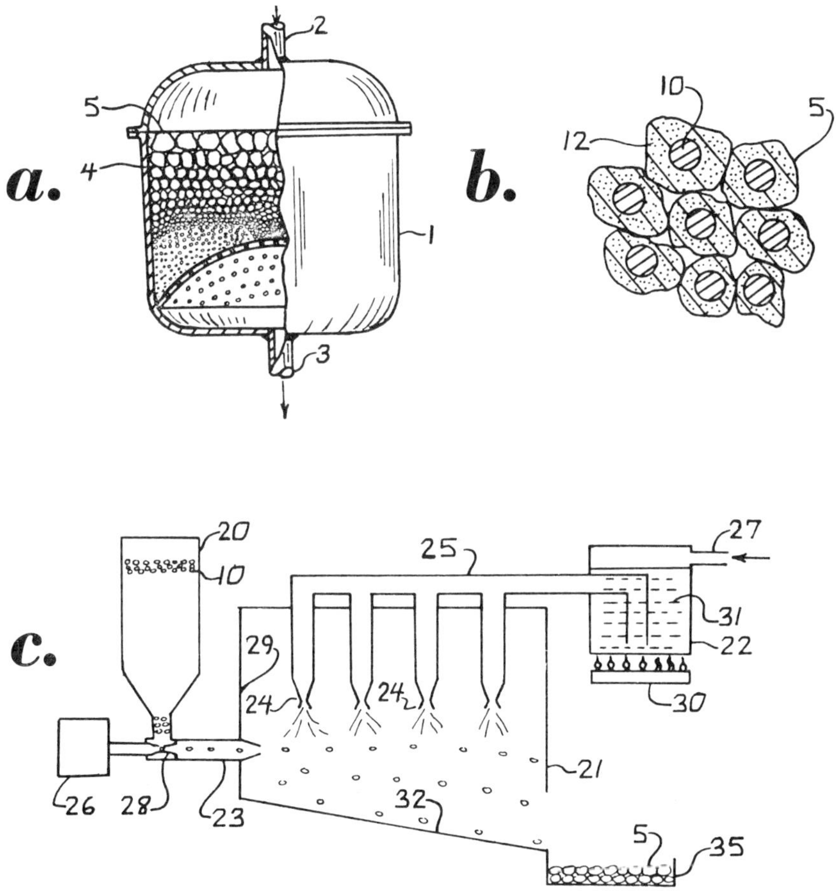

Source: U.S. Patent 3,704,786

As an alternative to encapsulating every filter particle in bed **4**, the smallest particles can comprise merely uncoated cores **10**. Since these particles will necessarily have the highest specific gravity in the bed, they will sink to the lowermost portion thereof. Similarly, the largest particles in the bed can be plastic or other material of approximately the same specific gravity with no core. Although thermoplastic material is preferred for the outer coating because of its ability to be applied to the core, where no core is employed both thermoplastic and thermosetting materials are suitable.

Figure 16c is a view of the preferred coating means for effectuating the formation of the filter media particles. The particles **5** can be formed by using this apparatus. Core material **10** is contained in hopper **20** which is connected at its lower end to a nozzle **23** by means of a venturi **28**. Nozzle **23** is disposed through the wall **29** of a chamber **21** in which particles **5** are coated. A source of high pressure air **26** connected to venturi **28** is adapted to provide sufficient air velocity to project particles **5** from the nozzle **23** into chamber **21**.

Thermoplastic coating material **31** in its liquid state is stored in sealed reservoir **22**. A plurality of spray nozzles **24** disposed at the top of chamber **21** communicate with reservoir **22** by means of manifold **25**. A gas burner or the like **30** disposed beneath reservoir **22** by means of manifold **25**. A gas burner or the like **30** disposed beneath reservoir **22** is provided to heat the thermoplastic material **31** and thereby maintain its liquidity. High pressure air introduced by means of line **27** pressurizes the reservoir to provide sufficient motive force for the thermoplastic material to be sprayed into chamber **21** through nozzles **24**.

In operation, large quantities of particles **10** are projected into chamber **21** simultaneously with the spraying therein of liquid thermoplastic material. As the particles **10** are subjected to the thermoplastic spray, they become coated. The core particles are maintained at a relatively cool temperature, so that the coating material solidifies almost immediately upon contact. The thickness of the coating built up on the cores is dependent upon the length of time they are in contact with the spray. Those particles which are projected across the entire chamber will generally have thicker outer coatings than those particles which fall immediately upon entering the chamber.

Although the core particles **10** are all approximately the same size, they are preferably of irregular shape. It is this factor together with the interference of one particle with another that determines at what point within chamber **21** a particle will fall out of contact with the thermoplastic spray. The falling point of a particle within the chamber is generally random. Accordingly, the overall thicknesses of the thermoplastic coatings on the cores are randomly distributed.

As the coated particles fall they are discharged from the chamber via inclined lower surface **32** to container **25**. The particulate filter media in container **25** can be placed directly into a filter housing as shown in Figure 1.6a. Since the larger particles have a lesser specific gravity than the smaller particles, the force of gravity and the initial filter flow tends to arrange the particles in a gradient of decreasing size from top to bottom.

FIBROUS FILTERS

NONWOVEN FABRIC

Sewn Fabric Filter

C.A. Lee; U.S. Patents 3,506,133; April 14, 1970; and 3,535,191; October 20, 1970; both assigned to Appleton Wire Works Corporation describes a nonwoven filter fabric formed by sewing a layer of substantially straight and parallel laterally independent rack filaments together by sewn filaments stitched in closely spaced parallel rows transversely across the rack filaments.

A nonwoven filler is sewn to the rack filaments preferably using different sewn filaments on each side of the filler to provide different qualities for the two sides of the resulting fabric. The rack filaments, sewn filaments and filler are then bonded together, as with a resin. The filler is composed of randomly disposed filler filaments providing a predetermined permeability, porosity and void opening which give the resulting fabric qualities desirable for its use as a filter medium, the rack filaments providing tensile strength.

Figure 2.1a is a plan view of a piece of sewn fabric; Figure 2.1b is a cross-sectional view of the fabric taken along section **2–2** of Figure 2.1a; Figure 2.1c is a cross-sectional view of the fabric taken along section **3–3** of Figure 2.1a; Figure 2.1d is an illustration of one form of apparatus for making the fabric; and Figure 2.1e is an illustration of a modified form of apparatus for making the fabric in endless form. In the process as illustrated in Figures 2.1a, 2.1b and 2.1c, the fabric **10** is formed of a plurality of relatively straight and parallel rack filaments **12** lying substantially in a sheet or layer in which is disposed a filler **14** which is sewn to the rack filaments **12** by sewn filaments **16** and **18**.

Only a portion of the fabric is illustrated in Figure 2.1a. The fabric ordinarily takes the form of a sheet, that is, it is thin relative to its width and length. However, although the rack filaments are described as being relatively straight or parallel and lying in a sheet or layer and forming a sheet-like fabric, this includes the formation of an endless belt even though the latter requires that the sheet itself be curved and rot entirely planar. As shown, the filler **14** is nonwoven and is preferably formed of filler filaments **20** of predetermined diameter and length depending upon the particular application to which the fabric is to be put.

FIGURE 2.1: SEWN FABRIC FILTER

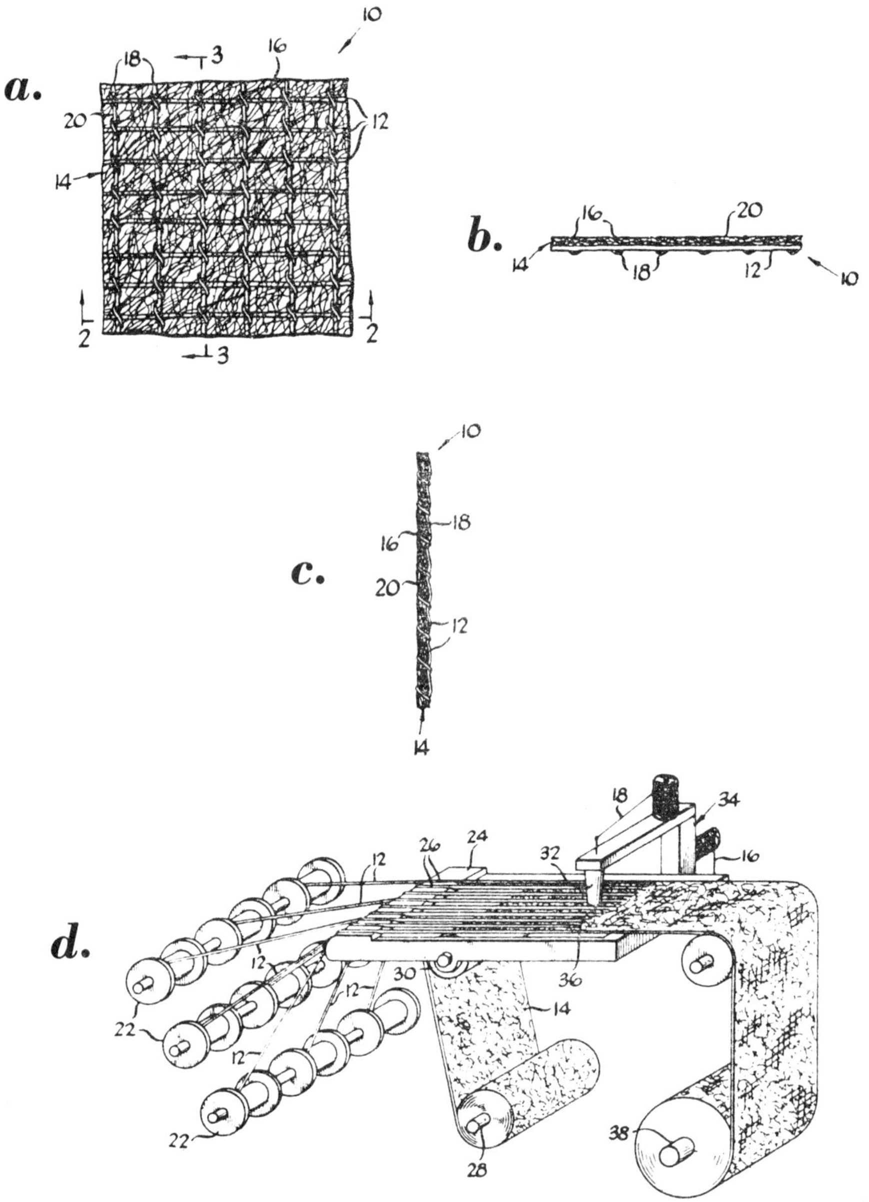

(continued)

FIGURE 2.1: (continued)

e.

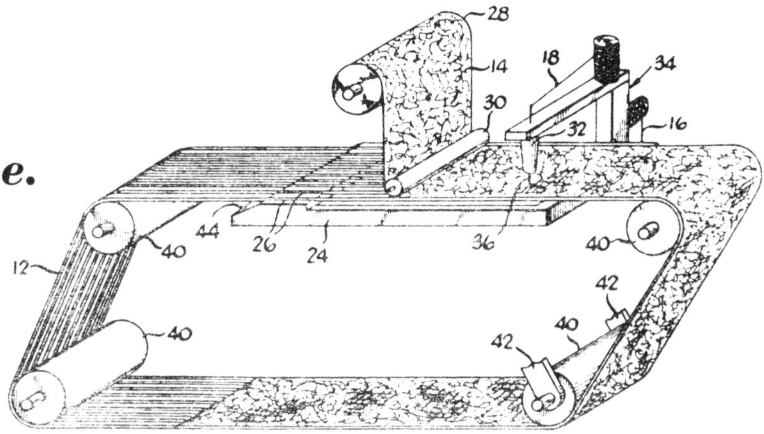

Source: U.S. Patent 3,506,133

In general it is preferable that the filler filaments be long relative to the space between rows of sewn filaments **16, 18** in order that filler filaments bridge the gap between sewn filaments and add their strength to the fabric.

In the fabric illustrated in Figure 2.1a, a lock stitch was employed to sew the filler to the rack filaments; however, other stitches can be employed. In using a lock stitch, it is frequently desirable, as shown, to utilize different threads on the respective sides of the rack filaments. As illustrated in Figure 2.1a, the sewn filaments on the filler side of the rack filaments are relatively fine whereas the sewn filaments **18** on the other side of the rack filaments are relatively coarse.

Further, the sewn filaments are, as shown, relatively straight, providing certain advantages in some applications. Two-sidedness may be provided by stitches other than the lock stitch. For example, a two filament chain stitch can be used with different sewn filament materials on the respective sides of the fabric. Of course, in many instances it is quite satisfactory to use the same material and filament size on both sides, in which case a single filament chain stitch can be used as well as any of the two filament stitches.

In Figure 2.1d is illustrated one way in which the fabric of Figures 2.1a, 2.1b and 2.1c can be made. The rack filaments **12** may be supplied by respective reels **22**. The rack filaments are passed over a rack **24** which may have a plurality of slots **26** or other means for guiding the relative position of the respective rack filaments. The filler **14** may be supplied, as shown, from a supply drum **28** in the form of a sheet. This sheet is directed, as by a roll **30**, to place the filler sheet **14** in contact with the rack filaments **12**. A sewing machine head **32** is mounted on a bracket **34** which is moved by motive means not shown so as to traverse the rack **24** transversely of the rack filaments **12**. The sewing machine head carries sewn filament **18** to the needle **36**. At the same time, a bobbin is mounted on bracket **34** beneath the rack **24** to supply the sewn filament **16** so that as the sewing machine head **32** traverses the rack filament sewn filaments **16** and **18** lock together around the filler and rack filaments, thus sewing the filler to the rack filaments. Preferably, the traversal of the sewing machine head and its stitching actions are programmed so that the stitches occur between rack filaments **12** rather than at a rack filament, thus avoiding piercing the rack filaments and reducing their strength. Piercing of the rack filaments may also be avoided by using a needle too dull to pierce the filaments. This is particularly effective when the rack filaments are monofilaments, i.e., single rods.

The rack filaments and filler are advanced after each traversal of the sewing head by a predetermined distance to provide the desired spacing between rows of stitches. The completed fabric is then taken up on a take-up reel **38**. Conventional braking means is applied to the respective reels **22** to resist the advance of take-up reel **38** and, therefore, keep the rack filaments under suitable tension so as to keep them substantially straight and parallel in the rack during the sewing operation.

After the filler **14** has been sewn to the rack filament **12**, it is frequently desirable and sometimes necessary to treat the fabric further. In many applications, it is desirable to treat the fabric with suitable resin acting to bond the various filaments together. If the fabric is to be used for filtering, care must be taken not to block or blind the passages through the fabric.

For some other uses, sufficient resin may be added to block all of the openings including the holes made in sewing. The needle holes may also be reduced or eliminated by shrinking the filler about the needle holes. This may be achieved by applying a controlled amount of heat, which heat may be applied during the curing of the resin.

There are various alternative ways of making the fabric. For example, the rack filaments may be looped around pegs at the end of a rack, and the fabrics sewn as individual sheets rather than in long strips as described in connection with Figure 2.1d. Alternatively, it is frequently desirable to make an endless belt as for a Fourdrinier wire. In such instances, the fabric may be made by apparatus as shown in Figure 2.1e. In this case, the rack filaments **12** are disposed endlessly about a plurality of rolls **40**.

At least one of the rolls is adjustable, as by movement of brackets **42**, to make the loop of rack filaments **12** the appropriate length and to maintain appropriate tension in the respective filaments. In disposing the rack filaments endlessly, it may be desirable to splice them individually, in which case they are preferably spliced at different positions along the length of the endless belt in order that the belt will not be particularly weakened at the same place for each rack filament. Alternatively, the rack filaments may be formed by a single strand wound helically around the rolls so as to minimize splicing.

In the arrangement shown in Figure 2.1e, the filler **14** is supplied from the roll **28** to the top of the rack filaments. Other than as occasioned by these structural modifications, the apparatus of Figure 2.1e may be operated substantially as described above in connection with Figure 2.1d, and the fabric may be finished in a similar manner. A sloping forward surface **44** may be formed in the rack **24** so as the previously sewn part of the endless fabric approaches the sewing head **36**, the sewn filaments ride against the surface **44**, lifting the rack filaments out of the slots **26** and permitting the fabric to advance until the entire fabric is sewn.

In addition to the advantages of the fabric as made with the apparatus shown in Figure 2.1d, the endless belt made in this fashion has an additional advantage over woven belts. It does not shrink in the process of manufacture. That is, the rack filaments are held straight during the process of manufacture and are not deformed by sewing, whereas in the process of weaving, the warp and weft yarns are bent around each other, thus shortening them. In the manufacture of the fabric, the rack filaments **12** may be formed into endless loops of length in the sewing process.

Fabrics made as described above can be used in a number of different applications, many of them requiring quite different properties for the fabric. For each application, there are particular desirable properties for the fabric. These properties may be provided by appropriate selection of the rack filaments, sewn filaments and filler. For example, as a filter medium the permeability, porosity and mean and largest void openings of the fabric should be different for different filtering operations. These properties can be effectively controlled by control of the filament diameter and filament length of the filler filaments **20** and by the thickness of the filler **14**.

The strength of the fabric is largely determined by the number, diameter and distribution of the rack filaments and by the particular material from which the rack filaments are made. For certain purposes, it is desirable that the fabric have more strength in certain areas, such as at the edge or in the center, in which event the rack filaments can be concentrated in those areas, or filaments of different diameter used. It is also within the scope of this process to lay strength filaments across the rack filaments and sew them to the rack filaments to strengthen the fabric in the cross direction.

The abrasion resistance of the fabric may be controlled by control of the sewn filament diameter, the material of which the sewn filaments are made, and by spacing of the sewn filaments and the sewing stroke. At the same time, other surface qualities of the fabric may be achieved. For example, finer sewn filaments may be used to provide a smooth fabric.

Example 1: With the above criteria in mind, the example of a filter medium can be considered. In particular, a filter medium designed for filtering paper mill effluent will be considered. In such filters, it is desirable to remove as much of the solids as possible while handling a substantial volume of water. For example, in one operating mill it is desirable to filter 50,000 gallons of effluent per minute while removing about 100 tons per day of solid.

It has been the practice to utilize virgin pulp as the filter medium. Typically, such filters removed about 0.1 pound of solid per square foot per minute while passing about 2 gallons of filtrate per square foot per minute at a pressure differential of about 6 inches of mercury. Typically this left in the filtrate about 0.25 to 1.4 pounds of solids per 1,000 gallons of filtrate.

A fabric designed pursuant to this process has proved to retain 0.15 pound of solids per square foot per minute while passing 6 gallons of filtrate per square foot per minute at the same pressure drop of 6 inches of mercury. At the same time, this filter medium of the process passed only about 0.05 to 0.06 pound of solids per 1,000 gallons of filtrate. Thus, the fabric of the process in the design tested, removed 50% more solids while passing three times as much water and left the filtrate with less than a fourth to a twentieth as much solids.

The filter medium thus tested was made with filler **14** made from very fine filler filaments **20** formed of Dacron polyester material. These filaments were prepared in a sheet form known as Spunbonded. The diameter of the filaments **20** was about 1.25 inches. The filler was used in sheets about 10 mils thick in which the fibers were bonded by resin.

The filler **14** was placed on rack filaments **12** made of Dacron of 660 denier distributed uniformly with 16 filaments per inch. The sewn filaments **16** over the filler were 5.5 mil monofilaments of nylon. The sewn filaments **18** on the other side of the rack filaments were substantially heavier being 440 denier of Dacron. The rows of stitches were uniformly distributed at 12 rows per inch, and the stitches were placed discretely between the rack filaments at 16 per inch. All of the material in this filter fabric was nonabsorbent of water.

The improvement in filtering properties lies both in increased permeability and in smaller void openings. This is partly occasioned by the use of fine, long filler filaments. That is, the finer fibers permit a larger total void space (permeability) while making the individual void spaces smaller. The filter therefore permits more water to pass through while stopping smaller particles.

Example 2: Another example of fabric made according to the process is a Fourdrinier wire or forming fabric. Such fabrics are formed as endless belts with the rack filaments **12** running in the belt direction to provide the necessary tensile strength to drive the fabric around the papermaking machine. The filler **20** is designed to provide the drainage necessary for the particular papermaking machine and, together with the top sewn filament **16**, provides a smooth upper surface on which to form the paper web.

At the same time, the sewn filament **18** inside the loop of the fabric **10** is made abrasion resistant to preserve the rack filaments from wear as the fabric passes over the machinery, as for example, over the usual suction boxes. As in the case of the previous example, all of the material is nonabsorbent of water.

A typically Fourdrinier fabric is made as follows according to the process: The filter material is made of Dacron filaments prepared in the Spunbonded sheet form. The rack filaments are Dacron filaments of 220 denier distributed uniformly at 32 filaments per inch. The top sewn yarn is 220 denier Dacron and the inside sewn filament is 150 denier Dacron. The sewn filaments are stitched at rows uniformly distributed with 32 rows per inch and with the stitches discretely between rack yarns at 32 stitches per inch.

Such Fourdrinier fabric provides a more rugged fabric than has previously been formed by weaving with filaments of comparable weight. At the same time, the drainage properties are improved by use of the nonwoven filler, and the use of fine filler filaments and fine upper sewn filaments provides a smoother surface upon which to form the paper web than has been previously achieved with Fourdrinier fabrics of comparable drainage rate and strength.

Varying Density Cartridge Filters Using Spray Spinning

According to a process described by *C.M. Vogt, J.W. Soehngen and J.C. Polise; U.S. Patent 3,904,798; September 9, 1975; assigned to Celanese Corporation* polypropylene or other fiber-forming polymer is melt extruded as a continuous molten stream which is attenuated by a plurality of gaseous streams into a fine filament and projected onto a rotating collection surface, such as a mandrel.

During collection there is changed, one or more times, at least one of (1) temperature of the filament-forming material, (2) the speed of extrusion, (3) the speed of rotation of the collector, (4) the distance between extruder and collector, or (5) the weight of an idler roll on the collector, thereby to form an annular layer whose density differs relatively from the layer immediately preceding it.

The resulting cylindrical, self-bonded, nonwoven structure has a predetermined, density profile. For uses such as filtration, the fluid to be filtered enters from the low density side so that solids will deposit throughout the filter more or less uniformly rather than building up so much in the first-contacted layer that it becomes fully clogged before the balance has had any appreciable solids buildup.

Figure 2.2a is a schematic illustration of one extrusion and collection apparatus. Figure 2.2b is a schematic plan view of the preferred extrusion apparatus and process. Referring to Figure 2.2a, a fiber-forming, thermoplastic polymer, preferably a polyolefin, is fed to an extruder **10** provided with an adapter section **12** to which a gas, such as steam or air, is supplied. While extrusion temperatures may be anywhere above the melting point of the polymer, it has been found that best results are obtained by heating the polymer to at least 150°C, and preferably from about 250° to 350°C above the softening point of the polymer being extruded.

For example, polypropylene having hereinafter defined characteristics will generally be heated to temperatures of from about 325° to 400°C. Polyethylene, on the other hand, will be heated to from about 350° to 450°C. A hot, molten stream of polymer **16** is discharged through a nozzle **14**. It is to be understood that nozzles having one or more polymer orifices may be used. Also, a plurality of nozzles per collector may be employed. However, there must be at least two planar gas streams per polymer orifice. The attenuating gas orifices **18** are of an elongated rectangular cross section, to emit substantially planar gas streams **17**.

The gas streams act on the polymer stream **16** in convergence region **20** to form an attenuated filament **22** wherein it cools and partially solidifies while moving toward a mandrel **24** on which it is collected as an annular cartridge **26**. The mandrel is ordinarily rotated at a speed sufficient to provide a moving surface of from about 25 to 125 feet per minute by a motor drive. Mandrel **24** is in surface contact with roller **28**, which acts as an idler roll and whose bias against the mandrel can be adjusted; the extent of the bias will affect how tightly the tacky filament packs against previous layers on the cartridge **26**.

Both the mandrel and the roller are reciprocated laterally by a traversing mechanism **30** whose throw determines the shape of the cartridge; the throw may be of constant length or may change in the course of package buildup to produce a particular shape as may be needed for acceptance in a receptacle of predetermined corresponding shape.

FIGURE 2.2: VARYING DENSITY CARTRIDGE FILTERS

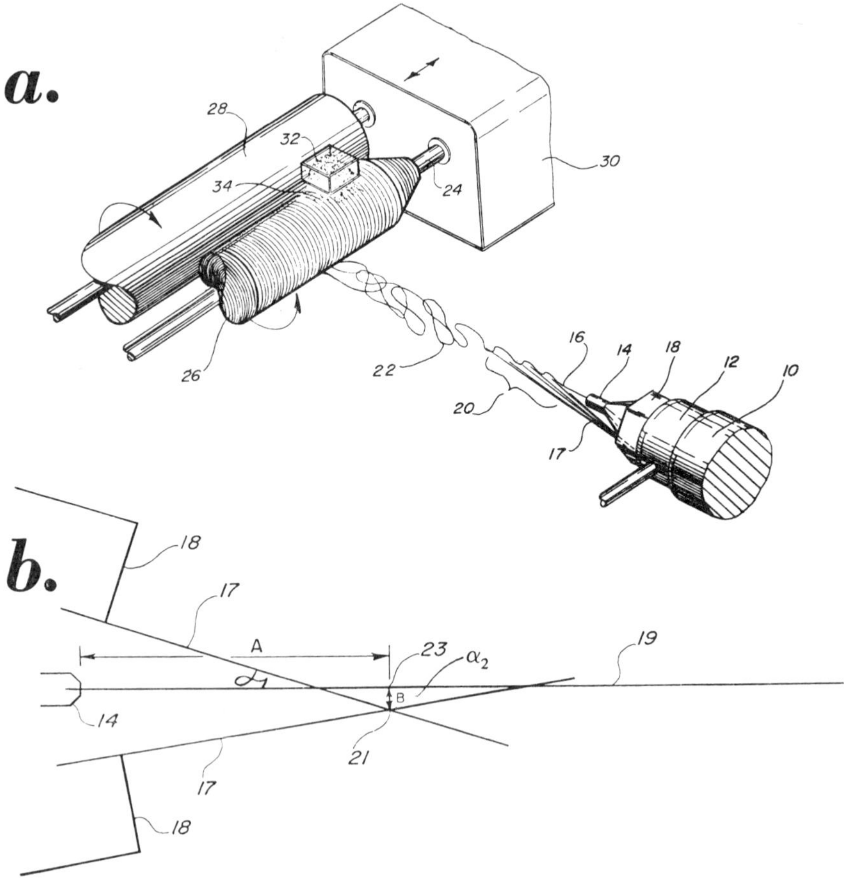

Source: U.S. Patent 3,904,798

The force of the attenuating gas on the polymer stream causes the polymer to attenuate greatly, e.g., from 10 to 500 times, based on diameter ratios, and possibly fibrillate to a slight degree to produce a substantially continuous fiber. Some turbulence and resultant whipping about of the polymer stream occurs, consequently, a generally random, stereoreticulate structure of fiber results as the material impinges on the collector.

Since the polymer is still in a somewhat molten or tacky state when it strikes the collector, some sticking together occurs at the points where fiber intersects. For brevity, this sticking will be referred to as interfiber bonding, although it is to be understood that this bonding will ordinarily result from an individual fiber looping about and sticking or bonding to itself.

For best results, the surface of mandrel **24** should be from about 6 to 48 inches, preferably 10 to 30 inches from polymer exit nozzle **14**. With greater distances the spray pattern is difficult to control and the resultant web tends to be nonuniform. Shorter distances result in a web which contains too great a quantity of shot, i.e., beads of nonattenuated polymer, which undesirably affects subsequent processing, web uniformity and surface area. The bias of roller **28** against cartridge **26** may be varied during operation, or the spacing between the cartridge and nozzle **14**, the rate of polymer extrusion, the force and quantity of gas, or the like.

A hopper **32** is positioned above the mandrel **24** and powdered material **34** such as carbon is dropped onto the cartridge during all or part of the cycle. Since the cartridge traverses and rotates, the carbon can be uniformly distributed even though the hopper is stationary. Substantially all of the fibers comprising the cartridge will have diameter falling within the range of 0.5 to 50 microns, and under preferred conditions, within the range of about 1 to 20 microns. The density at any given part of the cartridge filter is less than 50% of the density of the polymer substrate. Thus when polypropylene having a density of 0.9 g/cc is spray spun, the density at any given part of the cartridge filter will be less than 0.45 g/cc. Preferably the density will range from 0.1 to 0.5 g/cc.

In Figure 2.2b there is schematically shown a top view of the preferred apparatus for producing cartridge filters. A plurality of converging substantially planar gas streams **18** (which correspond to planar projections of the gas outlet passages) issue from substantially rectangular gas outlet passages **17**. The axis **19** of the nozzle 14 corresponds to the direction in which the polymer stream is extruded. The gas jets **17** are positioned along side the extrusion nozzle **14** in such a manner that the gas streams are directed substantially in the direction of flow of the polymer extrudate along the nozzle axis **19**. The planes of the gas streams (and projections of the gas outlet passages) intersect at a point **21** which is at a distance **B** measured perpendicularly from intersection point **21** to the nozzle axis **19**.

The distance **B** is at least equal to the diameter of the extrudate stream at a point **23** along the nozzle axis in juxtaposition to the point of intersection **21**. Preferably **B** is at least 0.06 inch, most preferably from about 0.2 to 2.0 inches. The point **23**, which defines the perpendicular distance from the nozzle **14** to the intersection point, is a distance **A** of at least 2.0 inches from the point of extrusion nozzle **14**, preferably from about 2.5 to 7.0 inches. The attenuating gas jets **17** are positioned alongside the extrusion nozzle such that the planes of the attenuating gas streams **18** intersect the nozzle axis **19** (also the axis of the extrudate stream) at an angle (α_1 and α_2) less than 45° to more than 5°, preferably from about 10° to 40°, to project the extrudate stream away from the extrusion nozzle.

Example 1: Isotactic polypropylene having a molecular weight of about 350,000, a density of about 0.9 g/cc, a melting point of about 160°C and a degree of crystallinity of about 61% is spray spun through an orifice having a substantially circular cross section of about 0.016 inch diameter in accordance with the Example of U.S. Patent 3,543,332, i.e., three steam passages discharging steam at 350°C and 40 psig at a skew angle of 2° and an angle of convergence of 12° toward a filament of 0.03 inch diameter polypropylene extruded at 350°C at a rate of 2.5 lb/hr. The steam passages are 0.082 inch in diameter and are set off 0.290

inch from the axis of the polypropylene orifice where the steam exits. The polypropylene is extruded as an expanding cone having a diameter of about 2 inches just prior to collection on a metal drum. The extrudate is collected on the metal drum having a diameter of 1 inch and rotated at a speed of 100 rpm (a surface speed of about 7.7 meters per minute), the drum being unheated and initially spaced 6 inches from the extruder outlet. The product is collected as a 26 inch wide band with a weight of 4 kg pressing down on the outer rotating surface to densify the structure.

After 10 minutes of extrusion and collection, the extrusion nozzle is moved back so it is 8 inches from the collector, the extrusion rate is reduced to 2 pounds per hour, and the weight on the collection roll surface is reduced to 2 kg. After 10 minutes the nozzle is moved back 10 inches from the collector, the extrusion rate is reduced to 1.5 pounds per hour and the weight is reduced to 1 kg. After another 10 minutes the extrusion rate is reduced to 1 pound per hour, the weight is altogether removed and extrusion is continued at the same 10 inch spacing between extruder and collector for 22 minutes more; extrusion is then discontinued.

For test purposes, the annular cylindrical mass is removed from the collector and a thin disc sawed off one end. Filaments in the disc range in denier from about 1 to 30 and in tenacity from about 0.3 to 1 gram per denier. The product is substantially free of shot.

Individual density measurements were made on the various sections of the filters outward along the radius. Several slices of approximately 0.3 inch each were taken off one end of each blank. Three concentric rings were cut from each slice with a special set of cork borers. The difference between the inside and the outside radius of each ring is approximately ¼ inch. The volume of each doughnut ring section was calculated from the various measurements on each section. Densities of the rings were calculated from their weights and calculated volumes. The density of the innermost, central and outermost rings was about 0.287, 0.230 and 0.220 gram per cubic centimeter, respectively.

Example 2: Using an apparatus such as shown in Figure 2.2a, polypropylene having a density of about 0.9 is spray spun at a melt temperature of 355°C onto a 1 inch diameter collection mandrel 6½ inches away from the nozzle tip. Initially, the idler roll is weighted with 8,000 g. After 15 minutes collection, the weight is changed to 6,000 g and after another 15 minute collection period the weight is changed to 4,000 g. The weight is finally changed to 2,000 g after another 15 minute collection period and collection is continued for an additional 28 minutes. The final filter is 2¾ inches in diameter and has apparent densities of 0.348, 0.265 and 0.210 g/cc respectively on the outside, middle and inside thirds of the cross sections. The filter is quite strong and exhibited good self-bonding.

Example 3: A series of filters are prepared at 354°C by spray spinning polypropylene having a density of about 0.9 onto a 1 inch diameter collection mandrel. The distance between the tip of the spinning nozzle and the collection mandrel is changed during the filter cartridge's preparation in a programmed manner. Also, the weight being applied to the idler roll is changed in a programmed manner along with collection times. Results obtained on several filter cartridges are summarized in the accompanying table. It can be seen from these runs that the apparent density of any section of the filter can be varied to any desired value. In addition, the reproducibility of the process is excellent as observed in the last two cartridges.

Ex. 1 Run	o.d. (in)	1st W*	1st D**	1st T***	2nd W	2nd D	2nd T	3rd W	3rd D	3rd T	4th W	4th D	4th T	Total Collection Time (min)	Filter Weight (g)	Apparent Density (g/cc) 4th Outside	Apparent Density (g/cc) 3rd & 2nd Middle	Apparent Density (g/cc) 1st Inside
A	2.75	8	4	15	4	6	15	2	8	15	1	10	30	75	528	0.229	0.280	0.363
B	2.75	8	7	15	4	9	15	2	11	15	1	13	12	57	405	0.213	0.240	0.290
C	2.75	8	5	15	4	6	15	2	7	15	1	8	38	83	598	0.335	0.317	0.332
D	2.75	8	6	15	4	8	15	2	10	15	1	12	16	61	450	0.217	0.258	0.285
E	2.75	8	6	15	4	8	15	2	10	15	1	12	16	61	450	0.212	0.254	0.283

*W = weight on idler roll, kg.
**D = distance from spray nozzle to collection roll, inches.
***T = time for collection, minutes.

The spray spun cartridge filters of this process are also useful as absolute filters, carburetor air filters and filters for cryogenic fluids such as air, nitrogen, oxygen, hydrogen, and the like, both during production and use where the temperatures are less than about 0°C and usually below about -100°C. Such filters can easily be formed or cut into whatever sizes or shapes are desired, including annular cylinders, and because they can yield under strong pressures, they can be peripherally clamped, if necessary, to provide a tight seal and prevent liquid bypass.

Aromatic Polyamide Fibers

According to a process described by *I.J. Gusman and C. Strack; U.S. Patent 3,937,860; February 10, 1976; assigned to J.P. Stevens & Co., Inc.* a high temperature resistant filtration material or felt is produced which does not have to be calendered to achieve the desired air permeability, thereby assuring permanence of product density during use and continued high filtration efficiencies. Additionally, the product has good flex strength and, thus, can be utilized for a relatively long time.

The layers of the high temperature resistant filtration material or felt are needled together by one or more passes through a needle loom. All the layers can be assembled prior to the first needling operation or the layers can be added in stages in a multistep needling operation. The two outer layers of the composite are formed of nonwoven fabric or batting of an aromatic polyamide known as Nomex, which is composed of recurring units of the formula

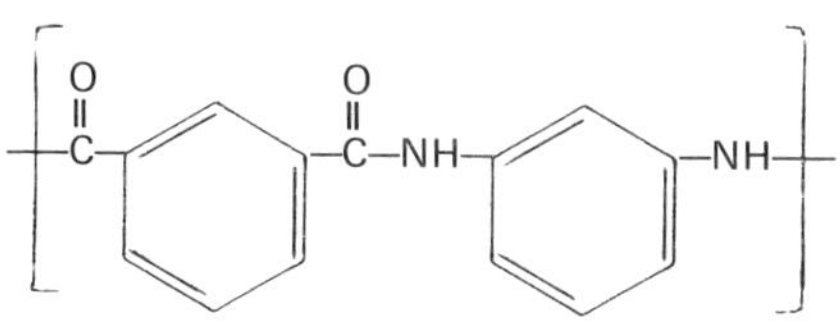

The fibers should be of fine denier, not exceeding five, with the denier of the fiber preferably being in the range of 1.5 to 3. The weight of each Nomex layer should preferably be in the range of 1.5 to 10 oz/yd^2. The nonwoven glass layer should be composed of fine denier fiberglass having a maximum denier of about 5. The weight preferably should be in the range of about 1 to 10 oz/yd^2. The woven fabric or scrim layer is made of yarn of either Nomex polyamide or glass. Preferably, the weight of the fabric is within the range of 0.75 to 10 oz/yd^2.

Example: The nonwoven Nomex materials used for the outside layers were carded cross-laid webs of 2 denier Nomex fibers having an average length of about two inches. Each of the webs weighed about 3.5 oz/yd². The Nomex woven fabric or scrim was made of single ply yarn which was spun from 2 denier, 2 inch fibers and which had a yarn number of 8.5 based on the cotton system (American hank). There were 18 yarns per inch in both the warp and fill directions.

The glass woven fabric or scrim was made of continuous glass filaments of diameter size G of electrical grade E glass and had a yarn number of 37 (glass system). There were 14 yarns per inch in the warp direction and 13 yarns per inch in the fill direction.

The nonwoven glass fabric or batt was composed of fine denier glass fibers lightly bonded with a thermosetting resin. A filtration material was made using 5 passes through a needleloom. In the first pass, a layer of the nonwoven Nomex was needled into a layer of the Nomex scrim using the following needleloom settings: strokes/minute = 582; penetrations/inch² = 380; needle penetration depth, ¾ inch; manufacturing speed, 12 ft/min (take-up roll advance).

The scrim side of the composite obtained from the first pass was covered with a layer of the glass batting which in turn was covered with a layer of the nonwoven Nomex material and this assembly was needled together by a pass through a needleloom using the same settings as in the first pass. The product was then further consolidated by three more passes through a needleloom at the following settings:

Needleloom Settings	3rd Pass	4th Pass	5th Pass
Strokes/minute	650	650	650
Penetrations/inch²	550	550	550
Needle penetration depth, inch	⅝	⅜	3/16
Manufacturing speed, ft/min (take-up roll advance)	9.6	9.6	9.6

A second product was then made by the above procedure except that the above-described glass scrim was substituted for the Nomex scrim. The properties of the two products are set forth in the following table:

	Product 1	Product 2
Nomex (top and bottom layers), oz/yd²	3.5	3.5
Nomex scrim, oz/yd²	3.5	-
Glass media, oz/yd²	1.5	1.5
Glass scrim, oz/yd²	-	4.5
Thickness, inches	0.092-0.097	0.087-0.091
Mullen burst, psi	345-375	210-255
Air flow, cfm/ft², 0.5" H_2O Δ pressure	27-30	24-26
Thermal stability, % shrinkage 24 hr at 450°F		
Machine direction	1.0	0.0
Cross direction	1.0	0.5
Tensile strength 2" strip, lb		
Machine direction	134	102
Cross direction	134	167
Elongation at 100 lb, %		
Machine direction	13	78
Cross direction	18	33

The filtering efficiencies of the two products are set forth in the following table.

Dust Particle Size, microns	- - - - -Filter Efficiency, %- - - - -	
	Product 1	Product 2
0.5	95	84
1.0	99	99
2.0	99	100
3.0	100	-

Note: These samples were all tested with a ΔP of 3.6" of water.

Twisted Strand Fabric

C.A. Butler and D.R. Lazzaro; U.S. Patent 3,510,005; May 5, 1970; assigned to The Lindsay Wire Weaving Company describe a porous structure which is comprised of a plurality of polymeric strands which are each twisted a predetermined number of turns per unit of length in a generally circular direction. The strands being of a predetermined transverse dimension and having a generally noncircular shape are disposed in contiguous side-by-side relationship to provide spaced points of contact. An elongated absorbent element containing a predetermined amount of adhesive material is disposed longitudinally immediately adjacent the strands and in overlying relationship relative to the points of contact therebetween.

Upon application of pressure, a predetermined amount of the adhesive material contained within the absorbent elements is exuded or released and deposited at the spaced points of contact, thereby to bond the elements together into a unitary, porous structure. Moreover, by a relative correlation between the construction and arrangement of the individual twisted strands and the mode of bonding the strands together, there is achieved optimum hydraulic characteristics in the porous structure without impairment to its strength characteristics. Furthermore, by this relative correlation, there is provided the optimum desired permeability characteristics in the porous structure resulting from an effective and selective control of the porosity, specific surface and tortuosity in the structure.

Referring to Figures 2.3a and 2.3b, there is illustrated one form of the porous structure, designated generally at **1**. As shown, the structure of **1** includes a plurality of twisted monofilaments or strands **2** comprised of polymeric material, such as plastic or the like. The strands are disposed in a generally common place and are attached together in substantially parallel, side-by-side relationship to provide a nonwoven, single layer, porous structure.

Prior to being attached together, the individual strands are each twisted about their longitudinal central axis a predetermined number of turns for a given unit length in either a clockwise or counter-clockwise direction. As best illustrated in Figure 2.3b, it is preferred that certain strands in the layer be twisted in one direction while selective other strands are twisted in the opposite direction. By this arrangement, adjacent strands are twisted in the opposite direction relative to one another while alternate strands are twisted in the same direction relative to one another across the transverse width of the structure, as shown by the arrows.

In this example, the shape of the individual strands **2** prior to twisting is generally that of a flat ribbon-like configuration in cross-direction. Moreover, and as best illustrated in Figure 2.3b, each strand, in effect, includes a pair of oppositely disposed splines **4** and **6** which extend radially outwardly to the outer peripheral edge **8** of the strand. The amount of twisting and the particular shape of the strands may vary in each case to achieve the necessary structural and hydraulic characteristics for a given application.

FIGURE 2.3: NONWOVEN TWISTED STRAND FILTER FABRIC

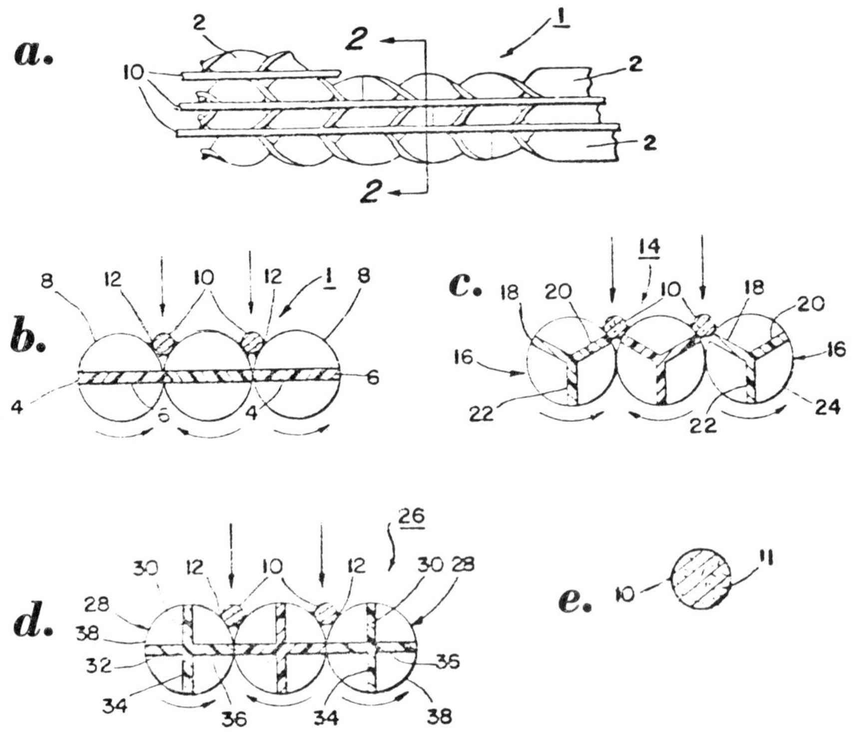

(a) Fragmentary top plan view showing one form of the nonwoven porous structure
(b) Enlarged vertical section view taken along the plane of line **2–2** of Figure 2.3a
(c) Enlarged vertical section view showing a modification of the porous structure
(d) Enlarged vertical section view showing another modification of the porous structure
(e) Enlarged section view of one of the absorbent elements used in making the porous structure of Figures 2.3a and 2.3d, and showing the element saturated with an adhesive material

Source: U.S. Patent 3,510,005

Any polymeric material which may be readily formed and twisted into the desired strand configuration may be used for making the strands of the porous structure. It has been found, however, that some thermoplastic materials are particularly suitable for making such a porous structure, particularly when utilized in making endless belts for papermaking machines.

Preferred among the common thermoplastic materials are polycarbonate and nylon which have the property that when expanded under the proper temperature conditions, the molecular chains, of which the plastic is comprised, become oriented in the direction of expanding, thus greatly increasing the strength characteristics of the material. Furthermore, it has been found that the structural orientation of such material, such as by twisting, is retained so that the material does not return to its original shape prior to twisting.

In accordance with the process, the porous structure **1** of Figure 2.3a is achieved by bonding the individual twisted strands **2** together by the use of absorbent elements **10** disposed to extend longitudinally between adjacent pairs of the twisted strands. Preferably, the absorbent elements **10** have a substantially reduced transverse dimension compared to the transverse dimension of the twisted strands and are preferably comprised of an absorbent material, such as soft unmercerized cotton thread or the like. The absorbent elements are saturated with a liquid adhesive **11** (Figure 2.3e), such as solvent-cement, which is held mechanically within the interstices of the material.

In application, the absorbent elements which contain a predetermined amount of adhesive material are disposed to extend longitudinally between adjacent twisted strands, as shown in Figure 2.3a. Thus positioned, a slight pressure may then be applied to the absorbent elements, as shown by the arrows. This pressure somewhat deforms the absorbent elements at the areas of contact adjacent the twisted strands. Such deformation automatically causes the release and deposit of a sufficient amount of the adhesive, as at **12**, at the points of contact between the twisted strands to provide a positive bonding of the twisted strands.

In many cases, the pressure required to release the cement from the absorbent strands is produced merely by applying tension to these strands, though proper use of rollers or other devices to increase this pressure is sometimes beneficial. But it is extremely important that the strands **2** and the strands **10** be held immovable with respect to each other until the solvent has evaporated to a sufficient extent to form a bond between these strands. It is also imperative that no flexure take place during this time.

The rate of evaporation can be conveniently increased by application of heat such as by assembling the porous structure on a heated drum. This drum may be grooved to help hold the strands in the proper position to each other while the cement is setting. Any temperature between the room temperature and 200°F may be used with a solvent-cement, such as polycarbonate dissolved in methylene chloride, with the preferred temperature being about 140°F.

Further control of the absorbency of the adhesive material by the elements **10** may be achieved by surface pretreatment, such as by dipping or by spraying, with an adhesive material to control the porosity of these elements. The adhesive material may be a type, such as solvent-cement, similar to that used to saturate the interior of the absorbent elements just before the assembly of the porous structure. In certain cases, it has been found that in the absence of such pretreatment, too great an amount of adhesive was enabled to permeate the absorbent elements immediately prior to assembly with the result that an excessive amount of adhesive was released upon deformation which resulted in partial stoppage in the openings of the structure. On the other hand, it was found in some instances, that too heavy a pretreatment prevented adequate permeation of adhesive interiorly of the elements during assembly of the structure. This resulted in unreliable bonding of the twisted strands **2**.

In Figure 2.3c, there is shown a modification of the porous structure, designated generally at **14**, which is generally similar in arrangement to that illustrated in Figures 2.3a and 2.3b, except that the individual twisted strands **16** are of a different shape in cross section. In this form, the individual twisted strands each include three equispaced splines **18, 20** and **22** which extend radially to the other peripheral edge **24** of the strand.

In Figure 2.3d there is shown another modification of the porous structure, designated generally at **26**, wherein each twisted strand **28** includes equispaced splines **30, 32, 34** and **36** which extend radially to the outer peripheral edge **38** of the strand. Accordingly, though the one to four spline shapes of the strands have been illustrated, it is to be recognized that other numbers of splines or other noncircular shapes for the strands may be utilized in accordance with the process.

Referring to the table below, it will be seen that several important variables relating to the structural characteristics of the individual twisted strands and the mode of bonding the strands together effect the hydraulic characteristics of the porous structure, and particularly the extent of independence of such variables in relation to the relative permeability in the structure for a given application, such as in the production of a Fourdrinier belt for papermaking machines.

In the following table, this relative permeability is expressed in relationship to the air and water permeability of a Lindsay Wire Weaving Company 74/58 PBLC Fourdrinier Wire Cloth. The functional effect of each variable, with the other variables held constant, is discussed below. In the table, and looking from left to right, the first column is tabulated with reference to the cross-sectional shape of the twisted strands **2**. In this column, the shape is given particularly with reference to the number of projections or splines on the strand, as illustrated in Figures 2.3b, 2.3c and 2.3d.

Moreover, it can be observed that the four spline **30–36** shape of Figure 2.3d has less relative permeability to air and water compared to the three spline **18–22** shape of Figure 2.3c, while the latter shape has less relative permeability compared to the two splines **4** and **6** shape of Figure 2.3b. It has been found, for example, that the three and four spline shapes are not as effective as the two spline shape, such as for instance in Fourdrinier belts, due to reduced permeability and clogging.

Accordingly, it has been found that the two spline or twisted ribbon-shape of Figure 2.3b is preferred and provides extremely good results for this purpose. Furthermore, in the two spline shape, the relative permeability to water is only slightly less relative to the Fourdrinier woven wire while the relative permeability to air is much less than the woven wire which is particularly advantageous for papermaking purposes. Thus, in the process, the relative permeability to both air and water may be effectively controlled by variation in the number of projections or splines provided on the individual twisted strands.

Splines	Twist	Cement	Cotton	Relative Permeability Air	Relative Permeability Water
2	6	6/100	60	0.214	0.944
3	6	6/100	60	0.107	0.925
4	6	6/100	60	0.077	0.840
2	4	6/100	60	0.188	0.944
2	6	6/100	60	0.214	0.926
2	12	6/100	60	0.154	0.926
3	6	6/100	60	0.107	0.926
3	12	6/100	60	0.093	0.926
2	4	4/100	90	0.214	0.935
2	4	6/100	90	0.231	0.944
2	4	8/100	90	0.231	0.935
2	4	10/100	90	0.231	0.926
2	4	12/100	90	0.231	0.926
2	4	8/100	40	0.200	0.926
2	4	8/100	60	0.200	0.935
2	4	8/100	90	0.214	0.935
2	4	10/100	40	0.188	0.926
2	4	10/100	60	0.200	0.926
2	4	10/100	90	0.231	0.926
2	4	12/100	40	0.200	0.900
2	4	12/100	60	0.200	0.909
2	4	12/100	90	0.231	0.926

The second column is tabulated with reference to the amount of twist imparted to the strands, and more particularly, to the number of turns per inch of strand length. In this column, the strands have been given 4, 6 and 12 turns per inch of length. It can be observed that the number of turns per inch has relatively little effect on the relative permeability to water with the four turn per inch strand being the most permeable.

It has been found, however, that the tortuosity is greatest with a lower number of turns than with a greater number, but as the number of turns is further increased another variable becomes important. That is, though the tortuosity is reduced by further twisting, the twisted strands are foreshortened so that the tortuosity tends to decrease to the point wherein the reduction in permeability, due to reduction in porosity, is greater than the increase in permeability due to the decrease in the tortuosity of the passages in the porous structure.

Moreover, with an infinite amount of twisting the splines would be disposed in substantially perpendicular relationship relative to the longitudinal central axis of the strands so that the passageways between adjacent strands would be substantially linear, as in the case of a woven, monofilament construction.

In the third column, the tabulation is with reference to the composition of the adhesive material **11** (Figure 2.3e), and more particularly to the solids content of the cement in the solvent. The solids content is expressed in terms of the ratio of the amount of cement, by weight, dissolved in the solvent solution. As illustrated, the ratios expressed are the proportions of polycarbonate dissolved in methylene chloride. Moreover, it can be observed that various ratios or concentrations of the cement dissolved in solvent may be utilized so that sufficient amounts may be released and deposited at the spaced points of contact between adjacent strands. As shown, the composition may be in the range from about 4 to 12%, by weight, with the preferred range being about 6 to 8%, by weight, and with the preferred amount being about 6%, by weight, to provide optimum bonding of the strands without impairment to the hydraulic characteristics of the porous structure.

Thus, it can be seen that with the preferred adhesive composition there is substantially no difference in relative air permeability, and a progressively slight decrease in water permeability as the concentration of the solids content of the cement is increased.

In the fourth column, the tabulation is made with reference to the size or diameter of the absorbent elements **10**. Looking from top to bottom in the table, it can be seen that there are three groups of twisted strands which contain 8, 10 and 12% solvent-cements which exhibit no significant variation in permeability to water, except with No. 40 Cotton with a 12% solvent-cement. Moreover, it can generally be observed that the relative permeability to air increases as the size of the cotton decreases.

In addition to their effect upon the relative permeability, the above described variables also have an effect upon other related hydraulic characteristics of the porous structure, such as the porosity, specific surface, and tortuosity. Generally, the porosity of the individual strands, and therefore, of the porous structure comprised of such strands, may be determined by the ratio of the area of a circle circumscribing the extremities of the projections or splines to the cross-sectional area of the solid material of which the strand is comprised, as shown in Figures 2.3b, 2.3c and 2.3d. Moreover, it has been found that the greatest porosity would result from a twisted strand having two splines **4** and **6**, as shown in Figure 2.3b.

In this connection, it has been found that the shape of the strand, assuming that the cross section of the solid material in the strand is kept constant, affects specific surface, as well as porosity. Thus, a strand having a four spline shape (Figure 2.3d) has a greater specific surface as compared to a circular strand of the same cross section. Moreover, the circular strand would provide a structure having substantially no porosity.

Accordingly, the four spline strand shape would have a greater specific surface than the two spline shape (Figure 2.3b) even though the latter shape in this case would have a greater porosity compared to the shape shown in Figure 2.3d. Moreover, it can be seen that in general the number of projections or splines on the strand has a pronounced effect upon the specific surface of the porous structure.

Controlled Fluid Flow Resistance in Two Directions

M.W. Gilpatrick; U.S. Patent 3,930,096; December 30, 1975; assigned to Deering Milliken Research Corporation describes a composite fabric structure that has a high resistance to fluid flow in one direction and a substantially reduced resistance to fluid flow in the opposite direction.

Figure 2.4a is a top view of the fabric; Figure 2.4b is a cross-sectional view of the fabric on line **2–2** of Figure 2.4a; Figure 2.4c is a perspective view of the apparatus employed to manufacture the fabric shown in Figures 2.4a and 2.4b; Figure 2.4d is a graphic representation of the fluid flow in both directions through the fabric.

In the preferred form of the process, the fabric consists of an open mesh type of woven, nonwoven or knitted substrate to which is laminated a nonpermeable film material. It should be kept in mind that when fluid flow is discussed it refers to liquid as well as gaseous flow. Also, the composite fabric is not restricted to any particular fabric construction so long as it meets the requirement that the fluid flow in one direction is substantially greater than the fluid flow in the opposite direction.

Fabric **10** in Figures 2.4a and 2.4b basically consists of a base layer of a loosely woven 65% polyester-35% cotton fabric **12** to which is laminated at long narrow places **14** a nonpermeable film material **16** such as polyethylene. Then the film material is slit longitudinally at **18** to form a series of flaps **20** secured at the longitudinal edges by the strips **14**. Looking at Figures 2.4b and 2.4d it can be seen that when a fluid is passed upwardly from the bottom of Figure 2.4b to the top, that the flaps **20** will be forced upwardly thereby allowing a large flow rate of fluid to pass therethrough as shown by curve **22** on the graph shown in Figure 2.4d.

FIGURE 2.4: COMPOSITE POROUS FABRIC-IMPERVIOUS FILM ARTICLE

a.

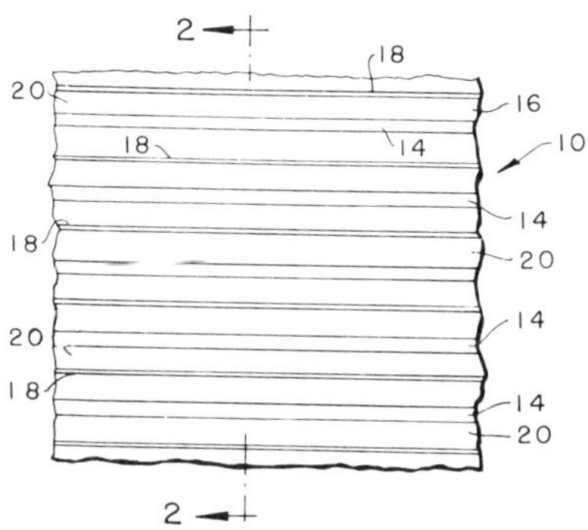

b.

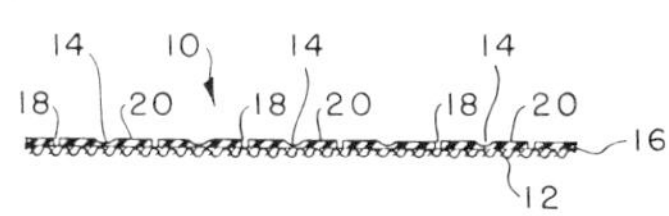

(continued)

FIGURE 2.4: (continued)

c.

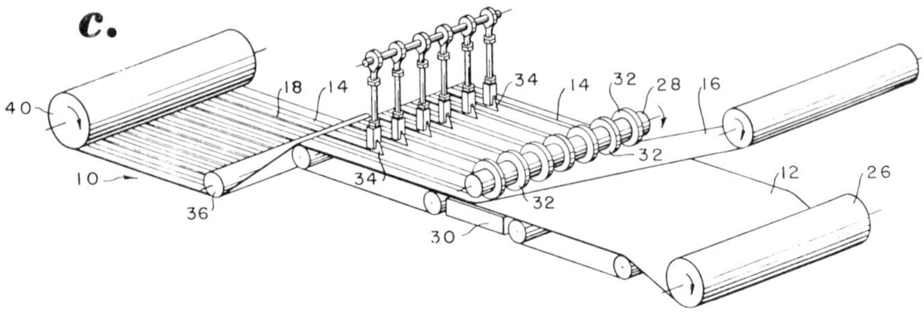

d.

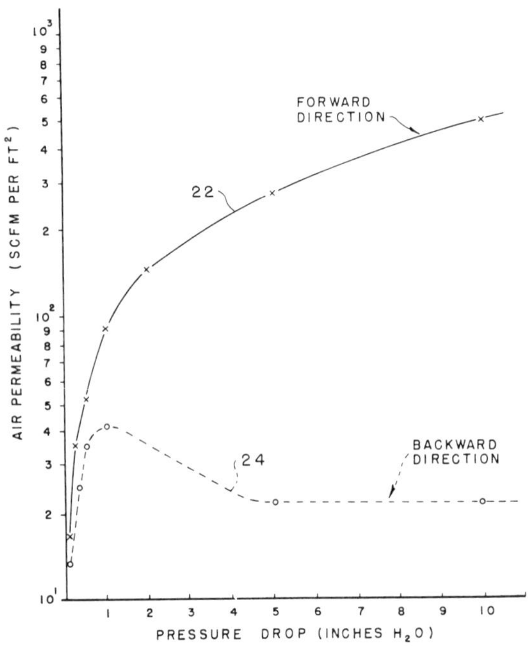

Source: U.S. Patent 3,930,096

Conversely, if the flow should be reversed, the pressure of the fluid will press the flaps against the substrate **12** to substantially cut down the flow of fluid therethrough, as evidenced by curve **24** on Figure 2.4d. The flow differential between the two will depend on the permeability of the individual laminated fabrics, fabric construction, etc.

Figure 2.4c illustrates the preferred method of constructing the above fabric. A woven fabric **12** is supplied from a supply roll **26** and is brought together with polyethylene web material **16** at a point between the finned pressure roller **28** and the contact heater **30**. The combination of the heat from the heater and the pressure of the fins **32** provide the laminated strips **14**. From the pressure roller the combined fabric is delivered to the slitting knives **34** where the film material is slit at **18** to provide the flaps **20**. Then the combined fabric is guided under idler guide roll **36** to the take-up roll **40**.

It can be seen that the fabric has definite use where it is desired to allow free flow of fluid in one direction but inhibited fluid flow in the opposite direction. Such fabric could be used as a check valve and in relief dampers.

GLASS FIBERS

Cellulosic, Asbestos and Glass Fiber Pads

R.K. Jones and R.W. Hess; U.S. Patent 3,796,659; March 12, 1974; assigned to Johns-Manville Corporation have found that liquids containing suspended particles can be filtered through fibrous filter pads comprising cellulosic fibers and asbestos fibers at increased flow rates and at acceptable levels of filtrate clarity by including glass fibers in a filter pad composition. Glass fibers, when substituted for a portion of the asbestos fibers present in conventional filter pads, in an amount of about 5 to 80% of the asbestos fibers originally present, result in a marked increase in liquid flow rates through a filter press with filtrate clarities at least as good as those obtained with similar fibrous filter pads not incorporating the glass fibers.

Increases in flow rates of over 100% as compared to conventional filter pads have been attained by incorporating glass fibers in the pad, with clarity levels better than those of the conventional pads. It is believed that this increase in flow rate without a deleterious effect on filtrate clarity results from the tendency of the glass fibers distributed throughout the pad to increase the pad's ability to resist compression.

In addition, it has been found that by incorporating glass fibers into filter pads in the amounts stated above, such pads possess a more uniform permeability than that possessed by prior pads formed of cellulosic fiber and asbestos fiber alone. It is believed that the glass fiber aids in uniformly forming the pad from the precursor slurry and thus results in a filter pad which is more uniformly permeable.

Figure 2.5a is an isometric view of one form of filter pad. Figure 2.5b is a sectional view taken on line **2–2** of Figure 2.5a. Figure 2.5c is a schematic illustration of the process and apparatus for forming a filter pad. Figure 2.5d is an isometric view of a filter press. Figure 2.5e is an exploded front elevation view of a portion of the apparatus of Figure 2.5d. Figure 2.5f is a perspective view of a filter plate shown in Figures 2.5d and 2.5e. Figure 2.5g is a graphic representation of the filtration characteristics of the filter pads of this process compared to other filter pads, wherein flow rate and turbidity are plotted against percent glass fiber substituted for asbestos fiber.

The filter pads may be produced by conventional fibrous filter pad forming processes modified so as to include glass fiber in the pad. For example, a slurry containing a mixture of fibers as the solids content and water may be drawn under a vacuum differential against a filter pad forming screen having a mesh size, for example, ranging from about 14 to 20 U.S. Standard Sieve Screen size. The solids content of the aqueous slurry may typically range from about 0.5 to 8% by weight and comprise a mixture of cellulosic, asbestos and glass fibers in the approximate amounts desired in the filter pad. White water from the slurry passes through the screen while the cellulosic, asbestos and glass fibers are deposited in random orientation on the screen. The deposited layer is thereafter removed from the screen and dried in a conventional manner.

FIGURE 2.5: FILTER PAD CONTAINING CELLULOSIC, ASBESTOS AND GLASS FIBERS

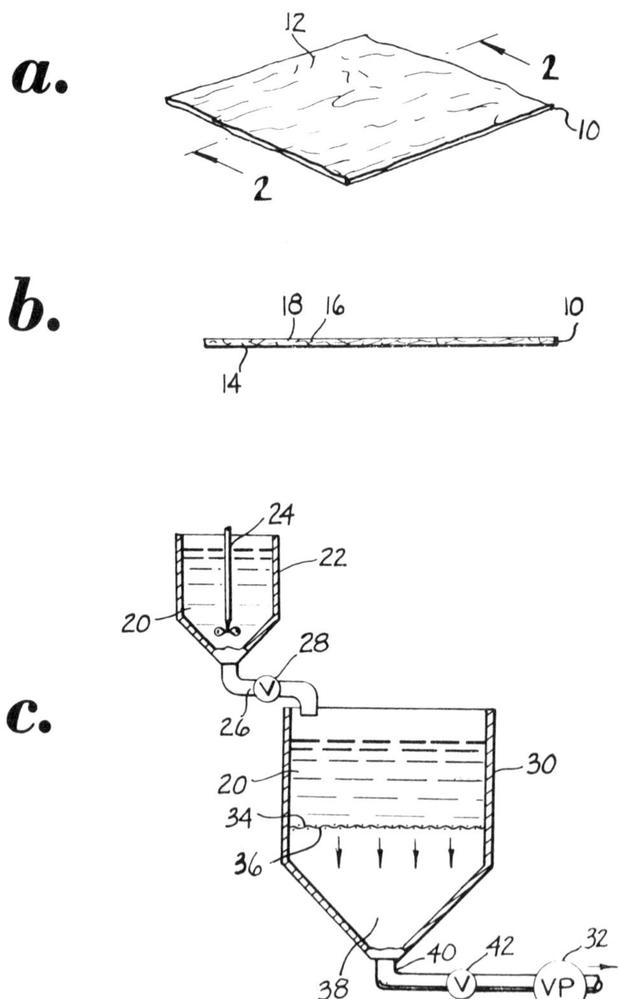

(continued)

FIGURE 2.5: (continued)

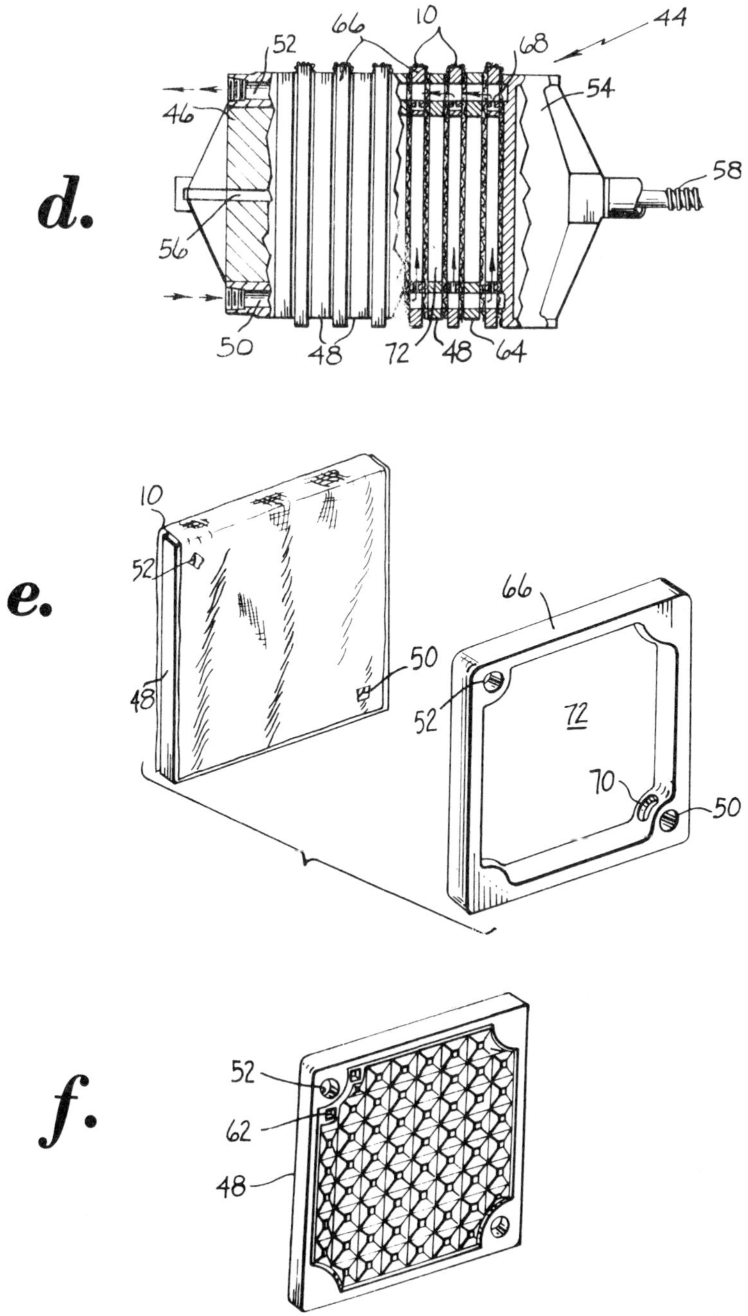

(continued)

FIGURE 2.5: (continued)

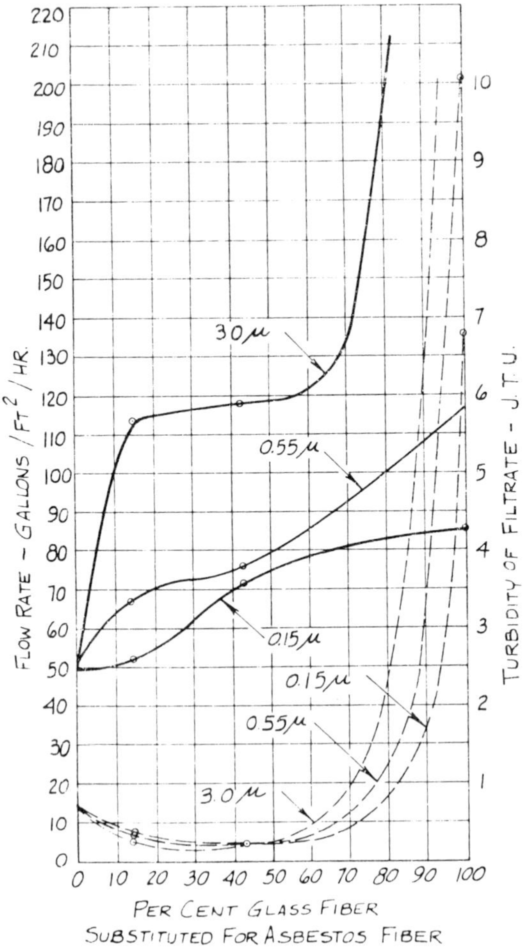

Source: U.S. Patent 3,796,659

Referring to Figure 2.5a there is illustrated a filter pad **10** comprising discontinuous individual fibers **12**. These fibers comprise glass fiber as well as asbestos fiber and cellulosic fiber of the type conventionally used in filter pads. The glass fiber preferred is a small diameter fiber of a diameter in the range of about 0.15 to 14 microns. The asbestos fiber suitable for use in the filter pad may include any crystalline hydrous magnesium silicate fiber which ranges from group 1 to group 9 Quebec Asbestos Mining Association grade. Chrysotile asbestos fibers are generally preferred. A cellulosic fiber generally of the type derived from wood which is either bleached or unbleached may be employed, although bleached cellulosic fiber is usually preferred. As illustrated in Figure 2.5a, the individual fibers of glass, asbestos, and cellulose are randomly oriented and relatively homogeneously mixed throughout a substantial portion of the pad.

In Figure 2.5b there is illustrated a section of pad **10**. The asbestos fiber **14** and cellulosic fiber **16** are of the same general character as is generally used in commercial filter pads. Obviously, however, the process of forming pad **10** differs from the process of forming pads from blends of asbestos and cellulosic fibers alone, in that glass fiber **18** is homogeneously mixed along with the cellulosic and asbestos fibers throughout the pad. The combined pad is substantially no thicker than a corresponding pad which employs a blend of asbestos and cellulosic fibers alone.

In Figure 2.5c, there is shown schematically a portion of an apparatus by which the product of Figure 2.5a may be produced. The fiber ingredients which form the solids content of an aqueous slurry **20** are combined in a tank **22** in a desired mixture and agitated by propeller or other type mixer **24**.

A predetermined quantity of the mixed slurry **20** is fed from tank **22** through conduit **26** and valve **28** into vacuum box **30**. The quantity of slurry transferred to vacuum box **30** generally depends on the size and thickness of the filter pad desired and upon the solids content of the slurry. In general, the solids content of the slurry will range from about 0.5 to 8%.

A vacuum of about 1 inch or more of mercury is then exerted for at least 2 seconds on the slurry within the vacuum box by vacuum pump **32**. The slurry is thereby drawn against screen **34** which has a wire mesh ranging from about 14 to 20 U.S. Standard mesh size and which is supported within the vacuum box by drainage member support **36**. The mesh size of the screen will of course depend on the type of fiber used to form the pad and can extend beyond a range from about 10 to 40 U.S. Standard mesh size depending on the diameter and length of the fiber employed. White water **38** from the slurry passes through the screen **34** and is discharged through conduit **40**, valve **42**, and vacuum pump **32** to waste. The asbestos, cellulosic and glass fibers deposited on the screen **34** form the filter pad **10** having fibers **14, 16,** and **18** randomly oriented and homogeneously mixed throughout.

In Figure 2.5d there is illustrated a portion of a plate and frame filter press in which the filter pad **10** of the process can be used. The apparatus shown in Figure 2.5d is entirely conventional and is operated in a conventional manner. Figure 2.5d is intended to illustrate but one of the many uses to which the pad may be put. It should be apparent, for example, that the pad could be employed in a chamber filter press or in a variety of pressure filtering apparatus. The pad need not have a rectangular shape but may be cut into discs or other suitable configurations shaped to coact with the apparatus with which the filter pad is employed.

As illustrated by Figure 2.5d, the filter press, shown generally at **44**, is provided with a fixed filtering head **46**, a plurality of filter plates **48** and frames **66** each provided with an inlet port **50** and a discharge port **52**, a loose filter head **54**, side support members **56**, and closing and tightening screw **58**. As noted above, the construction and operation of the parts of filter press **44** should be well understood by those skilled in the art.

In Figure 2.5e there is illustrated an exploded portion of the apparatus of Figure 2.5d. Filter press **44** comprises a plurality of plates **48** and frames **66**. The filter pad **10** can be positioned as illustrated over plate **48**. Loose filter head **54** is then moved toward fixed filter head **46** by the action of screw **58** until each plate **48** and frame **66** is tightly secured in abutting relationship adjacent side members **56** and between filter heads **46** and **54**.

Inlet ports **50** and discharge ports **52** form continuous closed pipes for inlet of the turbid liquid and discharge of the filtrate. Unfiltered liquid passing through port **50** and orifice **70** in frame **66** fills the cavity **72** between plate **48** and plate **64**. The unfiltered liquid then passes through filter pad **10**, which separates suspended particles from the liquid. As illustrated in Figure 2.5f, the filtrate is channeled along plate **48** into drainage orifice **62** and through conduit **68** to discharge port **52**.

Example: The filtration flow rate, permeability and particle size retention were determined for conventional filter pads consisting of asbestos and cellulosic fibers blended in various amounts. A sample in strip form was cut from each pad so as to extend in a direction generally radially outward from the center of the pad. The sample was cut into pieces, placed into a prefired and weighed crucible and dried overnight at 105°C.

The dried weight of the sample was measured. The sample was burned off over a Bunsen burner to carbonize cellulosic material present in the sample, and fired in an electric muffle at 1,800°F for one hour. The resultant asbestos residue was weighed. This weight was increased by 13% to correct for the water of crystallization removed. The corrected weight was used to calculate the percent of asbestos used in the composition of the pad.

Four discs were cut from the pad, one from the outer corner, one from exact center, and the remaining discs from each side of the center. Each disc was tightly sealed within a cylinder so that the surface of the disc extended across the cylinder's inside diameter and transversely of the longitudinal axes of the cylinder. To determine the permeability of the disc, deionized water was filtered through a filter pad having a controlled pore size of 0.22 micron.

The time was measured for the passage of 300 ml of the deionized and filtered water through each disc under a vacuum of 400 mm of mercury at 25°C. The measured flow times for the four discs of each pad sample were averaged and the flow time range (the difference between the highest and the lowest flow times) was determined. Percent range deviation was calculated by dividing the flow time range by the average flow time and multiplying by 100.

A standard test liquid was made up containing the following particles of artificial turbidity to determine filtration flow rate and clarity:

Particles	Mean Particle Size, microns
25%	2.0
25%	4.8
25%	7.0
25%	11.0

Each fraction of mean particle size solids weighed 1.3 g. The combined fractions, weighing 5.2 g in total, were treated with 25 mg of sodium tripolyphosphate in 20 ml of water to maintain the original size of the particles in a dispersed state. The slurry was homogenized to insure dispersion of the particles and diluted with water to 2,600 ml. The dilute slurry was used as the master batch for turbidity, and measured 58 Jackson Turbidity Units (JTU) when tested on a Hach Model 2100 turbidimeter.

For each filtration, the slurry was vigorously shaken and a representative sample of 200 ml was made up to 4 liters with a resultant concentration equivalent to 100 mg per liter. The 4-liter sample of 100 parts per million turbidity was used to conduct a constant pressure filtration in a stainless steel bomb filter. The four most permeable samples were filtered at a pressure differential of 5 psi and the remaining samples at 40 psi. Filtrate collected during the first 3 minutes was reserved for turbidity determination by means of the Hach turbidimeter.

Turbidity readings indicate the relative clarity of the filtrate, the clarity being inversely proportional to the measured turbidity. Each 4-liter sample was allowed to flow through each pad for a period of 21 minutes. The accumulative flow was plotted on log-log paper against time in minutes. The plotted points resulted in a straight line with a definite slope, which was extrapolated to give the total liquid flow at the end of one hour's elapsed time. The total flow was then expressed in terms of gallons per square foot per hour through each pad. The flow rate measured at 5 psi was converted to a 40 psi basis for comparison, using the following formula:

$$\text{Flow rate at 40 psi} = \text{Flow rate at 5 psi} \times \sqrt{40/5}$$

The results for this series of tests of conventional cellulosic-asbestos fiber pads are shown in Table 1 on the following page. Another series of tests were run to establish the performance of the cellulosic-asbestos glass fiber filter pads of this process. Individual filter pads of 500 g each were formed on a vacuum box from an aqueous slurry containing the fibrous solids. Control samples, containing 65% cellulosic fiber and 35% asbestos fiber, were made as the first and last samples. Glass fiber of an approximate length of $^1/_{16}$ inch and in the amounts of 5% and 15% of the total solids content was mixed with 65% cellulosic pulp, with the remainder of the solids content comprising asbestos fiber.

Pads comprising 65% cellulosic and 35% glass fibers were also formed. Pads with the above weight percentages were formed from three types of glass fiber: Micro-Fiber 102 (average diameter 0.15 micron), Micro-Fiber 106 (average diameter 0.55 micron) and Micro-Fiber 112 (average diameter 3.0 microns). The cellulosic pulp fiber and asbestos fiber were each of the type conventionally employed to form filter pads. The pads were tested for permeability, filtration flow rate and turbidity using the same procedures referred to above with respect to the conventional cellulosic-asbestos fiber filter pads. The results are shown in Table 2.

The column "Glass, % of asbestos" represents the percent of asbestos fiber originally present in Example 12 which was replaced by glass fiber. The results tabulated in Table 2 are shown graphically in Figure 2.5g, where flow rate and turbidity of filtrate are plotted against the glass fiber content of the filter pad, expressed as the percent of asbestos fiber replaced by glass fiber. Flow rate and turbidity readings for the control samples (samples 12 and 22) were averaged in preparing Figure 2.5g.

TABLE 1

Sample	Percent asbestos	Permeability: Averaged flow time (sec.)	Permeability: Flow time range (sec.)	Permeability: Percent range deviation	Flow rate: Operating pressure (p.s.i.)	Flow rate: Operating flow rate (gal./ft.^{2}/hr.)	Flow rate: Flow rate at 40 p.s.i. (gal/ft.^{2}/hr.)	Filtrate turbidity (JTU)
1	3.1	6.5	5	77	5	242	682	24
2	4.7	12.5	4	32	5	123.6	348	17
3	11.0	28.4	19	67	5	49.7	140	2.1
4	15.4	80	47	58	5	35	99	0.8
5	20.6	250	140	56	40	71	71	0.12
6	24.2	945	400	42	40	53	53	0.12
7	29.0	280	125	45	40	55	55	0.65
8	41.3	540	145	27	40	57	57	0.13
9	40.5	1,365	270	20	40	47	47	0.18
10	47.2	1,570	400	25	40	42	42	0.25
11	57.2	1,885	500	27	40	44	44	0.28

TABLE 2

Sample	Composition (percent)[1]: C	Composition (percent)[1]: A	Composition (percent)[1]: G	Glass, percent of asbestos	Average diameter fiber glass (microrons)	Permeability: Avg. flow time (sec.)	Permeability: Flow time range (sec.)	Permeability: Percent range deviation	Flow rate: Operating pressure (p.s.i.)	Flow rate: Operating flow rate[2]	Flow rate: Flow rate at 40 p.s.i.[2]	Filtrate turbidity (JTU)
12	65	35	0	0	-----	719	219	34.6	40	44.1	44.1	0.36
13	65	30	5	14.25	0.15	726	14	1.9	40	50.6	50.6	0.31
14	65	30	5	14.25	0.55	288	56	19.4	40	66.6	66.6	0.22
15	65	30	5	14.25	3.0	183	43	23.5	40	113.9	113.9	0.24
16	65	20	15	43	0.15	288	44	15.3	40	70.3	70.3	0.22
17	65	20	15	43	0.55	216	57	26.4	40	75.2	75.2	0.22
18	65	20	15	43	3.0	95	57	60.0	5	41.9	118.1	0.21
19	65	0	35	100	0.15	76	11	14.5	5	30.1	84.9	6.7
20	65	0	35	100	0.55	31	6	19.3	5	41.2	116.2	10.0
21	65	0	35	100	3.0	2.6	0.6	21.5	5	142	400	35.5
22	65	35	0	0	-----	442	58	13.1	40	52.6	52.6	1.0

[1] C = Cellulosic; A = Asbestos; G = Glass. [2] Gal/ft^2/hr

Table 2 and Figure 2.5g show the improved properties possessed by the filter pads of this process. It is apparent that for each of the glass fiber diameters tested, by substituting a small amount (e.g. 14.25%) of the asbestos fibers with such glass fibers, the filtration flow rate markedly increases without adversely affecting the clarity of the filtrate. The filtrate clarity actually increases with the increased flow rates. Filtration flow rates of more than twice that obtainable with cellulosic-asbestos filter pad control samples were obtained by substituting 14.25% of the asbestos fibers with glass fibers having an average diameter of 3.0 microns (sample 18). Significant flow rate increases were also measured for each of the glass fiber diameters tested at substitution levels of 43% of the asbestos fiber, again without adversely affecting filtrate clarity and again with an increase in clarity over the test samples.

When all the asbestos fiber in the filter pads was eliminated and substituted by glass fiber (samples 19 through 21) very large increases in filtration flow rate were measured, but the clarity of the filtrate decreased beyond normally acceptable levels. From the curve of turbidity vs glass fiber content of Figure 2.5g, it can be ascertained that by substituting up to about 80% of the originally present asbestos fibers with glass fibers, substantial increases in filtrate flow rates are obtainable without an adverse effect on filtrate clarity. The glass fiber should be present in an amount between about 5 to 80% based on the weight of asbestos fiber originally present. Glass fiber content can alternately be described as in the range of about 5 to 80% by weight of the combined weight of asbestos and glass fiber.

The results of Table 2 also indicate an increase in uniformity of pad permeability resulting from the presence of glass fiber. The column "% range deviation" indicates the extent of nonuniformity in flow time through four samples of each pad and hence is an indication of the extent of nonuniformity in permeability of the pad. The percent range deviations listed in Table 2 show in general a good uniformity of formation of the pad, with uniformity being controlled best by the smallest diameter glass fiber tested, 0.15 micron. When compared to the conventional cellulosic-asbestos fiber filter pads of Table 1, the glass-cellulosic-asbestos fiber pads of this process show in general a substantial increase in uniformity of permeability.

Zeta Potential Control Using Glass Fibers

According to a process described by *G.R. Machlan; U.S. Patent 3,542,674; November 24, 1970; assigned to Owens-Corning Fiberglas Corporation* all small particles including colloidal size particles are removed from a liquid suspension of the same by flowing the suspension through a bed of material having a pore size no less than approximately 10 times the smallest size particle to be removed, and the material of which has a zeta potential which causes the solids of the suspension to floc when under the influence of the zeta potential of the bed material.

In one form of the process, the liquid suspension is first treated to adjust the zeta potential of the suspended particles to a potential very nearly that at which the particles will floc. The potential provided by the treatment is such, however, as to still hold the particles dispersed throughout the liquid. The treated suspension is then filtered through fibers having a zeta potential in the suspension which is opposite in charge and magnitude from the zeta potential of the particles. When the treated particles approach the surface of the fibers, the dispersing zeta potential on the particles is counteracted by the zeta potential of the fibers to cause the particles to floc in situ and become attached to the fibers. The fibers can be mixed in with the suspension, but preferably are a fixed bed through which the suspension is passed. The preferred forms of fibers are glass fibers because of the durability, uniformity, and inexpensiveness of glass fibers.

Mats of the glass fibers can be made inexpensively to have extremely low densities and uniform pore sizes to provide high filtering capacity with very little pressure drop. Glass fibers have a negative charge in water, and where the particles to be removed are positively charged, the process is simply carried out by adjusting the zeta potential with acid and/or counter ions. Examples of positively charged particles are alumina, thoria, and ferric oxide. Positively charged particles can be deposited upon negatively charged glass fibers without charge reversal and by the adjustment of the zeta potential with pH control and/or counter ions.

Most suspended solids, however, are negatively charged. Examples of negatively charged particles are: organic materials particularly where they have undergone oxidation, colloidal silica, the aluminum silicates, clay, ferrous oxide, and various inorganic refractory materials excluding alumina, thoria and ferric oxide.

When the particles of the suspension are negatively charged, the expense of treating the suspension to be filtered can be greatly reduced or eliminated by using a filter bed of material that is positively charged, as for example chromium oxide particles, aluminum oxide particles, thorium dioxide particles, and preferably fibers of the same. The hydrous oxides of these materials are positively charged and under some conditions will cause negatively charged suspensions to floc without further treatment. In still other instances, it may be desired to only provide a coating of these materials on other base fibers, as for example stainless steel fibers, or glass fibers.

In those instances where the particles to be removed are negatively charged, the suspension can be treated with a charge reversal agent such as an aluminum hydroxide forming material or a ferric hydroxide forming material, as for example aluminum sulfate or ferric sulfate.

In one method of carrying out the process, the raw liquid containing the solid suspension is caused to flow upwardly through the filter bed. By so doing, optimum filtering conditions are provided requiring minimum backwash flows and providing optimum sanitary conditions both above and below the filter bed.

In another method of carrying out the process, particularly useful where the suspension contains large particles of debris which plug the surface of the filter bed, a thin layer of large size sand particles may be placed on top of the fibrous filter bed, and the suspension caused to flow downwardly through the composite bed. With this arrangement, a backwash flowing upwardly through the bed will expand the layer of large size sand particles to remove the troublesome debris.

Example 1: Colloidal size particles less than 1 micron in size of an acid ball clay was mixed into a suspension in a tank to provide a solid content of 500 parts per million. Circular discs of a phenol-formaldehyde bonded glass fiberboard $^1/_{16}$ inch larger in diameter than the inside diameter of the column were pressed into the column to provide an upper layer ten inches deep. Fibers of the phenol-formaldehyde bonded board had a diameter approximately 0.0006 inch, the board had a density of 10 lb/ft^3, and the fibers extended generally horizontally of the column. A lower layer of fibers was also used, and this layer was made from specially produced glass fiber mat weighing 2 lb/ft^3.

The fibers also had a diameter of 0.0006 inch and were bonded together by a phenol-formaldehyde binder. A 10 inch deep layer having a diameter $^1/_{16}$ inch larger than the internal diameter of the column was pressed into the column beneath the upper layer. The fibers in the lower layer extended generally in vertical parallel planes. The pH of the clay suspension prepared as above described was adjusted by the use of hydrochloric acid and/or sodium hydroxide to provide a predetermined pH, and this material was run through the column at a flow rate of 3 gal/min/ft^2 of cross section for one hour. Runs of various pH were made and the table below gives the percent of clay which was removed at various pH levels.

pH	Clay Removed, %
7.9	10.0
7.0	23.1
6.3	24.8
5.5	41.3
5.0	93.2
4.2	100.0

Example 2: The process of Example 1 was repeated excepting that slurries of an alkaline ball clay of various pH's were made and circulated through the column. The table below indicates the amount of the alkaline ball clay which was removed at various pH levels.

pH	Clay Removed, %
7.0	10.0
6.3	12.3
5.0	17.9
4.2	53.8

It will be noted that the alkaline ball clay could not be removed completely even at the low pH of 4.2. Another batch of the alkaline ball clay slurry of 500 parts per million and having a pH of 4.2 was made in which the pH was adjusted using sulfuric acid, and in which 5 parts per million of sodium sulfate were added. A hundred percent of this material was removed by the column.

Example 3: The process of Example 1 was repeated excepting that a slurry of 500 parts per million of colloidal silica was prepared and the pH adjusted with sulfuric acid. Various runs of this material were made at pH levels of from 4 to 8 during which there was no removal of the silica particles. A suspension of 500 parts per million of the colloidal silica was made and 15 parts per million of $Al_2(SO_4)_3$ plus 5 parts per million sodium sulfate were added and the pH adjusted to 4.2. Complete removal of the colloidal silica was obtained by the column. The 15 parts per million of aluminum sulfate was not sufficient to produce floccing and was used as a charge reversal agent to affect a charge reversal of the silica particles.

Coiled Package of Leached Glass Fibers

A. Marzocchi; U.S. Patent 3,692,186; September 19, 1972; assigned to Owens-Corning Fiberglas Corporation has found that it is possible to leach fibers completely and uniformly while in a coiled package provided: the fibers which are coiled into the package are devoid of film formers such as starch or other plastic resinous materials, and provided the fibers are coiled into an open package using low tensions and high crossover angles.

The method of producing the media comprises four major phases, i.e., the formation of a glass composition containing both siliceous and nonsiliceous components, the simultaneous formation of continuous fibers from the glass composition and application of a leaching agent capable of preferentially dissolving the nonsiliceous components; the separation of the nonsiliceous components to leave a continuous silica fiber; and the thermal treatment of the silica fiber to improve its strength and moisture resistance. A preferred method and materials for the preparation of silica fibers is provided by the following example.

Example: Fibers were prepared by flowing a molten mass of the following ingredients through minute orifices and attenuating the molten streams into minute filaments having an average diameter of less than 0.0006 inch: SiO_2, 69.6% by weight; Na_2SO_4, 2.0% by weight; and Na_2CO_3, 28.4% by weight. In the melting of the above batch composition, the sulfates and carbonates are vaporized and dispelled to yield fibers comprising approximately 80% by weight of silica and 20% by weight of sodium oxide. In view of the high silica content and low flux ratio, the batch is melted in a platinum furnace at temperatures of between 2700° to 2800°F and is introduced to the fiber forming bushing tip at temperatures in the area of 2300° to 2400°F.

The above formulation is highly suited for the process in that it yields glass fibers having a high silica content and free from boron and alumina. Since 20% sodium oxide is present, the acid applied at forming was applied in a ratio of 14% since only two thirds of the sodium oxide actually yields the sodium ion.

2,040 of the above described fibers are pulled from the bushing at a linear speed of approximately 4,500 feet per minute and are sprayed with a 15% aqueous solution of sulfuric acid by any of the methods and apparatus described in U.S. Patents 2,491,889, 2,846,348 and 2,906,470 with the spray contacting the glass fibers approximately three feet below the orifices from which the glass fibers were drawn and attenuated.

At approximately 10 feet from the spray applicator, the fibers are drawn together into a strand and coiled into a package by apparatus of the type shown in U.S. Patent 3,371,877, in which the tension is maintained below 10 ounces and a uniform crossover angle of 15° is used. Subsequently, the wound package of treated fibers was immersed in a bath of a 5% solution of sulfuric acid for a period of 30 minutes, for the purpose of removing the salts which were formed by leaching.

Following the acid immersion, the fibers were immersed in water for a final wash of 20 minutes. Prior to the water wash, it is beneficial to encase the package in a protective sock, such as a cotton sleeve, since the fibers are extremely weak at this phase. The leached and washed fibers were then oven treated at

1700°F for a period of eight hours. The basic glass batch may also be formulated to utilize or include metallic fluxes or glass formers, or combinations of such compounds, in place of the sodium oxide of the fibers yielded by the example.

According to the process, it has been found that it is possible to leach fibers when in a coiled package provided that film-forming and lubricant materials are not used, and provided that the filaments are wound into a package at unconventionally low speeds and low tensions, using an unconventionally high crossover angle of the fibers. The traverse of the coils must be such that successive coils or wraps are spaced apart from each other with the distance between coils being more than five times the width of the strand, preferably more than 10 times, and will usually be more than 20 times.

The strands preferably have 2,040 filaments, or more, each 0.003 inch in diameter and have a strand width of approximately one-eighth inch. In some instances, the strand may have 204 or 408 filaments and be correspondingly less wide. The spaced apart overlying coils must have a crossover angle of at least 4°, and may have a crossover angle up to approximately 25°. Preferably the crossover angle should be between 5° and 20°, and most preferably should be between approximately 6° and 18°. Where the coiled packages are made using a constant way-wind, the crossover points can be made to align across the length of the package.

By causing the gain to be greater than the distance across a strand taken circumferentially around the package, spaces between rows of aligned crossover points are produced to provide predetermined strand spacings that are greater than the spacings between the filaments of the strand. The spacings are also greater than that between adjacent strands of prior art wraps where adjacent coils are in touching engagement.

In addition, by increasing the crossover angle, the strand sections between crossover points are caused to have greater radial separation, and the porosity of the package is thereby increased. This porosity, or build, occurs in channels, and is sometimes called honeycomb. Where low porosity is desired, the gain is caused to be equal to or less than the distance across a strand taken circumferentially of the package. It has been found that not only do the filaments not break after being leached, but that the coils in all of the regions of the package are leached uniformly.

The coiled packages will have uses in and of themselves because of their extremely large surface areas, which after heat treatment will be active. The coiled packages are useful because they have a fixed shape and can be handled as a single piece, and they are also useful by reason of the fact that they are generally tubular in nature as produced. Because the structure is uniform and has been made in a manner wherein it is devoid of channels, uniform flow can be had through the package between its outer and inner cylindrical surfaces. The packages can be made of a size suitable for use in dryers and ion exchange equipment.

Referring to Figure 2.6, the cartridge **10** is generally cylindrically shaped and has an outside diameter of six inches, an inside diameter of 1½ inches, and a length of 12 inches. The cartridge **10** was produced according to the procedure outlined in the above example, excepting that it was fired after leaching by raising its temperature gradually 400°F. Thereafter it was held at this temperature for two hours followed by gradual cooling. This procedure drove off free moisture without destroying the porous nature of the fibers and made the cartridge suitable for absorbing moisture. While the cartridge can be installed in various types of casings and used for various purposes, it is shown installed in a casing **12** of an air dryer.

The casing comprises a section of 10-inch pipe **14** having a pipe cap **16** at its lower end. A pipe coupling **18** is provided at the bottom center of the pipe cap and a hole **2** is provided through the cap for use as an inlet connection to the dryer. The upper end of the pipe section has a diverging flanged section **20** thereon, for reasons which will later be explained, and the top flange **22** thereof has a flat sealing surface for abutment by an O-ring **26** that is supported in a recess **28** in a quick disconnect cover **30**.

FIGURE 2.6: COILED FILTER CONTAINING LEACHED GLASS FIBERS

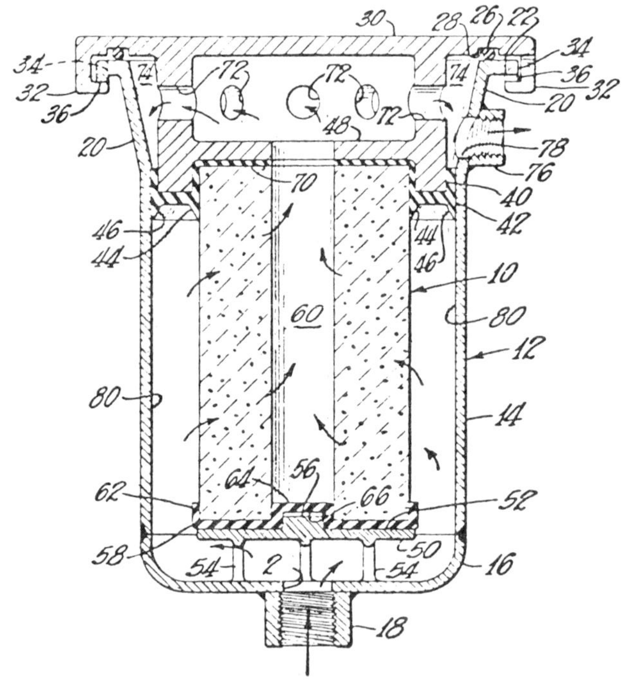

Source: U.S. Patent 3,692,186

The quick disconnect cover **30** has a plurality of downwardly extending fingers **32**, and the flange **22** has a corresponding plurality of recesses **34** to receive the fingers which allow the quick disconnect cover to be installed axially upon the flange section **20** with the fingers passing through the recesses **34**. The bottom portion of the flange **22** has ramps **36** adjacent the recesses **34**; and upon turning the quick disconnect cover, the fingers are caused to ride downwardly over the ramps **36** to pull the cover into tight engagement with the sealing surface of the flange **22**.

The quick disconnect cover includes a tubular projection **40** of a shape to fit between the upper end of the cartridge **10** and the sidewall of the casing **12**. A rubber sealing ring **42** of generally U-shaped cross section is slipped over the bottom end of the annular projection **40** to provide a seal between the cartridge and the sidewalls of the casing. The sealing ring **42** includes inner and outer lips **44** and **46**, which when pressure is exerted there against, deflect against the sidewalls of the cartridge and the casing to effect a seal therewith.

The tubular projection is provided with an annular flange **48** for the purpose of abutting the top edge of the cartridge and thereby holding the cartridge in firm engagement with a support **50** positioned in the bottom of the casing **12**.

The support has a generally flat upper surface **52** and legs **54** by which it is supported from the bottom cap of the casing, and includes a centering projection **56** which assures that the lower end of the cartridge will, at all times, be properly positioned in the casing.

Before installation in the casing, a rubber cap **58** is slipped over the bottom end of the cartridge to effectively close off the bottom end of the cartridge and prevent fluid flow from the inlet **2** directly to the center passage **60** of the cartridge. The cap **58** may include an annular lip **62** on its outer edge to help retain the cap **58** during insertion into the casing **12**, and also includes a center boss **64** which fits snugly into the lower end of the center passage **60**.

The boss includes a recess **66** into which the centering projection **56** fits snugly, so that the lower end of the cartridge is secured against movement. An annular rubber cap **70** may also be provided on the upper end of the cartridge between the cartridge and the flange **48**, and the tubular projection **40** is provided with a plurality of openings **72** to allow fluid flow from the central passage **60** to flow laterally through the openings **72** to the plenum **74** provided between the tapered walls of the flange section **20** and the tubular projection **40**.

A pipe coupling **76** is welded to the outside of the flange section **20**, and an opening is provided through the flange section to form the outlet connection for the casing. The casing **12** may be made of any suitable material, and as shown, is made of steel that has glass lining **80** on the inside and over the top surface of the flange **22** to form the sealing surface.

After insertion of the cartridge **10** into the casing **12** as above described, the rubber sealing ring **42** is inserted on the lower end of the tubular projection **40**, and the quick disconnect cover **30** is lowered axially into position over the top end of the cartridge.

It will be seen that the tapered surfaces of the flange **20** allow the outer lip **46** to be centered and guided, and the subsequent turning of the cap when tightened into position allows the legs **44** and **46** to assume a natural sealing position.

Microporous Fibrous Sheet

D.B. Pall and C.A. Keedwell; U.S. Patent 3,573,158; March 30, 1971; assigned to Pall Corporation describe a process for forming microporous sheet material useful as filters, which is substantially uniform in thickness, porosity and density, from a slurry of particulate material.

A porous sheet material also is provided, with one or a plurality of layers, each composed of a microporous multistratum layer which may optionally be adherent to a porous substrate and having a high voids volume, a pore size of less than 25 μ, and substantially uniform in thickness, porosity and density.

An apparatus is provided useful for preparing such material, such as a Fourdrinier machine, including one or more barriers such as doctor blades positioned to control deposition of slurried particulate material upon a support so that the deposition of deflocculated and uniformly flocculated material is uniform, and large nonuniform flocs cannot be deposited on the support.

Where the microporous material is employed to remove bacteria and other microorganisms from a fluid, the microporous layer should have a pore diameter within the range from about 0.03 to about 0.8 μ.

Filter units and elements comprising the microporous material of this process are capable of absolutely removing from fluid particles as small as 25 μ in size and even particles of from 10 down to 0.03 μ and smaller. At the same time, a relatively deep layer can be provided, that protracts the life of the microporous filter by serving as a type of depth filter, providing a high voids volume for retention of particles.

Figure 2.7a is a schematic view of an apparatus for forming microporous fibrous sheet materials substantially uniform in thickness and density and porosity. Figure 2.7b is a cross-sectional view of Figure 2.7a along the lines **2–2**. Figure 2.7c represents a cross-sectional view through a multilayer tristratum microporous material, showing the fibers of the first two microporous strata extending generally in the plane of the layer, and the proportion of fibers of the layer extending at an angle greater than about 30° to the plane in the top stratum of the layer, and the high voids volume there, due to the wide spacing of the fibers.

The apparatus as shown in Figures 2.7a and 2.7b comprises an endless Fourdrinier wire mesh belt **10**, which is driven by two spaced-apart rolls **12** and **14**, over a plurality of guide rolls **15**. The belt travels between closely abutting side frames **11**, which engage the edges of the belt in a fluid-tight seal, and with the belt as the moving bottom constitute a reservoir to retain a slurry of particulate material on the belt to the desired depth.

A head box **16** for delivering a slowly flocculating slurry of particulate material to the wire belt is positioned in close proximity to the roll **14**, which is the breast roll, and the belt. The head box includes a paddle wheel **2** which creates a shearing action at the head of feed passage **3** through which slurry is led to the belt. The passage has a long horizontal end portion **4** or apron which transfers the slurry to the belt, and transit time through this is long enough to permit flocculation to begin. By the time the slurry reaches the belt, it is partially flocculated.

FIGURE 2.7: PRODUCTION OF MICROPOROUS FIBROUS SHEETS

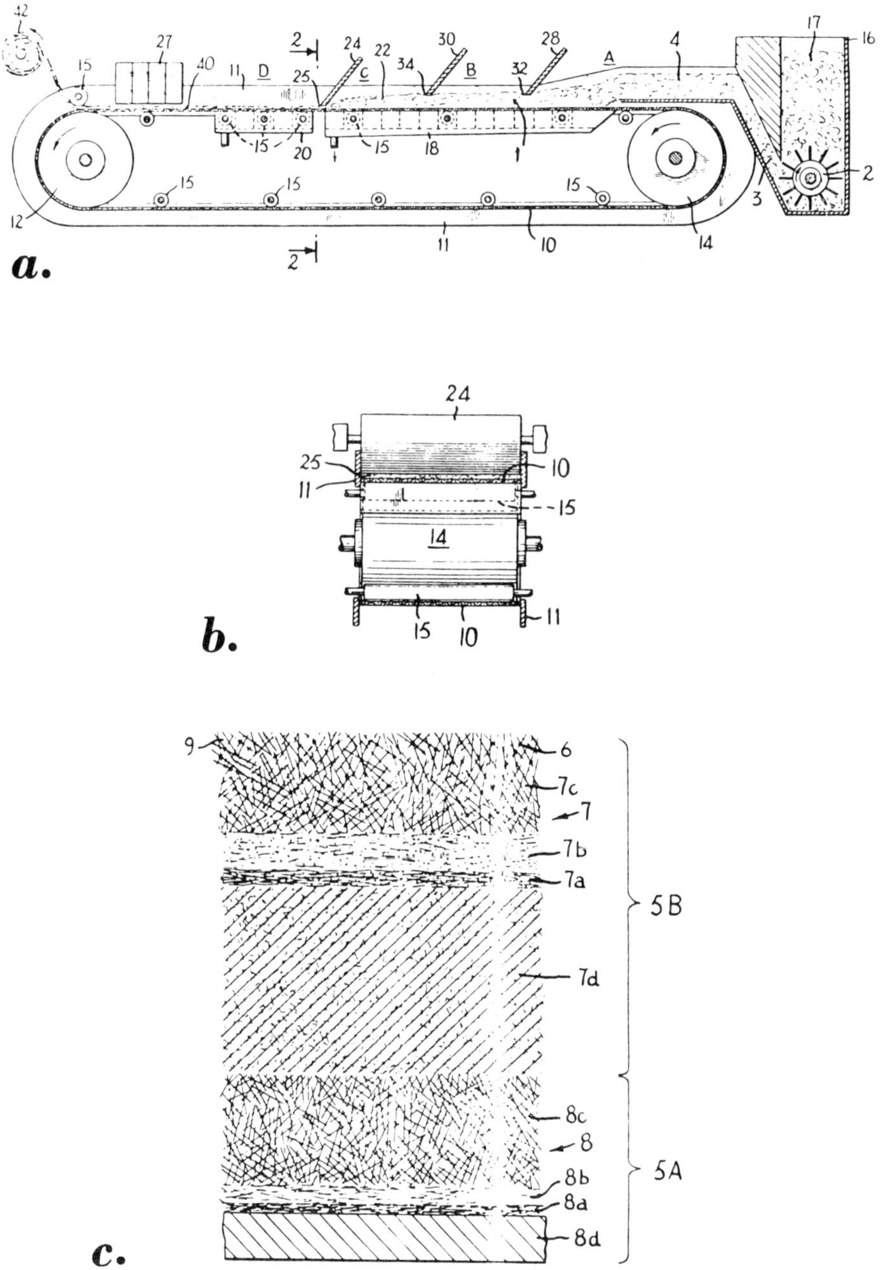

Source: U.S. Patent 3,573,158

A series of gravity drainage boxes **18** and vacuum or suction boxes **20** and a series of dryers **27** are disposed in close proximity to the wire belt. Immediately prior to the vacuum boxes and between the gravity boxes and the vacuum boxes, a primary deflocculating blade **24** in the form of a stainless steel blade having a thickness of about 0.010 to 0.015 inch is adjustably disposed above and across the wire belt to define a gap **25**, from about 0.01" to about 0.5" high, and inclined in the direction of flow at an angle to the wire belt within the range from about 30° to about 45°.

The gap is adjusted so as to restrict fluid flow, and build up a predetermined head or depth of slurry **1** on the belt in the relatively quiescent zones **A, B, C,** of gravity drainage, so as to control gravity settling at a desired rate there, and also reduce the depth of the slurry on the wire belt over the vacuum box to the desired depth, and break up large flocs and clumps of particles in the slurry before the slurry is drawn down on the support over the vacuum boxes.

Disposed between the primary deflocculating blade and the head box **16** in the quiescent zones are auxiliary deflocculating blades **28** and **30** which are disposed above and across the wire belt to define gaps **32** and **34**, respectively. The gaps defined by the blades are from about 2.5" to about 0.2" high, and the blades are inclined with flow at an angle to the wire belt within the range from about 30° to about 45°. The auxiliary deflocculating blades are especially important in the case of a slow draining slurry, to break up large flocs in the slurry in this zone.

In carrying out the method, using the apparatus as shown in the figures, feed stock or slurry **17**, such as an aqueous slurry of asbestos fibers, is fed from the head box onto the wire mesh belt **10** in zone **A** where it is retained by the blade **28**, within the reservoir created by sides **11** of a depth of, for example, about 3", and then passed through gap **32** into zone **B**, where depth is controlled by blade **30** to, for example, about 2", initially, and then through gap **34** into zone **C**, where depth is controlled by blade **24**, to, for example, about 1.5", and finally through gap **25** into zone **D** over the vacuum boxes.

In zone **A**, about 5 to 10% of the slurry fluid drains quickly by gravity through the wire mesh belt, to form the initial stratum. The slurry is only partially deflocculated, if at all, at this stage. Thereafter, flocs present in the slurry which are larger than gap **32** are broken up, as the slurry passes through this gap into zone **B**. In this zone, another 5 to 10% of the slurrying fluid of the dispersion, which has now become from 80 to 100% flocculated, gravity-drains through the wire mesh belt forming the first portion of the intermediate stratum **22**. It then passes beneath the blade **30** into zone **C**, and by the end of this zone, from about 15 to about 75% of deflocculated particulate material in the flocculated slurry, which is substantially free of large flocs, has gravity-settled on the mesh belt to form an intermediate stratum.

The supernatant slurry of particulate material above the initial layer is passed through the gap **25** beneath blade **24**, and as the slurry flows through the gap and enters zone **D** the turbulence on the head box side breaks up large flocs in the slurry and reduces the depth of the slurry to the approximate width of the gap. The fibrous material distribution in the slurry over the vacuum boxes is now uniform.

Immediately after, while the slurried flocculated fibrous material is uniformly distributed, and before large nonuniform flocs can form, the remaining supernatant slurrying liquid is drawn off over the vacuum boxes **20**, through the initial and intermediate strata, and through the mesh belt, and the remaining flocculated fibrous material is laid down as the top stratum of the layer, thereby forming a tristratate fibrous sheet material **40** substantially uniform in thickness and density. The sheet is then passed under the dryer **27**, and stripped from the mesh belt, and rolled up at **42**.

The barrier means or deflocculating blade or blades are ordinarily a thin plate of wood, glass, metal such as stainless steel, or plastic, and have a straight edge disposed across the wire mesh belt and parallel thereto. These can be provided with adjusting screws to vary the height and thickness, as is well-known in the art. Such blades, often referred to as doctor blades, should be at least as wide as the support, to ensure that substantially all slurried material must pass under the blade. In addition, the length of the blades is dependent upon type and rate of drainage of the slurry, and width, pore size and speed of the wire mesh support. The barrier means can also be in the form of a bar, plate, roll, or wedge. All such forms are referred to generically as bar or barrier.

Example 1: The apparatus shown in Figures 2.7a and 2.7b was used in the following way to prepare a fibrous sheet of uniform thickness and density, suitable for use as a filter medium, by coating an alpha cellulose-hemp paper substrate with asbestos fibers.

The deflocculating blades **24**, **28** and **30** were stainless steel strips, 42" in length and about 0.010" to 0.015" in thickness. The primary deflocculating blade **24** was positioned between the vacuum box and head box about 9/10 of the distance from the head box, and disposed above and across the wire mesh belt **10** at an angle of 40° to the mesh belt, and defined a gap about 0.024" higher over the mesh belt. The auxiliary deflocculating blades **28** and **30** were positioned between the blade **24** and the head box about 8" from each other, blade **30** being about 8" from blade **24** and blade **28** being positioned at the beginning of the belt. Blades **28** and **30** were inclined at an angle of 40° to the belt and formed a gap of about 0.375" above the belt. The wire mesh belt contained openings 0.008" x 0.012".

A slurry of crocidolite asbestos was prepared by mixing 1,500 gallons of water with 75 lb of crocidolite asbestos fibers averaging 0.01 micron in diameter and 30 microns in length. The slurry of asbestos was poured into the head box, and was fed therefrom at a rate of about 9 gal/min onto the substrate, an alpha-cellulose and hemp paper, with 25% polyvinylidene chloride as a binder, weight 4 g/ft^2, thickness 0.005", supported on a wire mesh belt traveling at a rate of 9 ft/min. In zone **A**, the slurry depth was 1.5". A portion of the water drained quickly through the belt, and there was formed an initial stratum 0.001" thick on the substrate, from about 5% of the fibers present.

The slurry of asbestos on the substrate then moved beneath blade **28** through the 0.375" gap between the blade and the mesh belt, which broke up the few flocs that had formed, into zone **B**. In this zone, the slurry depth was 0.75". The slurry passed beneath blade **30**, where the large flocs were broken up, and into zone **C**, where the slurry depth was 0.375". A portion of the water continued to gravity-drain through the substrate in zones **B** and **C**, and about 15%

of the asbestos fibers in the slurry had gravity-settled on the substrate by the end of zone **C**, within about 25 seconds, to form an intermediate stratum about 0.002" thick.

The substrate, coated with the initial and intermediate strata and the remaining water and asbestos fibers of the slurry, was passed beneath the blade **24**, in the course of which the large flocs present broke up, and the combined depth of the water and asbestos slurried therein and the thickness of the initial and intermediate strata of asbestos plus the substrate was reduced to about 0.025".

In zone **D**, the slurry and initial layer of asbestos were then passed over the vacuum box, where, before large flocs could reform, the remaining water was drawn through the substrate and mesh belt, depositing the remaining flocculated asbestos fibers on the initial layer of fibers in a manner such as to produce a top stratum thereon which had a substantial proportion of fibers extending across the plane of the layer at an angle of at least 30°. The fibrous coated substrate was of substantially uniform thickness, about 0.012".

Thereafter, the fibrous layer and substrate was passed under the dryers, dried, and then removed from the wire mesh belt. The final product was an asbestos-coated paper sheet having a maximum pore diameter of about 0.35 micron, an average pore diameter of about 0.1 micron; a water permeability of about 0.5 gal/min/ft^2 at an applied pressure differential of about 1 lb/in^2, and a voids volume of about 91%. The dried sheet had a substantially uniform thickness of about 0.014" with an asbestos layer about 0.009" thick; and a substantially uniform density of about 18 lb/ft^3, and was particularly suitable as a filter medium.

Microscopic inspection of a cross section of the material showed three microporous strata. In the first two strata, nearly all of the fibers were in planes approximating the plane of the layer, and in the top stratum a proportion of the fibers were oriented at an angle of 30° or more to the plane of the layer.

Example 2: The procedure of Example 1 was repeated, using glass fibers averaging 0.1 micron in diameter and 400 microns long, in place of the crocidolite asbestos fibers. The position of the deflocculating blade **24** was adjusted so that the gap between the bottom of the blade and the wire mesh screen was 0.06".

The slurry of glass fibers was prepared by mixing 1,500 gal of water with 12 lb of the glass fibers. The slurry of glass fibers was fed from the head box at a rate of 27 gal/min, onto the paper substrate on the wire mesh belt, traveling at 9 fpm, and allowed to gravity-settle through zones **A** to **C** for about 25 seconds, during which time about 50% of the glass fibers in the slurry gravity-settled on the paper, to form an initial stratum 0.001" thick, and an intermediate stratum 0.002" thick. The slurry then entered zone **D**, where the remaining liquid was drawn through, and the coated substrate was then dried.

The final product was a glass-coated paper sheet of substantially uniform thickness, about 0.008", and of substantially uniform density, about 15 pcf. The sheet had a maximum pore diameter of about 0.7 micron; an average pore diameter of about 0.4 micron; a water permeability of about 100 gal/min/ft^2 at an applied differential pressure of about 1 psi, and a voids volume of 95%.

Microscopic inspection of a cross section of the material showed three microporous strata. In the first two strata, nearly all of the fibers were in planes approximating the plane of the layer, and in the top stratum a proportion of the fibers were oriented at an angle of 30° or more to the plane of the layer.

When glass fibers averaging 3.0 microns by 5,000 microns were used in place of the smaller diameter glass fibers, a gap of 0.1" between the edge of the blade **24** and the wire mesh belt was employed. The slurry of glass fibers containing 0.1% by weight glass fibers was allowed to gravity drain through zones **A** to **C** for about 10 seconds, during which time 70% of the glass fibers in the slurry settled on the paper substrate to form an initial layer stratum and an intermediate stratum 0.001" and 0.005" thick respectively.

The remaining liquid was drawn through in zone **D**, forming after drying as the final product a sheet of substantially uniform thickness, about 0.010" thick, and of substantially uniform density, about 8 pcf. The sheet had a maximum pore diameter of about 15 microns; an average pore diameter of about 5 microns; a water permeability of about 3,000 gal/min/ft^2 at an applied pressure differential of about 1 psi; and a voids volume of about 97%. When the glass fiber sheet was held up near a source of light, substantially no light spots were observed in the sheet.

Microscopic inspection of a cross section of the material showed three microporous strata. In the first two strata, nearly all of the fibers were in planes approximating the plane of the layer, and in the top stratum a proportion of the fibers were oriented at an angle of 30° or more to the plane of the layer.

Silicone Resin Bonded Glass Fiber Filter Tube

A process described by *B.A. Head; U.S. Patent 3,972,694; August 3, 1976; assigned to Whatman Reeve Angel Limited, England* relates to an improved nonwoven fibrous filter tube composed of inorganic fibers, particularly glass fibers with a hardened silicone resin as a bonding agent. The filters may be used in steam filtration applications, have improved resistance to acids, and are characterized by a light white color and high hydrophobicity.

The process provides a filter tube composed of glass fibers, which filter tube is similar or better in filtration properties, e.g., flow rate and efficiency, as prior art filters with epoxy resin as a bonding agent, but of different physical properties which permit the filter tubes to be employed in demanding uses not possible by such epoxy resin tubes.

A filter tube is prepared by adding a volume of water to a beater, and while circulating the water, adding a quantity of the borosilicate glass microfibers to disperse the fibers. Filter tubes are formed by introducing a vacuum-type mandrel into the furnish to form a wet mass of nonwoven microfibers on the mandrel surface of the desired thickness. After formation, the tubes are then racked and dried at a temperature of from about 70° to 90°C.

The dried tube is then immersed in a silicone resin solution composed of approximately 7% by weight of a curable polysiloxane resin (I.C.I. R 282 resin) with 0.5% zinc catalyst as a curing agent in a solution of methyl ethyl ketone. The immersed tube is removed, the solvent evaporated, and then the tube is cured at 200°C for about 1 hour. The dried tubes are then cut to the desired length for use. The resulting dried filter tube of the process is composed of approximately 75% borosilicate glass microfibers and 25% silicone resin binder.

The tubes so prepared can be used for steam filtration applications, since they have good chemical resistance to steam. Steam of up to 100 psig (168°C) can be filtered for thirty 10 minute cycles during which flow is reduced from full (to atmosphere) to minimum, with only minor strength reduction (25%).

A comparison of the steam resistance of these tubes and standard epoxy resin tubes prepared in the same manner showed that after a steam cycle of 20 psi for 20 minutes, the strength of standard tubes was 30% of original, while the silicone tubes were 90% of original.

The superior resistance to steam makes the tubes superior for sterile air filters which must be sterilized by steam. In addition, the markedly increased hydrophobicity of the filter gives reduced condensate retention and higher flow capacities after steaming.

The reduced condensate retention of the tubes was demonstrated by steaming a standard tube and the tube of the same grade for 20 minutes at 20 psig. Immediately after steaming, air was passed through each tube at 80 lpm and the pressure differential across the filter plotted against time. The pressure differential across the tube was increased 67% by condensate retention and then fell steadily to 6.7% after 30 minutes. By comparison, the pressure differential across a standard tube was initially increased 217% and then fell to 83% after 30 minutes.

The silicone resin tubes show a marked increase in their resistance to degradation by strong acids, and, therefore, can be used for the filtration of such chemicals.

	Initial Collapse Strength, psi	Collapse Strength After 72 hr Soak in 10% HCl, psi	Collapse Strength After 72 hr Soak in 50% HNO_3, psi
Standard tube	29.5	19.5	9.5
Example tube	27	32	29

The extreme hydrophobicity of the tubes finds useful application in certain areas of filtration, such as coalescing or phase separation of emulsions, such as oil-in-water emulsions or removing oil droplets from a moisture-laden gas stream, and in other applications where hydrophobic filters are desirable.

OTHER FIBROUS FILTER MEDIA

Bonded Fibrous Product Using Treated Roving

A. Renjilian; U.S. Patent 3,834,547; September 10, 1974; assigned to Albany International Corp. describes an improved roving capable of being mechanically worked without first forming a yarn. The improved roving is in the form of a strand of staple fibers prepared in accordance with customary textile carding process. The formed roving is treated with a resin binder and, when cured, the fibers resemble a reticulated network.

The fibers are disoriented and cross over one another at a multiplicity of points to form bonding sites at the crossover point when the resin is cured, while maintaining the fibers spatially separated between the bonding sites. The thus treated roving is found to have good flexibility and abrasion resistance while providing the roving with unexpected tensile strength, increased elasticity, permeability, void volume and porosity even when the roving is subjected to compressive forces as compared with untreated roving subjected to the same forces.

Figure 2.8a is a schematic view of roving composed of a strand of condensed fibers; Figure 2.8b is a close-up view of the roving of Figure 2.8a showing the entangled fibers and numerous points of crossover or overlap between adjacent fibers; Figure 2.8c is a close-up view of a pair of fibers at a point of crossover or overlap showing a bonding of the fibers at the point of crossover; Figure 2.8d is a top plan view of a section of fabric woven in the filling direction; Figure 2.8e is a sectional view taken along the line **5–5** of Figure 2.8d; Figure 2.8f illustrates resistance to flow of fluids through filter media made in accordance with this process as compared with untreated roving and/or yarn as a function of varying flow rates; Figure 2.8g is a fragmentary perspective view illustrating treated roving wound about a perforated core which is useful as a filter cartridge; and Figure 2.8h is a top plan view of a section of fabric woven with treated roving both in the warp and filling directions.

FIGURE 2.8: BONDED FIBROUS PRODUCT

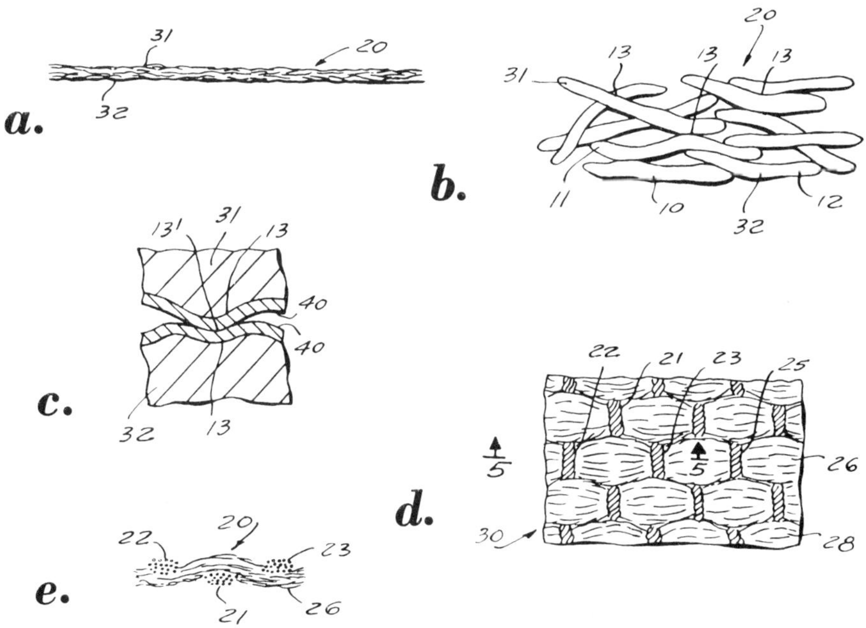

(continued)

FIGURE 2.8: (continued)

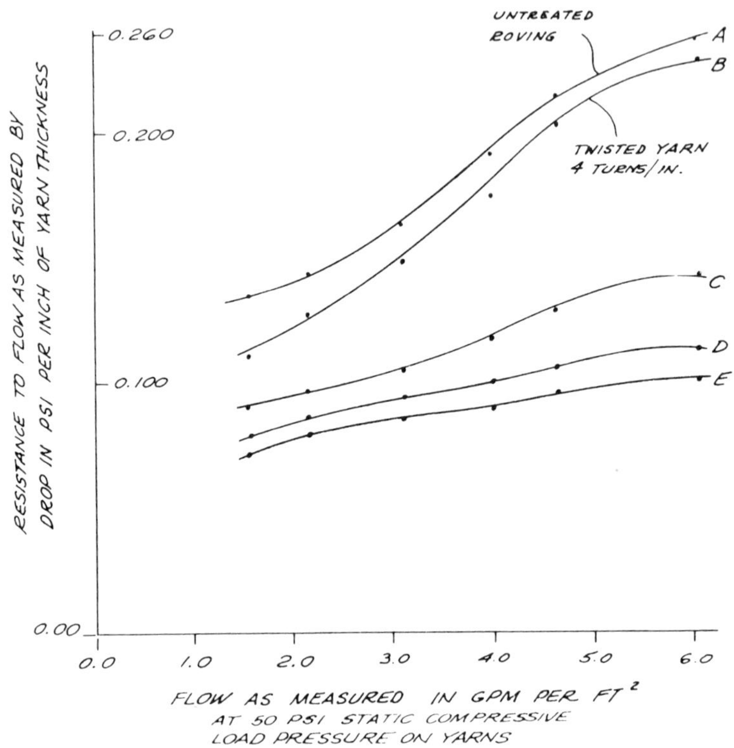

f.

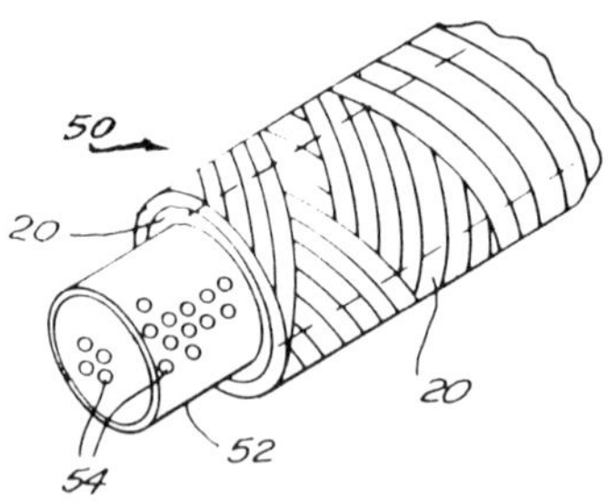

g.

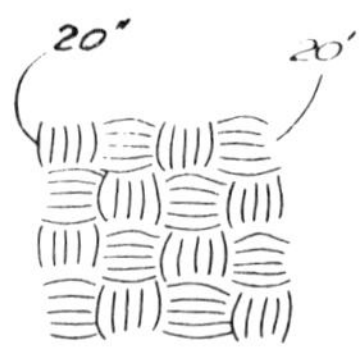

h.

Source: U.S. Patent 3,834,547

Figure 2.8a is a schematic view of roving **20** which includes a strand of entangled staple fibers such as **31** and **32** which form a condensed fiber reticulated network. These fibers cross over one another at numerous points **13** throughout the roving. Figure 2.8b is a close-up view of condensed fibers **31** and **32** having points of crossover or overlap **13**. Figure 2.8c is a greatly enlarged view of a bonding site at a point of crossover showing condensed fibers having adhesive coating **40** applied over their surfaces.

Figure 2.8d is a top plan view of fabric **30** woven from yarn **21, 22, 23** and **25** in the warp direction and roving **26** and **28** in the filling direction. Figure 2.8e is a sectional view of fabric **30** showing yarn **21, 22** and **23** in the warp direction and roving **26** in the filling direction.

The fibers within roving **20** are bonded at points of crossover due to the adhesive applied to the staple fibers **31** and **32** forming this treated roving which is not twisted, as is yarn **21, 22, 23** and **25** and is adaptable for weaving into fabric as shown in Figure 2.8d. It is to be noted, however, that adhesive **40** is applied to the fibers **31** and **32** respectively along the fiber lengths as well as at the point of crossover **13**. The adhesive improves the strength of this roving due to formation of bonding sites **13'** at points of crossover.

Figure 2.8g illustrates one form of a filtration device **50** in which a continuous strand of treated roving **20** is wound about a perforated core member **52** having perforations **54** uniformly distributed over substantially its entire surface. In operation the filtration device is inserted into a filter cartridge housing in which the fluid to be filtered is introduced through core **52**. The fluid passes through the perforations and is filtered by the treated roving by passing through the voids formed in the treated roving as illustrated in Figure 2.8b. The filtered fluid is then carried away after passage through the roving.

In Figure 2.8h a fabric is illustrated which is useful as a filtration medium, the fabric being formed of roving **20'** and **20"**. Roving **20'** is woven in the warp direction while roving **20"** is woven in the filling direction. Fabrics of the type illustrated in Figure 2.8h are useful as filter pads or other filter aids in removing particles from a fluid medium.

Example: Roving made in accordance with any customary textile carding process is wound on a spool forming a plurality of concentric layers of roving commonly referred to as a cheese. When a sufficient weight of roving has been wound about the spool, for example, 300 grams, a new spool is prepared and the wound roving is then ready for further processing in accordance with this process.

An adhesive material, for example, 5 parts of Elvamide 8061, a nylon that is soluble in lower alcohols, is dissolved in 80 parts of ethyl alcohol and 15 parts water. The solid nylon resin is dissolved by mixing, using moderate heat to facilitate solution of the solid resin. After the resin is in solution the temperature is maintained at room temperature but below the boiling point of the solvent in the resin solution to avoid distillation of the solvent.

The formed cheese is lowered into the resin solution and allowed to become impregnated. After complete impregnation the cheese is removed from the resin solution and the excess adhesive resin is removed, for example, by centrifugation techniques, squeezing or any other known means. After removal of excess adhesive solution the remaining solvent is removed as by air drying or by placing the cheese in a controlled temperature oven.

After drying has been completed the roving which initially weighed 300 grams increased in weight by 21 grams of 7% adhesive on dry weight basis. The adhesive carried by the fibers in the strand forming the roving is cured generally at the same time the solvent is removed. The fibers are bonded at their crossover points to form bonding sites so that the fibers remain spatially separated.

The roving made in accordance with this example is then adapted to be made into various products such as a fabric of Figure 2.8d without the necessity of first forming the roving into a yarn. In Figure 2.8f treated roving of the process was compared with untreated roving as a filtration medium. All tests were made at a static compressive load of 50 psi on yarns and roving. All of the roving and yarns were formed into pads having the same configuration, each pad being identical except for the treated roving which was made in accordance with the process herein. Each pad consisted of four layers of roving or yarn with each layer positioned at 90° with respect to the adjacent layer.

In the curves of Figure 2.8f the resistance to flow as measured by a drop in psig/in of yarn of roving thickness is shown graphically as a function of flow as measured in gpm/unit area (ft^2).

The results show that all yarns or roving exhibit an increase in resistance to flow as the rate of fluid being filtered increases. However, the resistance to flow increases at a greater rate for untreated roving and yarn (see curves **A** and **B**) than for treated roving (see curves **C, D** and **E**).

The untreated roving and the untreated but twisted yarn of curves **A** and **B** respectively exhibit a greater resistance to flow, that is, an increased pressure drop than the treated roving of curves **C, D** and **E**. Thus, it is apparent that the untreated roving and the untreated twisted yarn require greater work to be exerted on the fluid medium being filtered thus decreased force is required to filter a like volume of fluid through the treated roving of the process illustrated in curves **C, D** and **E**.

The roving of the process is capable of being employed as improved filter media for example by winding the roving about a perforated core and encasing the wound core into a housing fitted with an inlet and outlet opening. Thus, fluids such as liquids or gases may be filtered through the treated roving uniformly with high removal efficiency of particles entrained in the fluid while maintaining a relatively low pressure drop across the filter unit.

Also, increased flow rates of the fluid medium are obtained without loss of filtration efficiency. Further, no adverse filtration effects occur with the increase in temperature of the fluid medium being filtered. The roving is of high strength and does not require twisting. Also, the treated roving has improved physical characteristics of increased bulk density which maintains the relatively high percentage of voids, resistance to compression, increased abrasion resistance, permeability and flexibility.

Fiber Reinforced Plastic Filter Plate

W. Johnson, Jr.; U.S. Patent 3,497,065; February 24, 1970; assigned to GAF Corporation describes a process which provides a filter plate member formed from needled felt impregnated with in situ cured resin and having one or more fluid ports and drainage passages formed therein, the plate member also being provided with a recessed drainage area or zone communicating with the drainage passages. The drainage passage or passages extend axially or laterally from the fluid port or ports and may include elongated slots extending from face to face of the plate member. The thickened marginal portion of the plate member is of less density than the internal recessed portion to provide compressible gaskets. A groove extending around the recessed area may be provided in the thickened portion. The plate member, in its usual form, is provided with a recessed drainage area or zone on each surface of the plate member, and each such drainage zone communicates with the drainage passage or passages. The plate member is also preferably comprised of a plurality of sheets or laminations of the resin impregnated needled felt.

Figure 2.9a is a side elevational view of a fiber reinforced plastic filter plate; Figure 2.9b is a broken section taken along the line **2–2** of Figure 2.9a; Figure 2.9c is an enlarged section taken along the line **3–3** of Figure 2.9a; Figure 2.9d is a perspective view illustrating the method of making the plastic filter plate; Figure 2.9e is an enlarged view of the plug used in forming the drainage ports of the filter plate; and Figure 2.9f is a side view of the plug shown in Figure 2.9e.

FIGURE 2.9: FIBER REINFORCED PLASTIC FILTER PLATE

a.

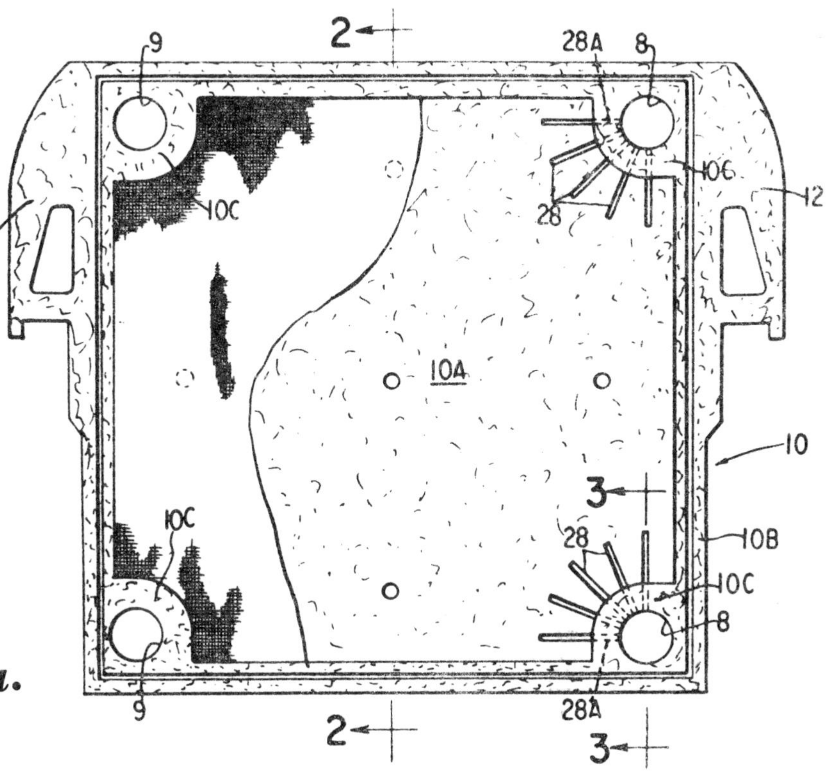

(continued)

FIGURE 2.9: (continued)

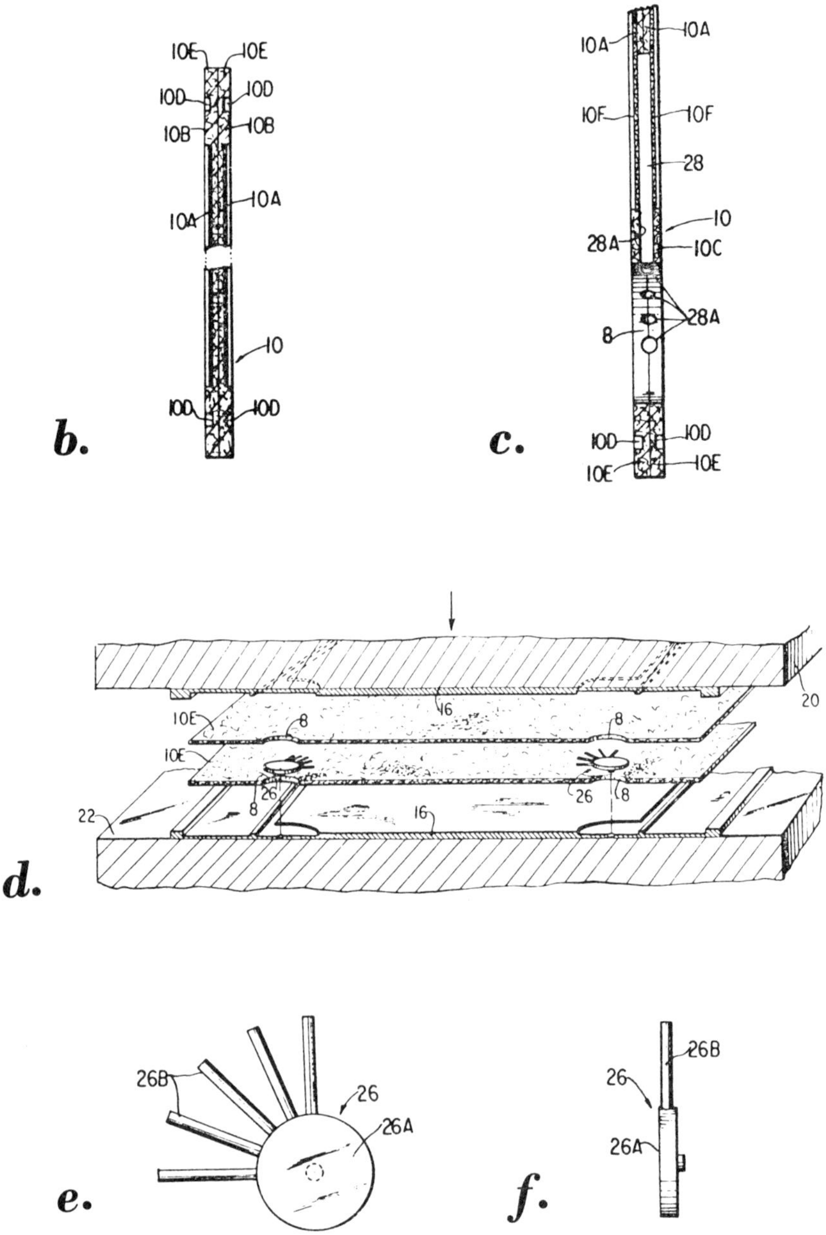

Source: U.S. Patent 3,497,065

Referring to Figures 2.9a, 2.9b and 2.9c, there is shown a filter plate **10** comprising a generally rectangular plate member provided with fluid discharge ports **8** and apertures **9** forming ports for incoming fluid. As is known in the plate and frame press art, a plurality of the plates are contained between spacer or frame members having openings into the inlet port or ports of the press and are maintained in assembled position by bolted rods or the like extending through ears or projections **12** on the plate members.

The plate member comprises a thickened marginal portion **10B** and a recessed area **10A** on each face thereof, providing drainage zones for fluids passing through a filter material extending across the faces of the plate outside of the recessed areas. The thickened portion is widened at each of the corners of the plate to accommodate the ports as indicated at **10C**.

A plurality of elongated finger-like slots **28** are provided in the recessed portions adjacent the discharge ports and extend in a radial outward direction from the ports. The slots also extend entirely through the plate, from face to face, at the recessed areas as shown in Figure 2.9c, so as to communicate with the drainage zones on each face of the plate. The slots also communicate with channels or openings **28A** provided in the thickened portions **10C** between the ports and the slots.

The thickened portion **10B** is also preferably provided with grooves **10D** which surround the recessed areas on each face of the plate. The plate is also preferably comprised of a plurality of sheets or laminations **10E** bonded together by the impregnating resin to provide a unitary plate structure as shown in Figures 2.9b and 2.9c. A lattice or screen-like member **10F** may also be contained within the recessed areas to assist in drainage of filtered fluids toward the drainage passages **28** and the ports **8**.

An important feature of the process, however, is the composition of the filter plate member. This member comprises a needled felt impregnated with in situ cured heat- and chemical-resistant resin. By needled felt is meant a batt of fibers which have been needle punched mechanically to cause certain of the fibers to be bent to extend transversely of other fibers. The felt material preferably comprises man-made fibers, such as nylon, Dacron and other acrylic, olefin or polyester fibers, or blends thereof with or without natural fibers such as cotton. The resin may be of a diallyl phthalatic type and constitutes at least 30% of the weight of the cured plate member, and preferably in the neighborhood of 65%.

The resin-impregnated needled felt is cured under heat and pressure to form a unitary, nonmetallic fiber-reinforced plastic filter plate having adequate strength for a filter press and yet having sufficient flexibility so as normally not to break under adverse pressure conditions, such as clogging on one side of the plate. The plate, including two laminations bonded together, has a cured or finished thickness at its marginal portion of about one-half inch, thereby permitting a greater number of plates to be used in a press of a given size. As a result of the composition and the method of forming the plate, its marginal portion has less density than its recessed or drainage portion, thereby inherently providing a compressible gasket means for the plate.

Another feature of the process is the method of forming a filter plate of the character above described. Referring to Figures 2.9d to 2.9f, and more particularly to Figure 2.9d, there are shown two mold press platens **20** and **22** having preformed molding plates or dies **16** attached. Two sheets or laminations **10E** of the filter plate **10** are positioned between the dies, each such sheet having fluid port openings **8** precut or preformed therein. The sheets are composed of needled felt impregnated with a resin which is heat- and chemical-resistant when cured and contains at least 1% of volatiles in the uncured state.

Interposed or inserted between the sheets is a plug member **26** shown in Figures 2.9e and 2.9f, having a central portion **26A** conforming to the shape of the fluid ports provided in the sheets. Extending radially outwardly from the plug member are a plurality or rods or fingers **26B** adapted to form the slots **28** in the sheets. The central portion of the plug member is adapted to fit within the fluid port openings of the sheets and the rod-like fingers extend between the sheets prior to a pressing or molding operation. The plug member including its rod-like fingers may be made of polytetrafluoroethylene or other suitable materials.

The dies are shaped or formed to provide the areas **10A**, the slots **28** and the grooves **10D** in the sheets when the platens are moved toward each other under pressure. At the same time, heat is applied to the platens and dies suitably to cure and set the resin of the sheets. The diameter or thickness of the rod-like fingers is substantially greater than the combined thickness of the depressed areas when the dies are closed, thereby to provide the slots in the sheets. At the same time, the channels or openings are formed in the thickened marginal corners **10C**. After a suitable curing time, which will bond all layers together, the platens and attached dies are opened, the sheets and the plugs are removed.

The method thus described provides a simple, facile and inexpensive means for forming fiber reinforced plastic filter plates. The plates, as above set forth, are flexible so as not readily to be broken under adverse pressure conditions and have a minimal thickness. At the same time, they are sufficiently strong, are impervious to filtered materials, and are heat- and chemical-resistant. They also provide an inherent gasket means.

Polymer Coated Fibers

H.M. Mikulski and H.E. Bixler; U.S. Patent 3,599,797; August 17, 1971; assigned to Met-Pro Water Treatment Corporation describe a filter which may be broadly classified as being of the molded cartridge type. The filter material used is a thermoplastic polymer coated fiber which may be a natural or synthetic fiber of the type described in U.S. Patent 3,121,698.

The filter has the following advantages:

(a) the filter can be constructed to have any given porosity;

(b) for any given porosity, the filter has a low density of 3 to 8 g/in^3; and

(c) with a pore size of one-half micron, the filter has a high pore volume of 80 to 90% and a high throughput of approximately 10 gpm at 2 psi pressure differential.

The coated fibers lie in substantially parallel planes generally parallel to the direction of flow so as to have a net fiber angle of approximately zero with the direction of flow. Since the fibers are coated with a polymer, the fiber substrate does not contact the fluid being filtered and therefore is not subject to attack by the fluid. In addition to being usable for filtering out solid particles from a gas stream, other fluids may be filtered. Thus, the filter can be used as a liquid separator for coalescing liquids such as water from perchlorethylene or separating high surface tension liquids from liquids such as gasoline.

If desired, activated carbon or other materials may be incorporated into the filter mass where adsorption is also desired. The method comprises placing fibers encapsulated in a polymer in a mold so that they occupy from 2 to 10% of the volume of the mold. Thereafter, the fibers are compressed by applying force in a direction generally perpendicular to the intended direction of flow. These steps are repeated until the desired filter mass is attained. Then the mold is placed into a pressure vessel wherein steam is generated to cause the polymer coatings to coalesce. After the filter mass has been subjected to the steam for a given period of time, it is removed from the mold and cooled.

If the filter is constructed so as to have zones of different porosity at different radii from the longitudinal axis, the thusly formed filter mass may be sealed at its ends in any convenient manner such as by applying a hot platen to the ends of the filter mass, thereby rendering them impervious to liquids. The ends may be subjected to a cutting step before the step of sealing the ends. The thusly constructed filter will be a self-supporting rigid structure which can withstand radial and axial loads in excess of 25 psi without substantial deformation. If higher load ratings are required, structural sections can be molded integrally with the fibers for added strength such as by use of reinforcing bars. The filter can be produced in any desired form including flat discs, hollow tubes, solid cylinders, etc.

The nature of the manufacturing process causes a gradation of the pore sizes. This gradation of pore sizes is in direct proportion to the density or amount of fibers per unit volume of the element. The manufacturing process can be varied so that the gradation of pore size can be with the larger size pores on the outer surface or on the inner surface as desired. Further, the size of the pores can be maintained substantially uniform.

Since the pore size can be so maintained, the filter is not dependent on tortuous path entrapment for its filtration efficiency. Thus, the filter can exhibit the best properties of a screen-type filter. That is, it will trap all particles larger than its pores while passing substantial portions of particles smaller than its pores. If the density varies at different radii from the longitudinal axis, the filter can perform as a depth filter whereby progressively smaller particles will be trapped until the minimum size pores are obtained between the inner and outer peripheries of the filter.

Electrically Conducting Filter Media

According to a process described by *F.E. Colvin, J.R. Mummert and F.G. Gilbert; U.S. Patent 3,933,643; January 20, 1976; assigned to The Carborundum Co.* conductive filter elements are made by treating nonconductive fibrous materials such as inorganic or organic fibers with resins containing finely divided carbon. The treated fibers are then fabricated into filter elements which are electrically conductive throughout. Strands of roving or yarn which are nonconductive may be joined by twisting together with strands of conductive roving or yarn to give conductivity to the finished multiple strand.

Conductive filter elements are also made from extruded fibers which are filled with carbonaceous material during extrusion. These elements permit grounding of static electrical potential which would otherwise accumulate in a nonconductive element. If desired, a controlled low potential direct or alternating electrical charge can be maintained on the filter element to enhance the separation of particles of opposite charge. Particulate filter media, such as that used in filter beds, may also be treated with carbon to form an electrically conductive filter mass.

Figure 2.10a is a perspective view of a sandwich construction having two sheets of porous fibrous material with conducting particles of conductive carbon disposed between the sheets, Figure 2.10b is a sectional view taken on the line 2–2 of Figure 2.10a, Figure 2.10c is a modification showing a porous filter sheet having conductive carbon fibers disposed throughout the sheet, Figure 2.10d is a plane view of a filter disc cut from the material of either Figure 2.10b or 2.10c, Figure 2.10e is a flat filter element adapted for air or gas filtration, Figure 2.10f is a top plane view with part broken away of a pleated liquid filter, Figure 2.10g is a horizontal sectional view through a cylindrical filter formed by wrapping a sheet of filter material spirally about a central core, Figure 2.10h is a top plane view of a molded filter, and Figure 2.10i is a lengthwise sectional view of a grounded filter cartridge.

FIGURE 2.10: ELECTRICALLY CONDUCTING FILTER MEDIA FOR FLUIDS

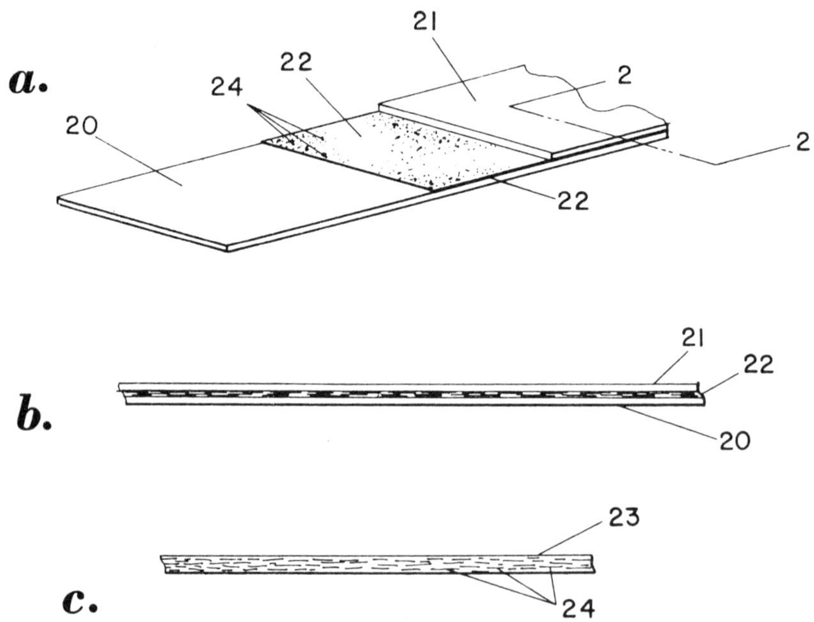

(continued)

FIGURE 2.10: (continued)

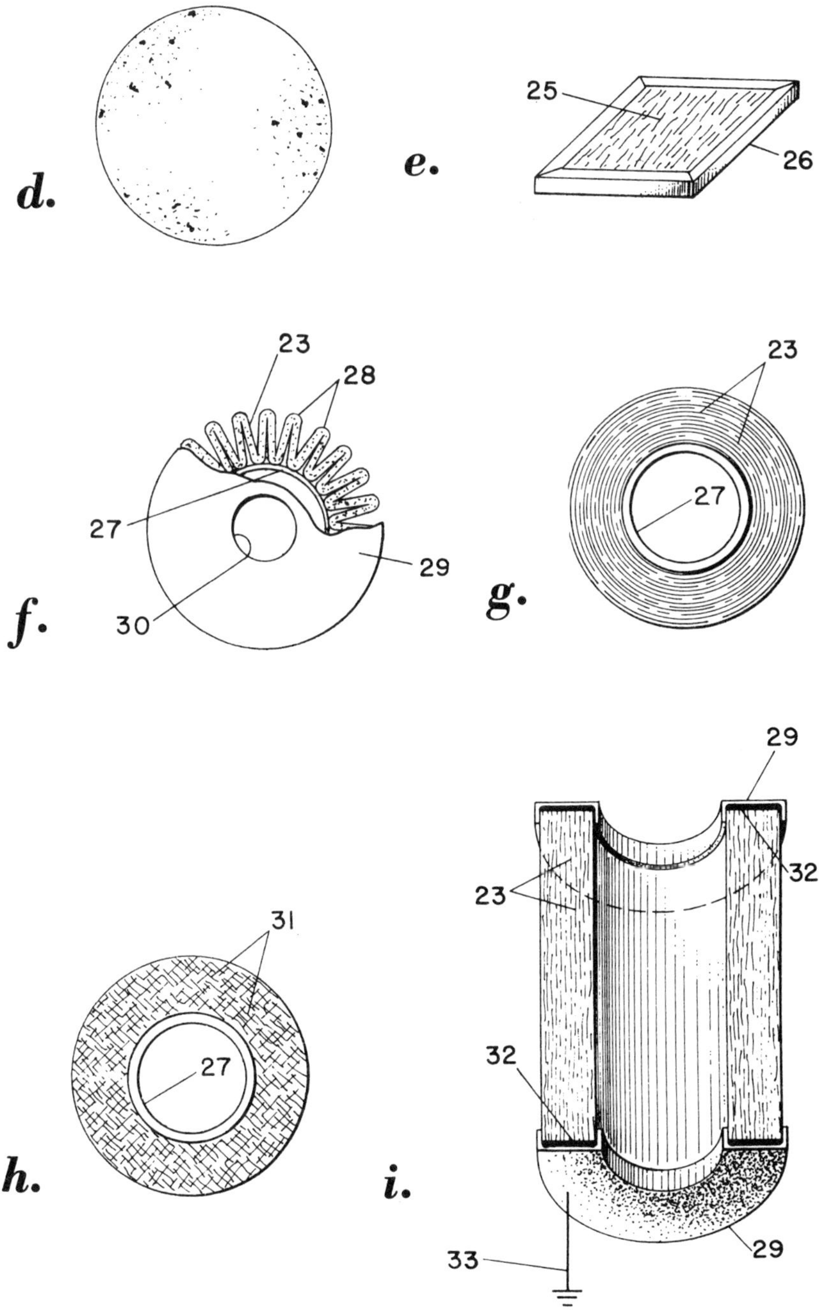

Source: U.S. Patent 3,933,643

One type of filter element construction is shown in Figure 2.10a in which two sheets of porous fibrous material **20** and **21** are assembled in a sandwich fashion, enclosing between them a layer of bonding resin **22** containing carbon particles or fibers **24**, the particles or fibers being so disposed as to furnish an electrically conductive path through the resin layer.

The porous fiber sheets may be formed of various fibers such as wood fibers, cotton linters or wool, as well as inorganic fibers made from alumina, silica, glass or asbestos. Synthetic fibers of nylon, Dacron, polyethylene or polypropylene may be used, also fibers drawn from phenolic, melamine or vinyl chloride resins. These materials may be blended, if desired, and the fibers may be water-laid or air-laid to form a sheet or batt. A cross section of the filter element is shown in Figure 2.10b.

Although the conductive carbon-containing part of the filter element is shown as a sandwich layer in Figure 2.10b, it is not restricted to this form but may be distributed throughout the filter element as shown in Figure 2.10c. In this preferred modification a porous fibrous filter element **23** has carbon particles or fibers **24** disposed throughout the element. The carbon particles or fibers are preferably mixed with the fibrous stock at the time the element is being formed. Resins or latexes such as melamine-formaldehyde, phenol-formaldehyde or similar melamine or phenolic resins may be used to bond the conductive material to the fibers. Regenerated cellulose available in powder form for solution preparation or other cellulosic solutions may be used, depending on the end use of the filter element.

Filter elements as shown in Figures 2.10b and 2.10c may be employed in various types of filters. One such filter is shown in Figure 2.10d in which the element of either Figure 2.10b or 2.10c is cut to form the disc shown, which may be secured in the desired filtering position to act as a surface type filter. Another modification is shown in Figure 2.10e in which the element **25** is made of a more loosely felted fiber batt, adapted for air or gas filtration, with the fibrous element being supported in a metal frame **26**.

The filter elements as shown in Figures 2.10b or 2.10c may also be employed to form a pleated filter element as shown in Figure 2.10f where the filter element **23** is pleated as indicated at **28** and the pleats are arranged in the form of an annulus about a perforated center tube **27**. The filter construction thus formed is preferably provided with end caps **29** which are bonded to the opposite ends of the element, such end caps having a center hole **30** to receive the usual filter support.

A similar type of construction is shown in Figure 2.10g in which the filter element **23** is wound spirally upon itself around a perforated center tube to form a filter cartridge. Alternatively, the cartridge may be formed by winding a yarn or roving of a conductive fiber or mixture of fibers in a honeycomb pattern on center tube **27**, or the cartridge may be formed from a molded or vacuum formed cylindrical mass of fibers **31** as shown in Figure 2.10h.

In the constructions described, the carbon particles or fibers are distributed throughout the filter element in such a manner as to insure electrical conductivity throughout the element. The element is then grounded through the end caps as shown for a filter cartridge in Figure 2.10i. In this figure, a cartridge is shown with end caps **29**. These end caps are of any suitable metal and are joined to the filter elements **23** by a suitable sealing composition **32**. Since most sealing compositions are resinous in nature and nonconductive, they must be blended with a sufficient amount of finely divided carbon to provide an electrically conductive bond between the filter elements and the end caps. The end caps are then grounded as indicated by electrical connection **33**.

This ground is usually supplied by the metal filter housing in contact with the end caps so that only special cases would require a separate ground connection to the filter elements. The process is not limited to filter cartridges having end caps however, since most of the wound honeycomb and felted (vacuum formed) filter cartridges are made without end caps. These would be grounded through the metal ends of the filter housing. The conductive filter elements of the process therefore provide a method for the dissipation of static charges that may rapidly build up in the pores of nonconductive filter media. This eliminates the hazard of high static electrical charge accumulations that may cause sparking across the filter media and ignition or explosion of flammable filter fluids or fluid mixtures.

Fibrous Cellulose Acetate

A process described by *A. Caputi, Jr. and T.C. Wong; U.S. Patent 3,842,007; October 15, 1974; assigned to E. & J. Winery* relates to a form of fibrous cellulose acetate which resembles asbestos in fibrous structure and is useful as a filter medium for filtering liquids.

Asbestos which has a large surface area per unit weight and consists of very fine, fluffy filaments intermingled with larger fibers is an excellent filter medium for liquids. Because such asbestos has excellent filtering properties it has been prevalently used to filter potable liquids such as fruit juices, vegetable juices, wine and the like. However, asbestos has recently been suspected of contributing to or causing injurious effects upon human health and, accordingly, its users have been fervently searching for an adequate substitute filter material.

A new form of cellulose acetate has been found which has a fibrous structure resembling asbestos and filtering properties equal to or better than asbestos. Cellulose acetate is odorless and tasteless and, to the best of present knowledge, is completely nontoxic to humans and other life forms. It is presently used in large amounts for packaging food products. Various forms of it have been used for cigarette filters (U.S. Patent 3,420,245 reporting such use of hairy cellulose acetate particles of fairly high surface area).

The cellulose acetate is fibrous in form and comprises rod-shaped fibers intermingled with fine, fluffy filaments. In an aqueous suspension it looks similar to cotton fibers (light and fluffy). However, microscopically it more closely resembles asbestos; that is, it forms a random matrix of fine, fluffy filaments backboned by large fibrils. The diameters of these fibrils and filaments range from about 0.5 to about 50 microns and the surface area of the dried fibrous mass is in the range of about 35 to about 55 m^2/g, which is slightly greater than asbestos.

This form of fibrous cellulose acetate is made by precipitating cellulose acetate of a particular viscosity from a low concentration solution. All values of viscosity are as measured in accordance with ASTM D1343-69. Commerical grade cellulose acetate in either powder or flake form, preferably powder, with a viscosity (as measured by ASTM D1343-69) in the range of about 25 sec to about 45 sec may be used to make this fibrous cellulose acetate.

The following examples illustrate the fluffy cellulose of this process, its method of manufacture and its use as a filtering material in comparison to asbestos and in comparison to the cellulose acetate of U.S. Patent 3,420,245.

Example 1: 3 grams of cellulose acetate powder (38.3% by weight acetyl, viscosity 40 sec by ASTM D1343-69) was added to 100 ml acetone in a vessel with stirring to avoid lumping. This mixture was stirred until a homogeneous solution was obtained. 500 ml of water was then poured quickly into this homogenous solution causing a fibrous precipitate to form instantaneously.

The entire contents of the vessel were poured onto a 100 mesh wire screen and the precipitate retained by the screen was washed with running water until it was acetone-free. A 2% by weight aqueous slurry of the precipitate was formed and placed in a 15 cm diameter stirring vessel equipped with a 5 cm diameter propeller type stirrer and stirred at 2,800 rpm for 20 minutes. The slurry was then suction filtered to form a wet cake of cellulose acetate (78.9% moisture). The surface area of the dried cellulose acetate was measured at 50 m^2/g.

A small portion of this wet cake was suspended in a small volume of water. In this aqueous suspension the fibers of the precipitate appeared light and fluffy and resembled cotton fibers. This suspension was placed in a Petri dish and observed under a dissecting microscope. The precipitate appeared as a matrix of fine, fluffy filaments backboned by larger fibrils (very similar to the fibrous structure of asbestos). Filament and fibril diameters were measured microscopically and found to range between 0.5 and 50 microns.

Example 2: A fibrous cellulose acetate precipitate was formed using the method of Example 1 except that a solution of 2 grams of cellulose acetate in 100 ml acetone was employed. The fibrous structure of the resulting precipitate was essentially identical to the precipitate of Example 1.

Example 3: A fibrous cellulose acetate precipitate was formed using the method of Example 1 except that a solution of 4 grams of cellulose acetate in 100 ml of acetone was employed. The resulting precipitate had a fibrous structure essentially identical to that of Example 1.

Example 4: A cellulose acetate precipitate was prepared by the method of Example 1, except that acetone was replaced by 1,4-dioxane for dissolving the cellulose acetate. The precipitated cellulose acetate filaments and fibers were essentially identical to those of Example 1.

Example 5: A cellulose acetate precipitate was prepared by the method of Example 1, except that acetone was replaced by methyl acetate for dissolving the cellulose acetate. The precipitated cellulose acetate filaments and fibers were essentially identical to those of Example 1.

The filtering abilities of the precipitates of each of Examples 1 through 5 were tested using the following procedure. A standard turbidity curve was prepared by filtering 100 ml dry white wine through a 0.1 micron Millipore membrane filter, adding various amounts (1 to 10 ppm) of SiO_2 to portions of the filtrate, measuring the turbidity of the SiO_2-containing portions with a Beckman model DU Spectrophotometer with matched 1 cm cells at 500 nanometers and plotting the results (turbidity vs ppm SiO_2).

A suspension of each cellulose acetate precipitate in the same dry, white wine was made (0.3 gram cellulose acetate/100 ml wine) and the precipitate was coated uniformly onto a Whatman No. 1 paper, 9.0 cm in diameter, from the suspension. The coated paper was mounted on a Millipore filter apparatus, 3.8 cm in diameter. 100 ml of the same dry, white wine containing 1,000 ppm carbon particles (0.5 micron diameter) was then filtered through the coated paper and the turbidity of the filtrate was measured as above. (The filtrates of Examples 4 and 5 were measured by Tyndall light only as described below.) These measurements, reported on the basis of the ppm of SiO_2 required to give the same turbidity, appear in the table below. Each filtrate was also observed with a Tyndall light.

Example	Turbidity, ppm SiO_2
1	<1
2*	<1
3**	<1
4	No turbidity viewed with Tyndall light
5	No turbidity viewed with Tyndall light

*Flow rate was slow due to compaction of matrix indicating "tender" fibers

**Matrix was sandy when rubbed between fingers indicating fiber coarseness

The effectiveness of the cellulose acetate of this process for filtering various liquids (carbon added only as indicated) as compared to asbestos was determined using the procedure described above. For red wines, turbidity was measured with a Turner model 110 Fluorometer with a 1 cm square cuvette and a No. 25 (red) filter at a slit setting of 30X. These tests are reported in the following table.

	Turbidity, ppm SiO_2	
Test Liquid	Cellulose Acetate	Asbestos
Dry white wine	<1	<1
Dry dark red wine	<1	<1
Dry white wine plus 1,000 ppm carbon particles (size $\approx 0.5\ \mu$)	<1	<1
Concord wine (6% v/v solids)	0	0
Concord juice (20°B, 19% v/v solids)	0	0
Orange juice (13°B, 5% v/v solids)	<1	6
Pineapple juice (13°B, 5% v/v solids)	<1	5
Sucrose solution (66°B) plus 1,000 ppm carbon particles (size $\approx 0.5\ \mu$)	<1	<1
0.1% yeast solution	0	0
0.1% bentonite solution	0	0

As can be seen from the above table, on an equal weight basis, the fluffy cellulose acetate had the same filtering ability as asbestos. In fact, cellulose acetate clarified fruit juices better than asbestos.

The cellulose acetate of U.S. Patent 3,420,245 was also tried as a liquid filter material for comparison purposes. It was prepared according to the method described in U.S. Patent 3,420,245, passed through a 60 mesh screen and placed in water. It did not disperse and repeated attempts to mechanically disperse it were generally unsuccessful. Accordingly, it did not coat the filter paper support uniformly and was therefore ineffective in filtration tests.

As illustrated by the above filter tests, in using the fluffy cellulose as a filter it is taken from its stored form, e.g., a wet cake, and dispersed in a small amount of the liquid to be filtered with stirring. The dispersed cellulose acetate is coated onto a suitable filter support such as wire screen, cloth, paper or porous ceramic from the dispersion by pouring the dispersion uniformly over the support. The amount used will usually be approximately 24 grams (dry weight) per square foot in most instances.

The liquid that drains from the coated support (liquid to be filtered contaminated with the nonsolvent in which the cellulose acetate was stored) is discarded and filtration proceeds as usual. The cellulose acetate of this process may be reused by peeling it off the filter support, suspending it in water with agitation, allowing the precipitate to settle and siphoning off the hazy liquid. It is estimated that 70 to 80% of the solid contaminants may be removed in this manner. If further purification is desired, the cellulose acetate may be redissolved in acetone, the solution filtered and the cellulose acetate precipitated from the filtrate.

In the above filter tests it was observed that the flow rates of the liquids through the fluffy cellulose acetate filter were slower than the flow rates through asbestos. It has been found that blending linear fibers of cellulose (commercially available) with the fluffy cellulose acetate improved flow rates substantially with no measurable change in filtering power. If desired, the fluffy cellulose acetate of this process may be blended with other filter materials such as glass wool and diatomaceous earth to provide materials with varying flow rates and filter properties.

GENERAL FILTER CONSTRUCTION

METALLIC FILTER ELEMENTS

Ultrathin Metallic Filters

R.J. La Botz; U.S. Patent 3,515,605; June 2, 1970; assigned to Aerojet-General Corporation describes ultrathin metal members having accurately dimensioned fluid flow passages. In the method of making these metallic filters a layer of etchant-resistant material is joined to at least one layer of etchant-susceptible material having a thickness equal to the desired depth of flow passages of the members, the layers of material forming a unitary structure.

The etchant-susceptible material is masked to define a precise pattern of flow passages thereon whereby when the masked unitary structure is exposed to an etchant bath and passages are etched to expose the etchant-resistant material, an ultrathin metal member having a pattern of flow passages of precise dimensions is formed.

Referring to Figure 3.1, there are shown articles of manufacture in the nature of platelets having on at least one surface thereof, a precisely controlled pattern of elevated and depressed areas which are utilized as fluid flow passages. In the process, the depth of etch control is not obtained by controlling the length of time the workpiece is in the etch bath, nor by the bath and spray strength, but rather by constructing the platelet to have a surface which is impervious to the etchant at the desired depth in the workpiece being etched.

This is accomplished by using a sandwiched or layered construction, as shown in the figures, where two or more layers of material are metallurgically bonded together to form a single, integral sheet before being inserted in an etchant bath.

Referring to Figure 3.1a, there is shown a partial section of an article of manufacture **10** which comprises an upper layer of one material **12**, and a second material **14**. The upper layer material is of the same thickness as the desired depth of etch and is subject to attack by the chemical etchant. The bottom layer is bonded to the upper layer and is selected so as to be inert to the etchant. When the composite sheet is placed in the etching bath, the etchant removes all the unmasked portions of the upper layer of material, but etching stops when the etchant-resistant material is reached. Thus, the depth and uniformity of the resulting etched flow passages are controlled only by the thickness and uniformity of the top layer.

FIGURE 3.1: METALLIC FILTER

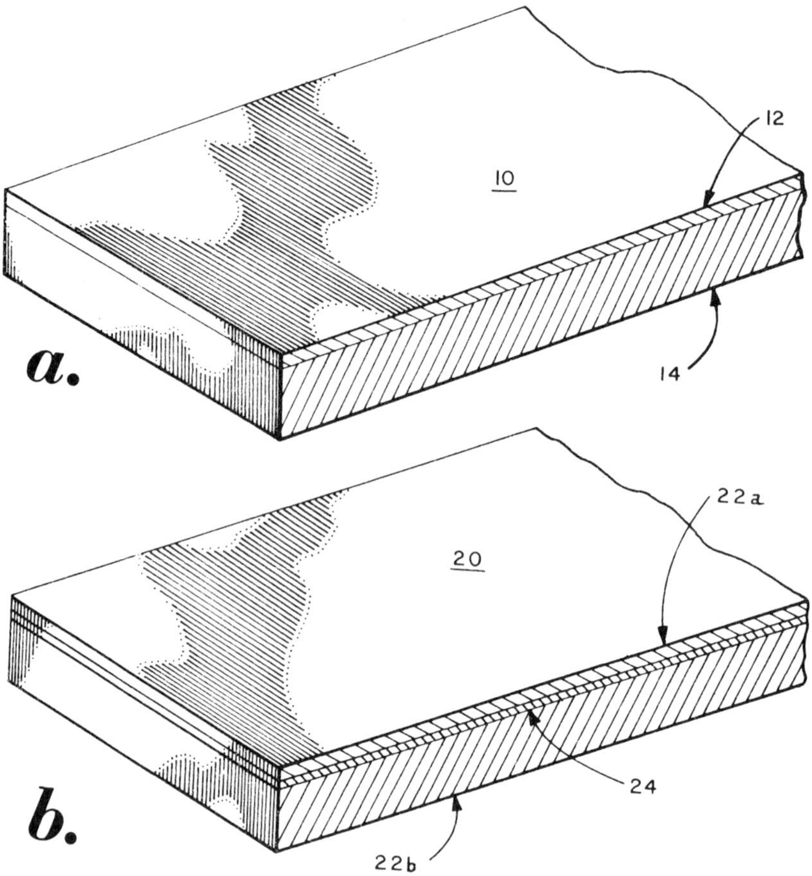

(a) Partial section view of a layered article of manufacture where the two layers are of different materials.
(b) Partial section view of a layered article of manufacture where the top and bottom layers are of the same material, but are separated by a thin foil of braze material.

Source: U.S. Patent 3,515,605

A typical example of materials which might be used in such a process are nickel and 347 stainless steel. A cladding of nickel equal in thickness to the desired depth of etch is placed on the stainless steel. After masking, the sheet **10** is placed in a hot nitric acid solution which rapidly etches the unmasked portions of the nickel layer **12**, but does not attack the stainless steel **14**.

After a suitable length of time, the piece is removed from the bath with only the masked portion of the nickel layer remaining on the stainless steel. Layered materials of the type described are available commercially. Layered materials may also be fabricated in a number of different ways. While some materials utilize diffusion bonding, the materials might also be brazed together with an extremely thin foil or plating of braze material applied to one of the mating surfaces of the layers.

In the example shown in Figure 3.1b, the upper layer **22a** and the lower layer **22b** are of the same material and they are bonded together with a foil of non-etching material **24**. This thin layer of braze material **24** is utilized to control the depth of etch. Thus, the entire surface of the lower layer is masked to withstand the etchant in the bath while the desired pattern is outlined on the surface layer of the platelet **20** by masking.

Where accurate plating techniques are available, it is possible to plate the upper layer **12** upon the lower layer **14** of the washer **10** shown in Figure 3.1a. While it is contemplated that the layer **12** will generally be of rolled material, where it cannot be rolled to the precise thickness desired because of the strictness of tolerances imposed, the top layer **12** may be made slightly thicker than desired and the final adjustment in layer thickness made by subjecting the entire surface to the etching process. In that case, if, for example, the top layer is 10% too thick, subjecting this layer to an etch process giving a 10% nonuniformity in depth of etch will result only in a 1% nonuniformity in the final etch.

Corrugated Metal Filter

A process described by *J. Atkins; U.S. Patent 3,620,375; November 16, 1971* relates to a filter which is useful for separating entrained matter, such as dirt particles or other forms of minute particles that may be carried by a flowing material such as air, liquid, or the like. The filter is formed having a metallic filtering element encased within a rigid frame.

The filter element is constructed having an inner core of a metallic filtering composition, formed from such materials as layers of expanded metal sheets laid one upon the other, sheets of porous or foraminous metal, or even an agglomeration of the turnings or shavings that are acquired from the standard operations of a lathe, the inner core being encased within a pair of outer coverings which are constructed of a more sturdy thickness of expanded metal so as to provide rigidity to the filter element itself.

The entire filter element is then exposed to a corrugating process wherein its surface area takes on a serpentine-like appearance which effectively increases the exposed surface area of the filter element. The significance of this type of arrangement in a filter element is that flowing material, such as air or a liquid, may more easily and quickly pass through a filter having a greater surface area within a frame of particular exterior dimensions than can pass through a filter of the same dimensions having a flat or planar filtering element. The entire filter element as formed, is then cut to the dimensions desired, and in the cutting process the severed edges of the metallic sheet material are pressure bonded together into an integral unit which retains its integrity and shape, and sustains the overlying relationship of the layers of sheet material even though the outer frame may not as yet have been placed around the element, or even when the element is removed from the frame.

As an example of the effective increase in the surface area of a filter subjected to a corrugating process, in a filter having dimensions of approximately 16 inches by 20 inches, and whose filter element is exposed to a three-quarter inch depth corrugating machine, when the filter element is once again expanded into a flat configuration, its 20 inch length can be expanded to approximately 36 inches, indicating that effective surface area of this particular size filter can perform just as efficiently as a filter having a flattened filter element which has a length of approximately 36 inches, instead of 20 inches.

Thus, a filter element formed in accordance with this process may be made of greatly reduced size but can perform just as effectively as the larger and more cumbersome filters. Furthermore, since the filter is constructed of metallic materials, it not only retains its shape for a much greater length of time, but it may be easily cleaned of its fine dirt or other dust particles accumulated during the filtering operation, and once again be reused just as effectively as a new filter. The filter may be repeatedly cleaned with any form of a variety of solvents and used almost indefinitely.

Sintered Filter Having Straight Holes

According to a process described by *T. Kimura, H. Hamamoto, A. Majima, Y. Awano, Y. Inaguma and M. Tomatsu; U.S. Patent 3,751,271; August 7, 1973; assigned to Kabushiki Kaisha Toyota Chuo Kenkyusho, Japan* a sintered metal, or ceramic filter having parallel, straight holes of uniform diameter and with smooth inner surfaces passing therethrough is formed by sintering a green compact about a large number of straight wires arranged parallel to one another.

Upon application of heat at a sintering temperature above the melting point of the wire material and less than the melting point of the sintering powders, the wires are absorbed into the pores of the sintering powders and leave holes of the same shape and size as the original wires in the resulting sintered filter.

Thin Porous Metal Alloy Foils

D. Lee; U.S. Patent 3,647,085; March 7, 1972; assigned to General Electric Company describes a process for preparing a body with apertures of small cross section useful as a filter. An alloy comprised of at least two phases in the solid state is treated to produce at least one phase in a fine form distributed in a matrix comprised of the second or other phases. The resulting treated structure is etched to remove the finely distributed phase to produce apertures or, if desired, recesses.

In one form of the process, a solid alloy comprised of at least two phases in the cast solid state is used. The alloy is cast and plastically deformed to substantially destroy its cast structure. The resulting alloy article is then heated to a temperature above its recrystallization temperature, which is the temperature at which the strained structure of the deformed alloy is replaced by a strain-free crystal structure.

The rate of recrystallization depends largely on the specific recrystallization temperature used. Generally, the higher the recrystallization temperature used, the shorter is the period of time required to complete the recrystallization. Generally, the greater the deformation of the alloy, the smaller are the phase particles produced upon recrystallization. Representative of the alloys which can be treated in this manner are Ag-Cu, Ag-Al, Al-Co, Al-Cu, Al-Mg, Au-Ge, Au-In, Be-Fe, Be-Cu, Be-Ni, Be-Zr, Bi-Mg, C-Fe, C-Mn, C-Ti, Co-Cr, Fe-Mn and V-Zr.

In another example, an alloy which undergoes phase transformation is used. Such an alloy may be a precipitating-type alloy, i.e., a solid solution alloy which cools from a single solid phase to precipitate a second phase.

There are a number of alloys which undergo solid state precipitation during cooling to produce two phases. These alloys undergo such precipitation within a certain composition and temperature range, i.e., below a certain temperature for the particular alloy composition. Such compositions and precipitation temperatures are known from phase diagrams in the literature. Representative of the solid state precipitating type alloys useful in the process are Cu-Co, Al-Mg, Ag-Zn, Cu-Al, Ni-Ti, Cu-Be, Ni-Be and Bi-Sn.

In still another form of the process, a solid alloy which has a two-phase structure but which undergoes martensitic transformation is used. The martensitic phase appears with a specific temperature range during cooling.

There are a number of alloys of certain composition which are comprised of at least two phases and which undergo a martensitic transformation during cooling. Such alloys and their compositions are known from the literature. Representative of such alloys are Fe-C, Ti-V, Fe-Ni, Au-Cd, Fe-Ni-C.

Example: A 94% titanium-6% molybdenum alloy button was cast in a vacuum by means of an electron beam. Each of the components was about 99.999% pure. The button was about ¾ inch thick. Two opposed periphery portions of the button were machined off to produce parallel sides. The resulting structure, i.e., workpiece, had a diameter of 2 inches and was about ¾ inch in height.

It was wrapped in titanium foil to prevent oxidation of the titanium and heated in a furnace having an atmosphere of purified helium. All subsequent heatings of the alloy workpiece were also carried out in an atmosphere of purified helium. When the workpiece attained a temperature of 1200°C, it was removed from the furnace and forged by means of a drop hammer to a thickness of 0.385 inch to destroy its cast structure. The workpiece was then heated in the furnace to a temperature of 800°C and was maintained at this temperature for 30 minutes to homogenize its structure and then water quenched to room temperature.

The workpiece was then heated to a temperature of 700°C which is above the temperature at which the martensitic transformation occurs, and hot rolled for about 20 seconds. This heating and hot rolling procedure was repeated two more times and then the workpiece was water quenched to room temperature. Its thickness by this procedure was reduced to 0.243 inch. It was then reheated to 750°C, hot rolled and water quenched to room temperature resulting in a thickness of 0.187 inch. It was then heated to 800°C, hot rolled and water quenched to room temperature resulting in a thickness of 0.138 inch.

The workpiece was then heated to 750°C, and maintained at this temperature for ½ hour to stabilize its structure. It was then rapidly cooled in air to room temperature.

A slice, approximately 0.050 inch in thickness was cut transversely from the workpiece. This slice was ground and polished on both faces to produce a foil 0.003 inch thick. This foil was then thinned, i.e., electropolished to a thickness of 5 microns using a solution of 94 ml acetic acid and 6 ml perchloric acid.

A micrograph of the foil is shown in Figure 3.2a. The thinned foil was then immersed at room temperature in an etchant comprised of 2 ml concentrated (49%) hydrofluoric acid, 2 ml concentrated (70%) nitric acid and 96 ml water until the molybdenum phase was substantially removed. The removal of the molybdenum-rich phase was determined by periodically removing a section of the foil from the etchant and examining it under an electron microscope.

The final etched foil was then placed in an electron microscope, and a transmission shadowgraph was prepared and is shown in Figure 3.2b.

The final thickness of the etched foil was about 0.5 micron. The holes formed by removing the molybdenum-rich phase were substantially uniform in size, i.e., about 5,000 A in diameter.

FIGURE 3.2: METAL ALLOY FILTERS

a.

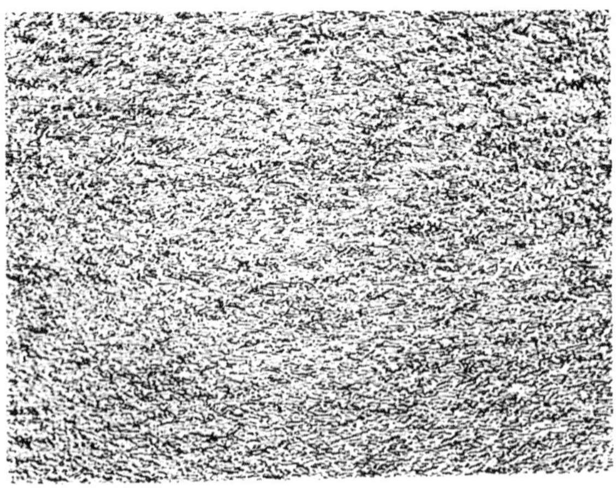

Photomicrograph (magnified 750 times) of a workpiece of 94% by weight molybdenum showing the molybdenum phase distributed in fine form in a matrix of the titanium-rich phase.

(continued)

FIGURE 3.2: (continued)

b.

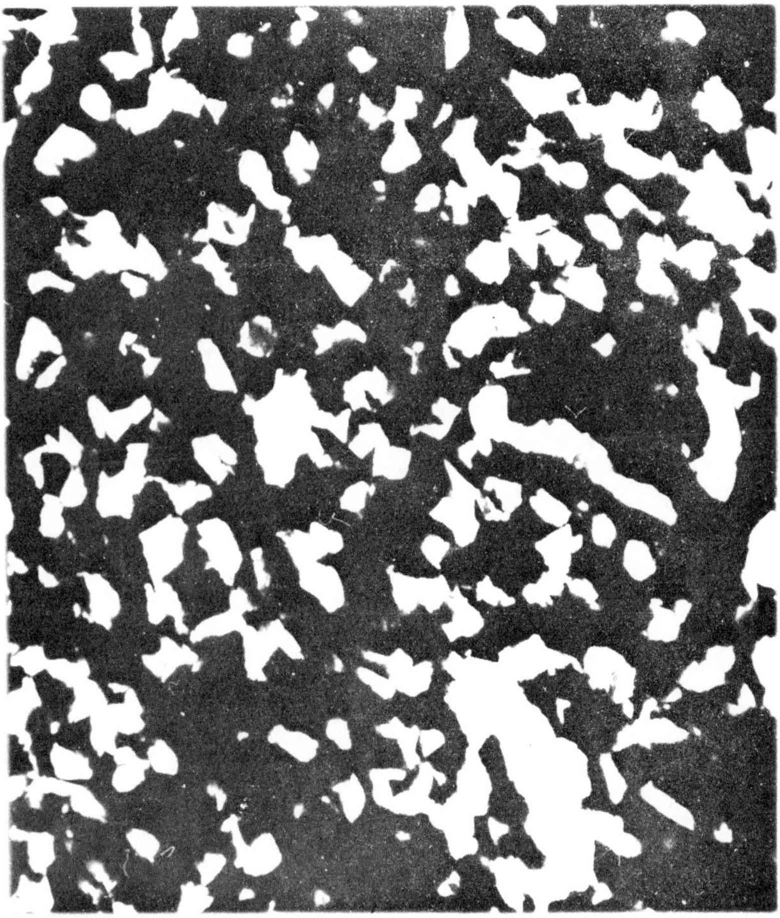

Transmission shadowgraph (magnified 10,000 times) of the work-piece of Figure 3.2a after the molybdenum-rich phase has been substantially removed to produce holes by etching as disclosed in the accompanying example.

Source: U.S. Patent 3,647,085

It appeared that the holes were substantially uniform throughout the thickness of the foil. The etched foil appeared to be suitable for use as a filter.

Since the porous etched solid can be produced in thin foil form having high tensile strength, it is especially useful as a filter for the separation of very fine materials, as for example in the purification of water. Such a filter allows good fluid flow since its thinness offers little drag or resistance for fluid to pass through. In addition, its high tensile strength would allow pressure to be applied to the fluid to still further increase flow.

If desired, composites can be formed for a wide variety of special applications by filling the holes or recesses of the etched material of the process with a foreign material, i.e., a material different from that of the etched material. For example, they can be filled with superconductive material or with iron particles to produce oriented, single-domain ferromagnetic sheet.

OTHER PROCESSES

Filter Leaf

C.B. Burkhart; U.S. Patent 3,679,062; July 25, 1972; assigned to AMBAC Industries, Inc. describes a filter leaf having a pair of spaced sheets disposed on opposite sides of a porous metallic spacer. At least one of the sheets consists of filter cloth through which a liquid flows to the spacer and then to a peripheral or central outlet in the leaf.

The spacer is desirably made from coarse woven wire cloth or expanded metal and is sprayed with a brazing material on one side. The spacer and sheets are positioned together in a brazing furnace with a metallic sealing ring surrounding the spacer. Brazing material flows to the areas of contact between the sheets and the spacer, to establish a number of metallurgically bonded connection between the sheets and the areas of contact between the sheets and the spacer to establish a number of metallurgically bonded connections between the sheets and the spacer and prevent movement of the sheets relative to the spacer.

Figure 3.3a shows a filter tank **10** in the form of a hollow cylinder oriented with its axis in a horizontal plane. The tank has an inlet **12**, a central axially extending collecting pipe **14** and an outlet **16** at one end of the pipe. A plurality of filter leaves **18** are positioned in axially spaced relationship along the pipe and each leaf registers with a separate opening in the pipe.

The construction of each filter leaf will be seen in Figure 3.3b. The filter leaf **18** includes a nonporous sheet **22**, a sheet **24** made from fine woven wire cloth and a spacer **26** made from coarse woven wire cloth. Sheet **22** is made from sheet steel or other metal. The filter leaf is of a generally disk-shaped or circular configuration, with a central opening **28** of a size permitting the filter leaf to be fitted snugly but slidably over the pipe **14**.

The space between the sheets **22** and **24** is sealed at the outer periphery of the filter leaf by means of an annular sealing ring **30**. The sealing ring flatly contacts the sheets on its opposite sides and is joined to the sheets by brazing material as well as by tack welding. The spacer **26** is joined to the sheet **24** by brazing material at joints **32** formed where the sheet contacts the spacer. Similarly, the spacer **26** is brazed to the sheet **22** at joints **34** formed at each of the areas of contact between the sheet and the spacer.

From the foregoing it will be seen that fluid is free to flow through the sheet **24** from the outer surface **38** thereof into the spaces provided between the individual wires of the spacer **26**. From this location the filtered fluid flows radially inwardly to the opening **20** with which the particular filter leaf registers. Gaskets **36** are provided on opposite sides of the filter leaf to prevent leakage of unfiltered fluid directly into the openings **20**.

FIGURE 3.3: FILTER LEAF UNIT

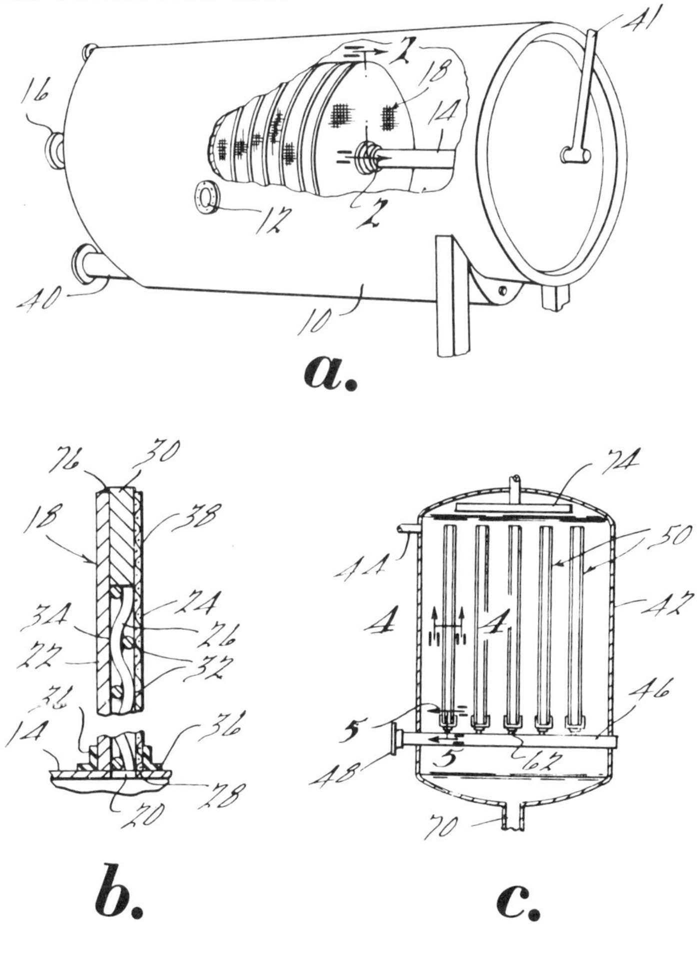

(a) Perspective view with parts broken away of a filter mechanism employing filter leaves.
(b) Broken sectional view of the structure illustrated in Figure 3.3a taken along the line **2–2**.
(c) Sectional view of another filter mechanism employing a modified form of filter leaf.

(continued)

FIGURE 3.3: (continued)

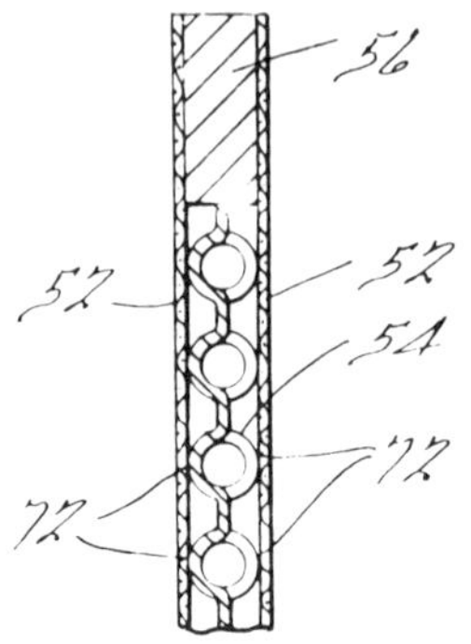

d.

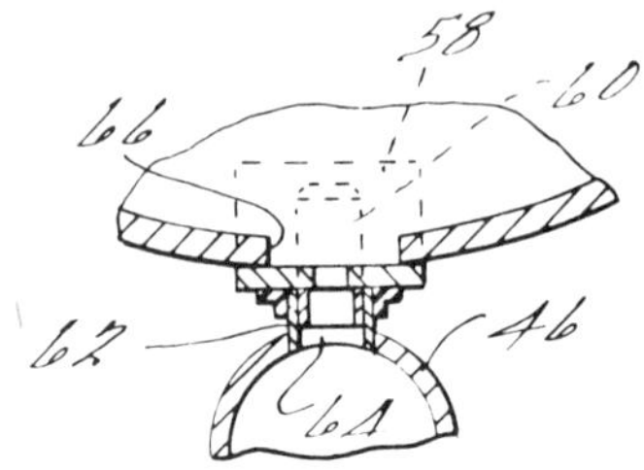

e.

Sectional views of the structure illustrated in Figure 3.3c taken along the lines **4–4** and **5–5**, respectively.

Source: U.S. Patent 3,679,062

It will be seen that the tank **10** can be drained through a drain outlet **40**. The cleaning of the individual filter leaves is aided by rotating the pipe **14** by means of a crank arm **41** at one end thereof. The filter leaf **18** may be used either as illustrated in the drawings or by applying a coating of diatomaceous earth or other particulate filtering material to the outer surface **38** of the filter sheet **24**.

The geometric shape of each filter leaf as seen in plan view is not critical. While the filter leaves are of circular configuration as seen in plan view, they may also be square shaped or possess a variety of other configurations. It will also be seen that the filter leaf has an outlet at its center.

The process can also provide a filter leaf having an outlet at its outer periphery. Such a filter leaf is illustrated in Figures 3.3c, 3.3d and 3.3e. The method of manufacturing the filter leaf as well as the filter leaf illustrated in Figures 3.3c to 3.3e will be explained after the description of the filter leaves.

Figure 3.3c shows a filter tank **42** having an inlet **44** and a collecting pipe **46**. An outlet fitting **48** is provided at one end of the collecting pipe. A plurality of identical filter leaves **50** are positioned in spaced locations along the collecting pipe. Each of the leaves includes a pair of spaced woven wire cloth filter sheets **52** disposed on opposite sides of a spacer **54**.

The spacer is made from expanded metal and is surrounded by a metallic sealing ring **56**. An outlet fitting **58** is secured to the outer periphery of the filter leaf and is provided with a hollow nipple **60** received within an upstanding sleeve **62** on the collecting pipe **46**. Each sleeve registers with an opening **64** in the pipe while the nipple is in communication with the spacer **54** through a gap or interruption **66** in the ring **56**. A gasket **68** aids in sealing the joint between the nipple and the sleeve.

The sheets **52** are brazed to the spacer **54** at joints **72** formed at all areas of contact between the sheets and the spacer. It will, of course, be apparent that the spacer **54** is porous in nature and provides a path for the flow of fluid from the space between the sheets to the nipple **60**. Cleaning of the filter leaf **50** is aided by a spray head **74** operable to direct jets of cleaning fluid generally tangentially along the outer surfaces of the sheets. The tank **42** may be drained through a drain **70**.

Manufacture of the filter leaves **50** is substantially the same as the manufacture of the filter leaves **18**. However, the process will be described with respect to the filter leaves **18**. Preferably, a light coating of nickel brazing material is applied to one side of the spacer **34**.

The brazing material is in a fine powdered form and is sprayed by means of conventional electrostatic spraying techniques. Next, the ring **30** is tack welded to the sheet **22** in the position illustrated in Figure 3.3b. The exposed surface of the ring **30** is also spray coated with brazing material. The spacer **26** is then held on the sheet **22** within the ring **30** and with the spray coated side of the spacer in contact with the sheet **22**.

The sheet **24** is then laid over the spacer **26** and ring **30** and is tack welded to the ring **30**. The tack welding of the ring to the sheets **22** and **24** leaves weldments at circumferentially spaced locations around the filter leaf. Such weldments serve only to position the parts properly during the subsequent brazing operation.

The assembly is then positioned in a brazing furnace with the sheets **22** and **24** arranged in a horizontal attitude and held against the spacer **26** by ceramic plates. A stack of filter leaves **18** may be positioned within the furnace with a separate ceramic plate between each filter leaf. The weight of the filter leaves and ceramic plates applies pressure to the opposite sides of each filter leaf to establish the desired contact between the sheets and the spacer **26**.

The stack of filter leaves is then subjected to a brazing temperature for a suitable length of time during which the brazing material will flow by wicking to all of the joints where the wires of the spacer overlap and to the areas of contact **32** and **34** between the spacer and the sheets. Extra brazing material is also laid at the outer periphery of the ring **30** as indicated at **76** to assure a good brazed joint between the ring and the sheet **22**. Of course, the sheet **24** is brazed to the ring.

The manufacture of the filter leaf **50** is essentially the same as that described for the filter leaf **18**. It is only necessary to add extra brazing material at the joints between the fitting **58** and the sheets **52** to insure a proper seal therebetween. As in the first embodiment, only one side of the spacer **54** is spray coated with brazing material. In this case, however, 10 to 20% of the pores of the sheet in contact with the sprayed side of the spacer are plugged with brazing material. The other pores of the sheet are left virtually unplugged by brazing material except at the joints **72**. A sufficient number of the pores of the screen **52** in contact with the sprayed side of the spacer remains unplugged to render the use of two porous sheets desirable for many applications.

It is to be emphasized, however, that the magnitude of brazing material applied to the spacer **26** or **54** should be very slight to avoid undesirable plugging of the pores of the screen **24** or the screens **52**. With a little experimentation, the optimum amount of brazing material can be determined. In various tested forms of the process illustrated in Figures 3.3a to 3.3e, the spacer **26** was made from different sized woven wire cloth.

In some cases the spacer had four wires to the inch with the wires possessing a diameter of 0.063 inch. In other cases the spacer had only one wire to the inch and this wire was 0.120 inch in diameter. While spacers of this size range work satisfactorily, it is believed that various other sized spacers would also perform well. Likewise a variety of fine woven wire cloth materials may be used for the sheets **24** and **52**, 24 x 110 Dutch weave and 60 x 50 twill weave materials having been found to operate satisfactorily. A variety of metal alloys might be used for the spacer and filter cloth although stainless steel was found to be entirely suitable.

Filter Candle

A process described by *J.P. Nommensen; U.S. Patent 3,975,274; August 17, 1976; assigned to Stamicarbon B.V., Netherlands* relates to filter candles of the type in which the fluid medium flows through a circumferential wall into the interior of the filter, the circumferential wall having a continuous slot or a plurality of slots therein of such narrow width that solid particles do not pass through.

Slotted filter candles of this general type are known from Swiss Patent 438,220 and French Patent 783,203 and 2,088,713. The known filter candles have the drawback that, if the slots are very narrow, e.g., in the order of 20 microns or smaller, they are difficult to clean by reversing the direction of flow of the filtrate because of the high filtration resistance. In most cases, it is necessary to dismantle them for cleaning, in order that they can be treated with chemical agents, for instance, with a dilute caustic solution if they should be contaminated by silicate components.

The object of this process is to provide a method which makes it possible to effect a proper cleaning action in slotted filter candles with a small slot width by reversing the direction of flow and in which the amount of medium washed back remains small.

According to the process, this can be achieved if use is made of a slotted filter candle in which the openings or slots between the sectional members which form the filter do not exceed about 20 microns in width, and the sectional members are fastened, at least in three points, to stay rods mounted approximately parallel to the axis of symmetry.

The whole of the connected sectional members is kept compressed during the filtration by means of a tension member. The filtration is interrupted at regular intervals and the direction of flow of the fluid medium is reversed and at the same time a short reverse pressure pulse of such strength is exerted on the inner surface of the bottom of the filter candle that the openings between the sectional members are enlarged by more than 5%, while the pretension is so adjusted that the pressure pulse virtually relieves the stay rods. The slot width is thereby enlarged during the pressure pulse, and particles caught in the slot are loosened and removed with the adhering filter cake. Each pressure pulse returns a controlled equal amount of filtrate during the fixed short time of the pressure pulse.

In Figure 3.4a, there is shown a filter candle made up from helical sectional members **1** formed by sectional wire wound into a helix so as to form slots between the members **1**.

FIGURE 3.4: FILTER CANDLE

a.

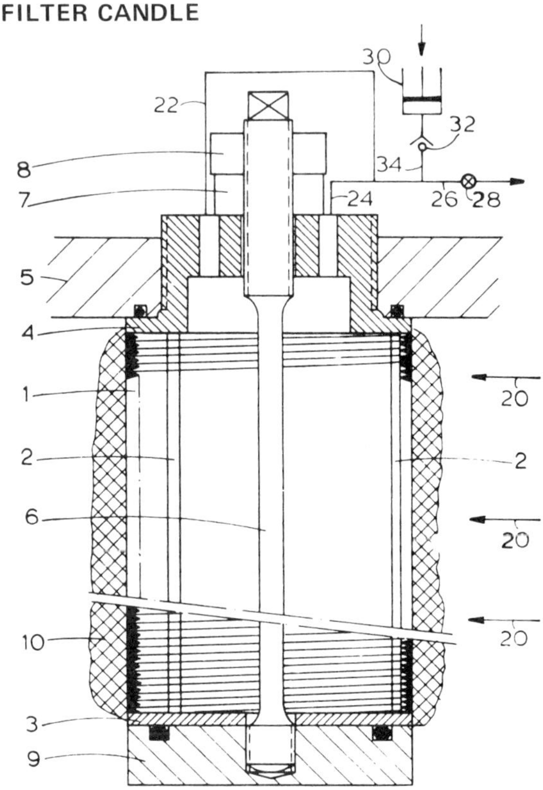

Vertical sectional view of a filter candle.

b.

Cross section of one of the sectional members of the filter of Figure 3.4a.

(continued)

FIGURE 3.4: (continued)

c.

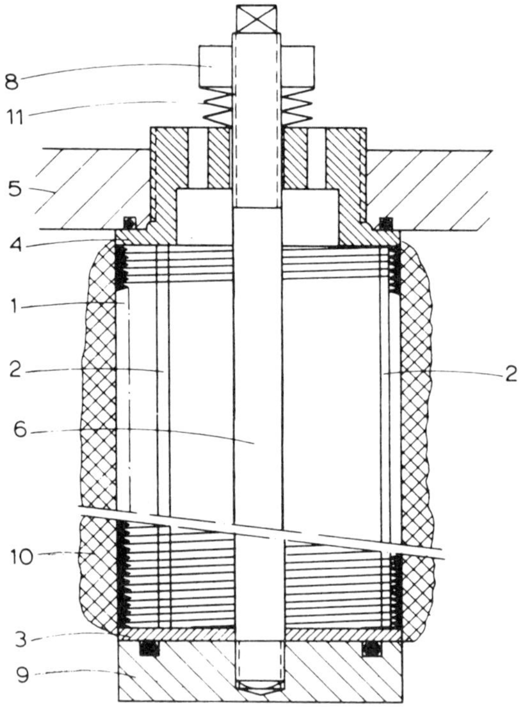

Vertical sectional view of a modified candle.

Source: U.S. Patent 3,975,274

The distance between these sectional members **1** is, for instance, about 7 microns. The sectional members are fastened to stay rods **2**, two of which are shown in Figure 3.4a. The distance between the stay rods **2** is usually about 15 mm, and the stay rods are disposed uniformly about the axis of symmetry.

The assembly of sectional members and rods is closed at the bottom by means of a bottom plate **3** and fixed at the top to a threaded connecting part **4**, which is screwed into a cover **5**. The assembly is compressed axially by means of a pretensioned central tension rod **6**, having a threaded upper end which is provided with a tightening nut **7** and a lock nut **8**.

The lower end of the rod **6** is fastened to a stay plate **9** which is mounted under the bottom plate **3**. The central tension rod **6** must be so dimensioned that, at the required tension, the limit of elasticity of the material is not exceeded during the fluid pressure pulse which is applied to the interior of the filter. It is moreover desirable that the tension rod is so shaped that fatigue fracture owing to the highly frequent pressure pulses can hardly occur.

Figure 3.4b is a cross section of one of the sectional members **1** of the slotted filter. By preference, these sectional members **1** slightly taper towards the inside of the filter, so that a kind of predistribution of the medium is effected before the slots. During the pressure pulse the liquid will then be pressed outwardly through the slots more easily, so that the filter cake will be dissolved better.

Figure 3.4c shows a variant of the tension member. In principle, the tension rod **6** has been maintained, but a spring has been fitted under the tightening nut **8**. The required tension is now produced by a spring, in this case the cupped springs **11**.

For a satisfactory performance of the filter candle it is necessary that the reverse fluid pressure pulse rapidly builds up and lasts for at most some tenths of a second, not more than about 0.4 second. Of course, this is only possible if the resistance of the filter is high, it being furthermore implied that the slot width must not be large. As a rule, a pressure plus of 3 bars is required to effect the required results.

According to the process the openings between the sectional members do not exceed about 20 microns and the sectional members are fastened, at least in three points, to stay rods mounted approximately parallel to the axis of symmetry, the whole of connected sectional members being kept compressed during filtration, by a tension member.

Means is provided for interrupting the filtration at regular intervals and reversing the direction of flow of the filtrate and at the same time exerting a pressure pulse on the inner surface of the bottom of the filter candle, the pretension being such that the stay rods are virtually relieved during the pressure pulse.

The medium may be either a liquid or a gas. The particles to be separated may be of the order of 1 to 10 microns in the filter described. If the medium is an incompressible liquid, the pressure pulse may exert pressure on the bottom by the liquid.

If a gas loaded with particles is filtered, it may be necessary, because of the compressibility of the gas, to promote the pressure pulse by mechanical means. To this end, pressure may be exerted for a short time on the cupped springs shown in Figure 3.4c, by external mechanical means during the reversal of the flow of the medium in the filter.

During a filtering operation fluid medium and dispersed particles flow in the direction of arrows **20** so that particles form a filter cake **10** on the outside of the filter. The fluid passes through the slots between the section members into the interior of the filter and then out through lines **22, 24** and **26**. Periodically the flow of fluid medium is stopped by closing a valve **28** in the line **26**.

Then a reverse fluid pressure pulse is applied to the interior of the filter by any suitable means such as a piston and cylinder illustrated schematically at **30**. A check valve **32** in the line **34** connecting the piston and cylinder **30** with the line **26** prevents flow into the piston and cylinder during filtration. The reverse pressure pulse is of sufficient magnitude to widen the slots between the filter members **1** and to cause a small back flow of fluid medium through the slots.

According to one example, use is made of a slotted filter candle with a diameter of 70 millimeters and a length of 600 millimeters. The number of stay rods of about 1 x 1 millimeter was 16 and the number of slots 540. The tension in the tension rod was about 200 kg (100 kg/mm^2). The slot width in the untensioned condition was 7 microns and in the tensioned condition 6.5 microns.

The slot widths were enlarged to 7 microns by exerting a pressure pulse for about 0.06 second. This filter was used to filter a catalyst with a particle size of $1\ \mu < d < 8\ \mu$ from an acid aqueous medium from a part flow of a reactor. When a pressure pulse was exerted every 30 minutes, the filter cake of catalyst particles could be made to recirculate. The filtration was carried out at a constant throughput of 100 liters per hour. One liter of filtrate was washed back every pressure pulse.

The application of the central tension rod at a very specific pretension is essential. It has been demonstrated by trial and error that without such a tension rod, after 500 pressure pulses, during about 250 operating hours, the slotted holes have become enlarged and nonuniform in slot width owing to plastic deformation of the thin stay rods. It it obvious that in case the tension stress is too great the slots are opened insufficiently during the back washing.

The pretension may also be effected with the aid of means fitted outside the filter. This will preferably be done when a number of filter candles is arranged in parallel. It will then suffice to use a joint bottom plate and a limited number of tension members.

Flow Path at Interface of Two Connecting Filter Members

J. Rosenblum; U.S. Patent 3,812,972; May 28, 1974 has found that a highly efficient filter comprised of a pair of nonporous filter members may be fabricated by forming at least one groove in at least one surface of the pair of members and then joining the members together in surface to surface contact so that the groove lies at the interface thereby defining at least a segment of a liquid flow path through the filter at the interface of the members.

In accordance with the process such a filter may be designed with a filtration capacity for a desired liquid of almost any viscosity by forming the filter with a predetermined number of liquid flow paths having predetermined dimensions, the number and dimensions of the flow paths being selected to accommodate the characteristics of the desired liquid. For example, a filter may be formed with a liquid flow path construction which enables the filter to be used in a process of treating raw sewage, separating sludge from the water.

In a preferred method of this process, the pair of filter members are metallic and a plurality of grooves are formed in at least one contacting surface of the pair of filter members by a chemical etching process, such as a conventional photochemical or electrochemical etching process. The grooves each represent at least a segment of a separate liquid flow path through the filter.

Chemical etching is preferred in forming such grooves since the grooves so formed are extremely precisely defined. Further, chemical etching enables the formation of a groove of almost any desired dimensions, e.g., 0.010 inch in depth, which is an important aspect of the method.

Further, the filter members are preferably affixed to each other by a conventional diffusion bonding process. Diffusion bonding is accomplished by joining two or more members constituted of identical heat deformable materials, such as plastics or metals, in a furnace having either a hydrogen atmosphere or vacuum.

Then, the pressure is increased to maintain a surface to surface contact between the members, and the temperature is elevated sufficiently to cause the molecules at the interface to diffuse and intermingle, thereby forming a permanent bond between the two members which remains when the temperature is reduced.

In one specific example of a filter of this process, a pair of plates, each plate having a multiplicity of holes, are mounted together in superimposed contacting relation. The holes are arranged in the same spatial pattern on each plate, and the plates are so aligned that the holes in one plate are completely out of register with the holes in the other plate.

A multiplicity of grooves are formed in at least one of the contacting plate surfaces, each such groove being positioned to provide a liquid flow path between a hole of each plate at the interface of the plates so that a multiplicity of tortuous liquid flow paths are defined through the filter.

The filters can be designed with a selected hole and groove construction (i.e., number, arrangement and dimensions) which will enable them to provide effective filtration of substantially any preselected liquid. Further, the specific construction of the holes and grooves also determines, in part, the maximum liquid flow rate of the filter.

More specifically, the greater the volume of liquid flow paths through the filter, the greater the liquid flow rate that can be handled before the filter breaks down. A balance must, of course, be struck between the viscosity of the liquid to be filtered and the rate of filtration in determining the number, arrangement and dimensions of the holes and grooves.

Disc Filter Element

A process described by *C.M. Radford; U.S. Patent 3,971,722; July 27, 1976* relates to filter elements having a generally planar configuration such as, for example, disc filter elements.

Filter elements of a planar configuration generally comprise a frame having support means secured thereover for a filter bag which, in use, encloses the frame and support. Such frame and support assemblies are generally fairly costly to manufacture or do not provide sufficiently large flow paths to provide for optimal flow conditions within the element. Another construction which comprises simply a sheet-like web of material having upstanding ridges thereon to act as a support for a filter cloth, also suffers to some extent from these disadvantages.

It is the object of this process to provide a filter element of the above type which is at least simpler to manufacture than many prior art elements, but which also provides for free flow of liquids within the covering filter medium.

In accordance with the process, there is provided a filter element comprising a pair of molded panels secured together to define an internal cavity, spaced supporting ribs on the outer surfaces of the panels, drainage perforations through the panels between the ribs and communicating with the cavity and a drainage outlet from the cavity.

Further features of the process provide for the two panels to each have an integral surrounding ridge with the ridges secured together to hold the panels in spaced relationship, for the panels to be additionally secured together by means of spaced lugs extending inwardly into the cavity, and for the two panels to be of identical shape and configuration.

Referring to Figure 3.5 a disc filter element **1** of conventional planar configuration is shaped as a truncated sector of a circle. The element comprises two identical molded panels **2** secured together by means of a suitable bonding agent which will depend on the material from which the panels are made. Of course, any other means of securing the panels together may be used such as mechanical means for instance, screws or nuts and bolts, but bonding is generally preferred. However, welding would be equally acceptable if it is possible.

FIGURE 3.5: DISC FILTER ELEMENT

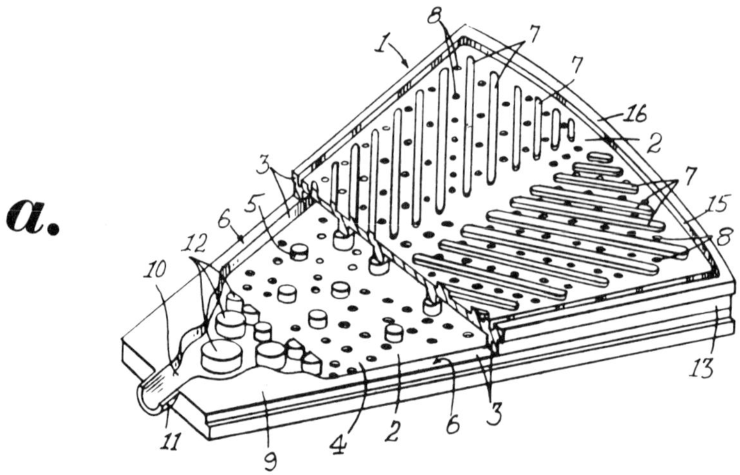

Partly broken isometric view of a disc filter element.

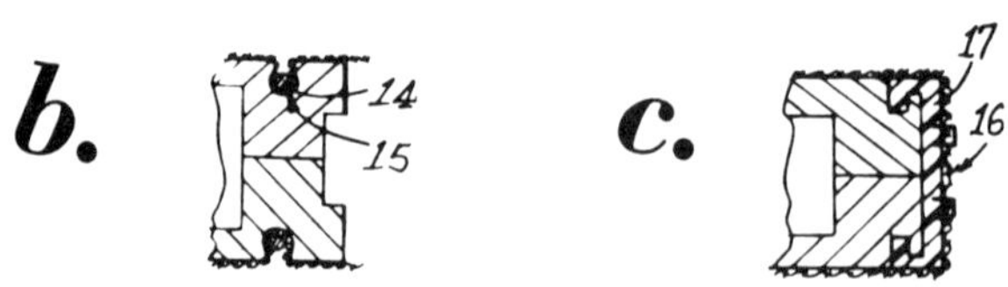

Cross-sectional views through the outer edge of the panel showing two alternative forms of securing filter media.

(continued)

FIGURE 3.5: (continued):

d.

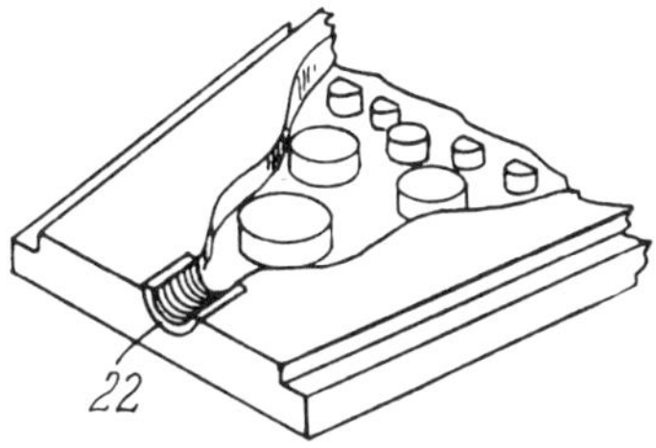

An alternative outlet from the element.

e.

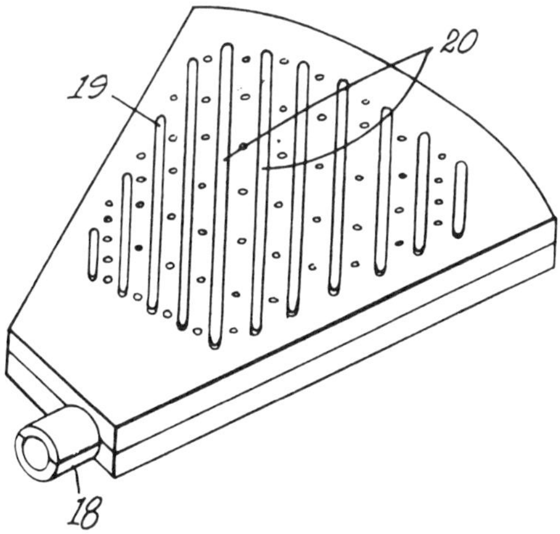

Isometric view of an alternative application of the process.

Source: U.S. Patent 3,971,722

Each panel has a surrounding ridge **3** with the ridges actually being bonded together to provide a cavity **4** between the panels. The panels also have lugs **5** on their inside surfaces with the free ends of the lugs being coplanar with the upper surface **6** of the ridges. These lugs are also bonded together when the panels are joined so that in use they maintain the panels in their correct space relationship whether a negative or positive pressure is applied to the interior of the cavity.

The panels are provided, in this particular case, on their outer faces with two series of parallel spaced ribs **7** each series being arranged on one side of the median radial line which divides the sector in half. The channels formed between the ridges are provided with series of perforations **8** communicating with the cavity.

At the narrow end **9** of the element the cavity is curved smoothly to a drainage outlet **10**. The outlet itself may be defined by two semicylindrical tubular extensions **11** one of which is molded integral with each of the panels. The extensions are arranged such that when the two panels are secured together a tubular connector pipe is formed, which is suitable for direct connection to the filter.

Alternatively, as shown in Figure 3.5d, semicylindrical screw threaded inserts **22** may be molded into the panels at the outlet so as to form a screw threaded socket when the two panels are secured together. An outlet tube having complementary screw threads may then be fitted to the socket. Furthermore, if required, the screw threads carried by the inserts could simply be molded into the outlet, thereby avoiding the use of inserts which are in general costly.

In the region adjacent the drainage outlet where the cavity narrows, is a series of baffles **12** of different shapes and sizes located to prevent direct blow back of filtrate into the cavity during the blow off cycle during filtration, and also to diffuse the blow off and spread it within the cavity. These baffles accept the wear from any abrasive particles present in the filtrate, thereby protecting the interior of the cavity itself.

Finally the panels are shaped to provide a groove **13** down each of the radially extending sides of the filter elements in order to fit into existing frames therefor where this is necessary. The presence or absence of these grooves will depend upon the construction of the filter with which the element is to be used. Two alternative methods of fitting material over the faces of the panels are considered to be particularly suitable.

The first, shown clearly in Figure 3.5b, is to locate independent sheets of filter material over each face and secure them in position by a caulking strip **14** forced into a caulking groove **15** around the periphery of the panel faces.

Alternatively a bag may be fitted over the element and secured to the outer peripheral edge **16** by means of staples. In order to facilitate this and protect the element, a disposable strip **17** of plastics material may be releasably secured to the element to act as a staple receiving pad. A type of double rebate arrangement is illustrated in Figure 3.5c for holding the pad in position but any other suitable arrangement can be used. In use, the channels provide for drainage of filtrate between the ribs **7**, the filtrate passing from there into the cavity. Substantially unobstructed flow is then provided through the cavity to the outlet.

This construction of the filter element allows it to be easily and swiftly constructed with a consequent saving in cost. Also, the lugs enable the panels to be made of a softer and thus more abrasion-resistant material such as a highly resistant grade of polyurethane since the lugs prevent collapse and/or bulging of the panels during the suction and blow off cycles respectively.

However, where required, the panels can be made of harder material and the lugs omitted. A preferred panel has been made of polyurethane having a Shore Hardness of about 90 and wherein the lugs were present. It is considered, however, that effective panels will also be made of suitable rubbers or the like.

The process is not confined in its application to the specific type of disc filter element described above. Thus it could also be applied, for example, to the so-called leading edge disc filter sector illustrated in Figure 3.5e. In this case the outlet pipe **18** is parallel to one radial edge of the sector **19** and in this case there is only one series of ridges **20** which extend in the general direction toward the outlet.

Integral In-Line Filter

A process described by *D.W. Voegeli; U.S. Patent 3,753,500; August 21, 1973; assigned to Minnesota Mining and Manufacturing Company* provides a low cost, corrosion-resistant in-line filter which can be made at low temperatures, in various configurations, and formed in housing or casings with diverse internal geometric configurations, in such a manner that the resulting filtration structure is an integral filter assembly.

The in-line filters described employ particulate material, granules, or beads such as glass beads, coated with a nontacky, powdered thermosetting resin, such as that described in U.S. Patent 3,175,935. The free-flowing coated beads are poured into a suitable filter housing to dispose a mass of the coated beads at a desired locus in the housing, and the so-disposed coated beads are consolidated therein to form a desired shaped body, the shaped body then being heated to cure the resin in situ within the housing.

Heating causes the contiguous beads to become bonded together in a cured thermoset resin matrix, thereby forming a solid, consolidated composite porous body which is also bonded to the contiguous wall of the housing in which it is disposed.

The integral in-line filters formed in these examples have been assigned an approximate micronic rating. The rating was derived by testing filters made from similar coated beads and under similar molding conditions. The test used to determine the bubble point and thereby the micronic rating was ASTM E128-61.

Example 1: Spherical glass beads of 100 mesh (149 micron diameter) were coated with Scotchkote 112 thermosetting epoxy resin, to the extent of about 16 grams of resin per 100 grams of glass beads. A three-eighth inch diameter steel rod was inserted as a dam into one end of a flare-to-flare brass fitting, having an internal diameter of three-eighths of an inch, the rod extending into the fitting to a point midway between its ends. Approximately 1.25 grams of resin-coated glass beads were introduced into the other or open end of the brass fitting and a second three-eighth inch steel rod was inserted as a ram into an open end of the fitting.

Using a single action hydraulic press, a pressure of 200 pounds was applied to the ram to consolidate the epoxy resin coated glass beads in the form of a disk. The steel rods were withdrawn with the consolidated body of coated beads disposed within the fitting in the form of a disk at the desired filtration locus and the assembly was transferred to a warm air oven and cured at about 190°C for one-half hour, forming an in-line filter comprising a porous, composite filter body integrally bonded in the brass fitting.

The in-line filter was connected to a compressed air line. A pressure drop of 23.5 cm of Hg across the filter body was measured at an air flow rate of 15 scfh with a pressure gage on the upstream side of the filter. The filter's bubble point when tested according to ASTM E128-61 was found to have a micronic rating of 50 microns. The strength of the bond between the porous filter element and the brass fitting was determined by measuring the force necessary to eject the filter element from the housing with a three-eighth inch diameter rod. A force of over 1,000 pounds was required to eject the filter element from the brass housing.

Example 2: A laminated, multiple porosity, in-line filter assembly was fabricated in a copper sweat fitting with a nominal internal diameter of one inch. The filter is useful to filter successively finer particles from the fluid being filtered without the filter being filled by a large number of particles caught or retained on one filter surface.

To form the filter, a steel plunger, which conformed to the internal dimensions of the sweat fitting, was inserted into one end of the fitting to about the midpoint of the fitting. About 5.0 grams of 24 mesh, about 2.5 grams of 65 mesh, and about 2.5 grams of 180 mesh resin-coated, soda lime silica glass beads (all three sizes of bead having been coated as in Example 1), were successively poured into the open end of the fitting and leveled in the internal cavity.

A second steel plunger ram was inserted into the opposite end of the fitting, sealing the fitting, and a pressure of 1,400 psi applied to the filter body in a single action hydraulic press. The steel plungers were withdrawn, and the assembly transferred to a warm air oven. The epoxy resin was cured at 190°C for one-half hour. The resulting filter assembly consisted of a laminated filter body having three different micronic ratings of 227, 89 and 39, the laminate integrally bonded to each other and to the inside of the fitting.

Example 3: A double laminated in-line filter assembly was fabricated in a copper sweat fitting using the technique of Example 2, the fitting having an internal diameter of one inch and a 45° shoulder. In this example, the filter comprised a 0.365 inch layer of 24 mesh epoxy-coated glass beads (710 microns) and a 0.295 inch layer of 33 mesh epoxy-coated glass beads (110 micron diameter).

The resulting filter had a laminated structure with micronic ratings of 227 and 176. A load of 1,250 pounds (1,650 psi) was necessary to eject the filter element from the housing. The ejection pressure was exerted from the smaller diameter of the housing to avoid any mechanical advantages being contributed by the shoulder of the housing and thus determined the strength of the bond between the filter element and housing.

Movable Two Frame System

A. Wehner; U.S. Patent 3,971,715; July 27, 1976 describes a device for sieving, sorting and filtering. The device comprises two frame systems movable relative to each other, each frame system comprising a number of carriers extending in the sieving direction and moving parallel to one another, the frame systems interengaging in such a way that one carrier of one system is always adjacent within a certain distance to one carrier of the other system.

Due to the fact that on each pair of adjacent carriers of the two systems there is attached a pair of flexible material mats, the pair of mats consisting of one perforated and one unperforated mat, and that the carriers are moved successively toward and away from each other by the relative motion of the frame systems, there are formed zones of continuously changing width at an alternately deep sag of the bottom mats of the mat pairs and at a simultaneously different large arching of the top mats of the mat pairs.

The mat pairs thus extend momentarily in an overstretched position parallel to each other which results in the formation of hose- or channel-shaped sorting decks, the sorting decks exerting, owing to their continuously changing cross section, a pressure or suction effect on the goods to be sieved, sorted or filtered. In spite of the fact that the carriers perform a swinging motion transverse to their longitudinal stretch, a significant effect of the device according to the process lies in the fact that the sieving, sorting or filtering goods fed into the hose- or channeled-shaped sorting zones of the inclined mounted device are transported in the direction of the longitudinal stretch.

Figures 3.6a through 3.6c illustrate the design according to the process of a sieving machine, which comprises two interengaged frame systems, each of the systems comprising a number of carriers stretched in the sieving direction and moving with a parallel spacing relative to each other. One carrier **1** of system **I** is always adjacent at a parallel spacing to a carrier **1'** of system **II** so that these adjacent carriers form slots of continuously changing width owing to the swinging motion resulting transversally to their longitudinal stretch.

To this end, the carriers **1** are moved over a swinging level **4** by a driving rod **3** whereas the carriers **1'** are moved over the swinging lever **4'** by the driving rod **3'**, the carriers **1**, swinging levers **4** and driving rod **3** representing the moving system **I**, and the carriers **1'**, swinging levers **4'** and driving rod **3'** forming the moving system **II**. The two moving systems are driven over a common crankshaft actuating the associated system **I, II** in a swinging motion with a shifting of 180°.

Elastic or flexible sieving mats **8, 8'** are fastened to the carriers **1, 1'**; due to the relative motion of the carriers, the sieving mats form adjacent zones of continuous change with an alternately differently deep sag curved arching or different tensioning of sieving mats. The relative motion of carriers **1,1'** are of such a magnitude that the sieving mats are momentarily overstretched upon attainment of their stretched length, which results in a momentary cross-sectional increase of the sieve holes and this guarantees, in turn, an absolute freedom from clogging.

Thus, mats **8, 8'** fastened to carriers **1,1'** form sieving channels extending in the longitudinal direction of carriers **1, 1'** and limited laterally by the carriers. This makes it possible to use instead of several narrow and long sieving mats a single continuous elastic or flexible sieve bottom extending over the whole width and length, which is then attached zone-like to the individual carriers by means of gibs.

Although carriers **1, 1'** of the two systems **I, II** perform a swinging motion transverse to their longitudinal stretch and with a shifting of 180°, an unobjectionable transport of the sieving goods to be treated is assured over the whole length of the sieving channels extending in the sieving direction, since the machine according to the process is mounted in an inclined position.

FIGURE 3.6: DEVICE FOR SIEVING, SORTING AND FILTERING

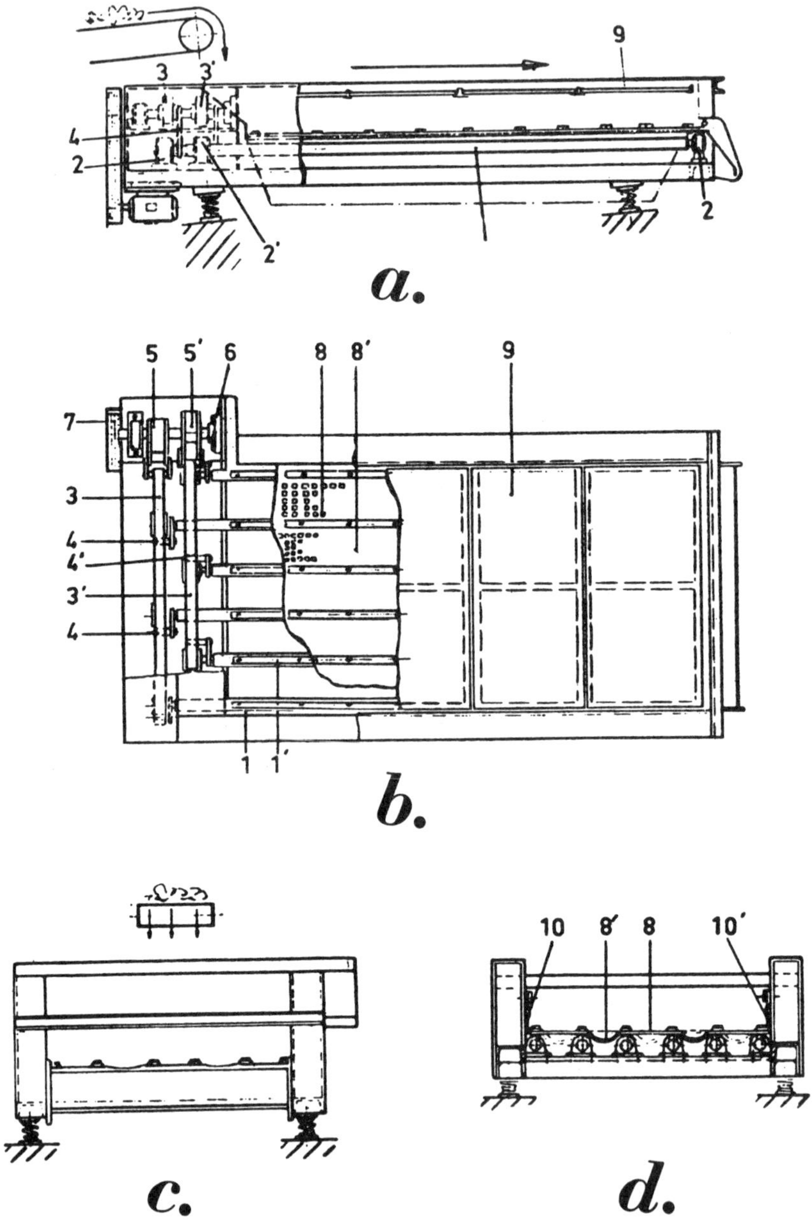

(a) Sieving machine.
(b) Top view of Figure 3.6a.
(c) Side view of Figure 3.6a.
(d) Section through the sieving machine of Figure 3.6a.

(continued)

FIGURE 3.6: (continued)

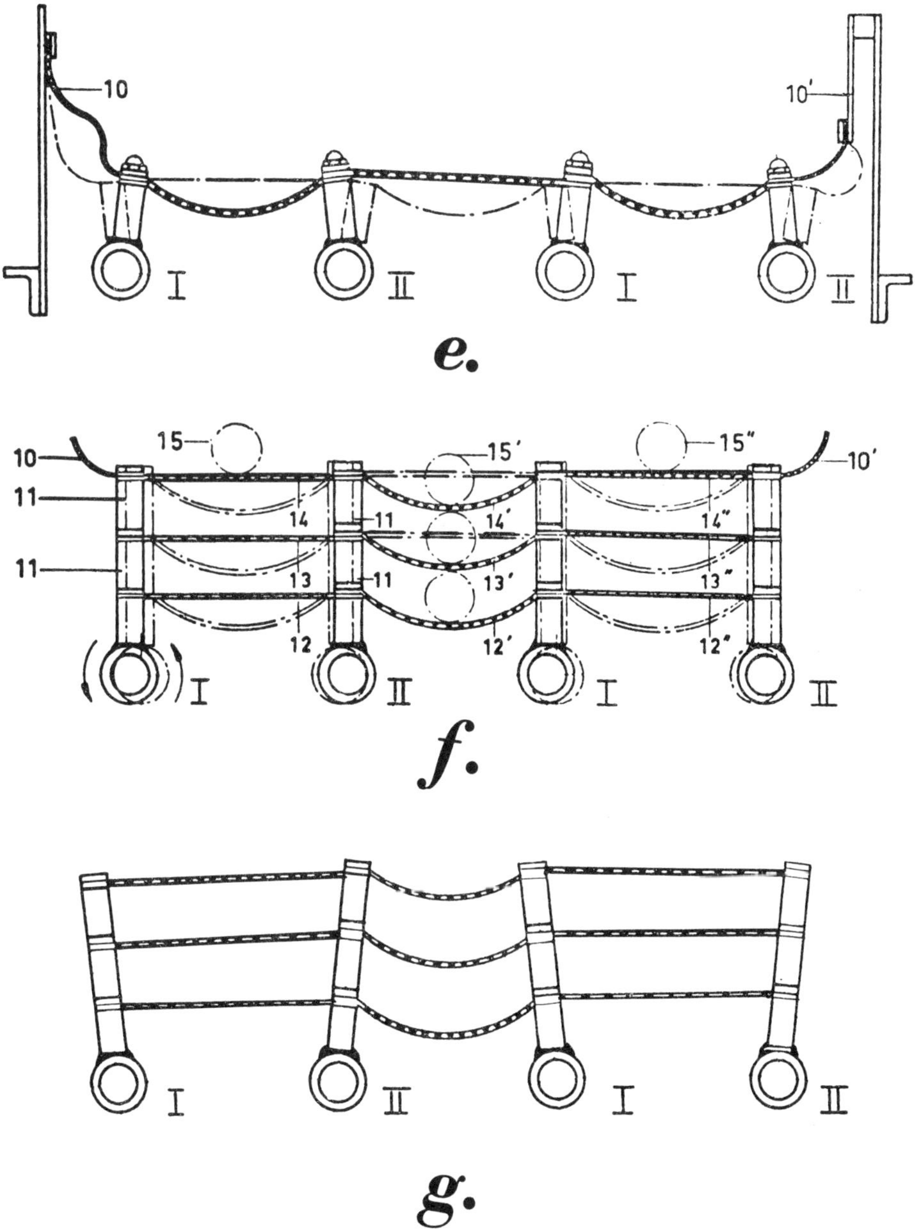

(e) Section of Figure 3.6d on a larger scale.
(f) Construction for multiple sorting.
(g) Similar to Figure 3.6f.

(continued)

FIGURE 3.6: (continued)

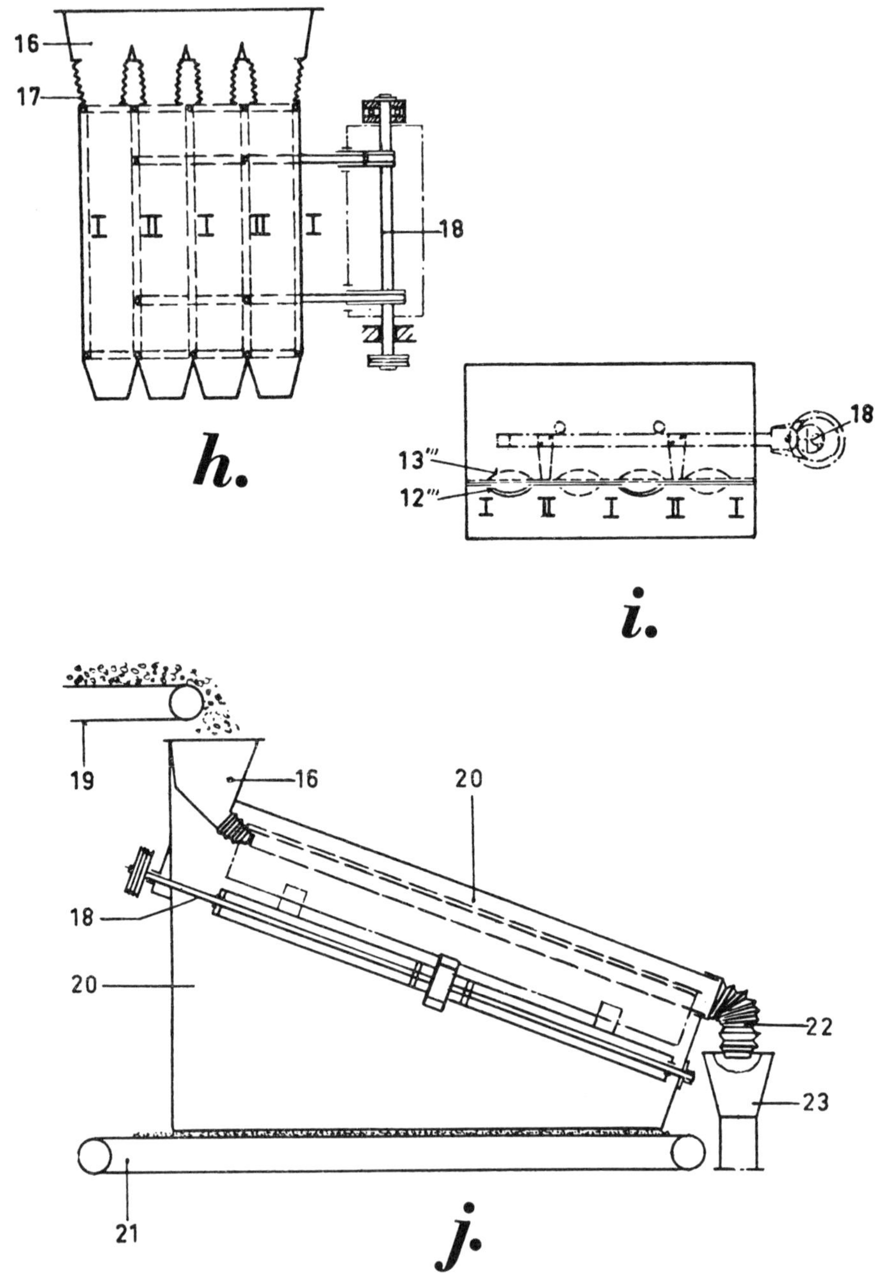

(h) General arrangement with hose-shaped zones.
(i) Top view to Figure 3.6h.
(j) Side view to Figure 3.6h.

(continued)

FIGURE 3.6: (continued)

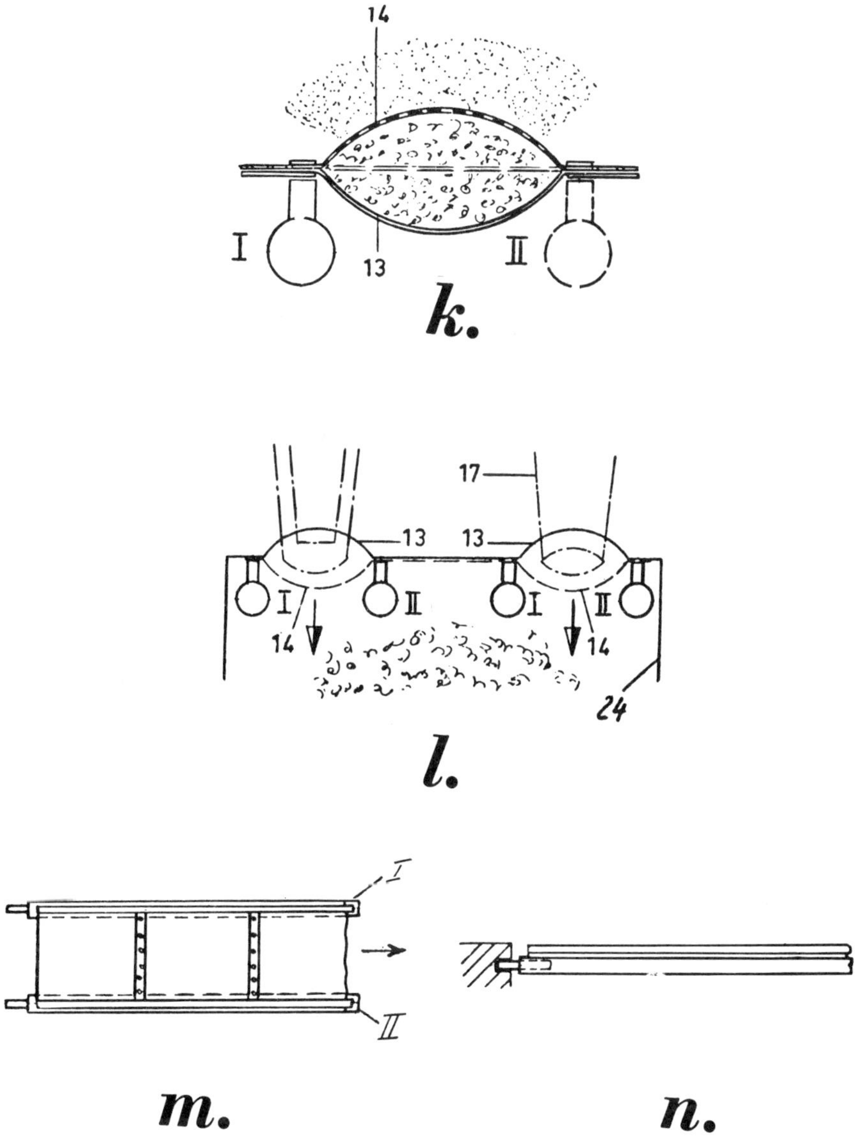

(k)(l) Section of a hose-shaped zone.
(m) Division into sections of a mat extending in the longitudinal direction of the carriers.
(n) Side view of Figure 3.6m.

(continued)

FIGURE 3.6: (continued)

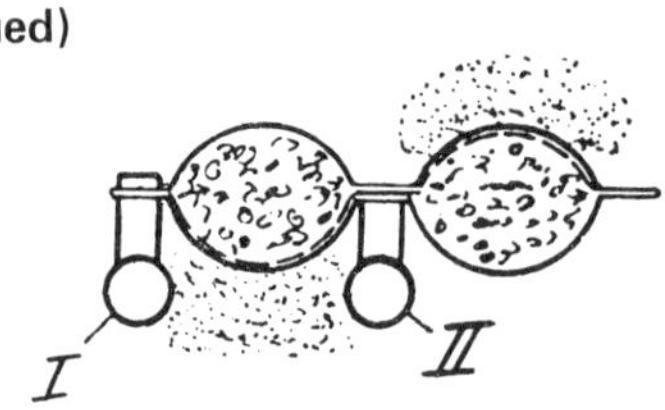

o.

Cross-section through two hose-shaped zones.

Source: U.S. Patent 3,971,715

Therefore during the tensioning and momentary overstretching of sieving mats **8, 8'** tossed up particles of sieving material fall on sieving mats **8, 8'** over an arched path directed toward the discharge and are tossed up again and fall again over an arched path on the sieving mats. However, in order to attain a higher sorting efficiency, a cover frame **9** made up of several sections is provided at a certain distance over the sieving mats, this distance being sized to ensure that upon tensioning of the sieving mats, the tossed up particles of sieving material will be bound against the cover frame in such a way that the particles of sieving material are deflected at a sharp angle from their normal arched path and fed to the sieving mats.

This results in the particles of sieving material being brought on the sieving mats much more often than would be the case under normal conditions. In this way, a substantial improvement of the sorting efficiency is attained by reducing the amount of fine particles in the overflow.

The sectional representation in Figures 3.6d and 3.6e illustrate the formation of sieving channels as a function of the position of carriers **1, 1'** relative to each other, on the one hand, and the trouble free lateral sealing effect, on the other hand. It is thus evident that during the first motion phase, which corresponds to one half of the motion cycle, mats **8** are, for instance, in a tensioned position, whereas the adjacent mats **8'** assume a concave sagging position.

During the second motion phase, the previously sagging mats **8'** are tensioned, while the adjacent mats **8** assume a concave sagging position. Figures 3.6d and 3.6e show, furthermore, the trouble free side seals **10, 10'** which can be achieved for example, by means of plain rubber or plastic covers loosely tensioned to the outwardly extending carriers or the side walls of the sieving machines, as illustrated by **10**.

Similarly as is illustrated by **10'**, it is possible to use bendable metal or hard plastic plates which may be attached, if required, over intermediate pieces to the outwardly extending carriers and/or the side walls of the sieving machine.

Figures 3.6f and 3.6g show the advantageous application of the sieving machine according to the process for a multiple sorting process or for increasing the sieving efficiency by means of several decks. Thus, for example, in the left hand longitudinal zone of Figure 3.6f a perforated mat **13** is arranged spaced by means of intermediate pieces **11** over a bottom perforated mat **12**, the perforated mat **13** being uncovered in turn by a sieving mat **14** spaced again by means of intermediate pieces **11**. In this way, it is possible to carry out simultaneously two similar sieving or sorting operations in a zone limited by two adjacent carriers of system **I, II** the central perforated mat **13** serving as transport mat for the sieving or sorting material which has passed through mat **14**, on the one hand, and at the same time as bouncing mat for the particles of sieving or sorting material tossed up by the bottom mat **12**, on the other hand. The increase in efficiency attainable in this way is evident.

In the central zone, these mats **12′**, **13′**, and **14′** with differently large sieve holes are arranged superposed and spaced by means of intermediate pieces **11**, the mats separating the material charged on the top mat **14′** into four fractions; the same applies to the right hand zone.

If, contrary to the representation, the top mat **14″** of the right hand zone is an unperforated mat and the material to be treated is charged on the central perforated mat **13″**, it is possible to carry out a dust-free sieving or sorting of the material into three fractions. It is apparent that in one zone either only perforated mats, or perforated and unperforated mats can be used, depending upon the intended use.

The carriers of systems **I, II** can be moved in this case also linearly or on a circular or partially circular path or they can carry out a rotary-tilting motion, depending on the type of sieving machine adopted.

Figure 3.6h shows a solution, where the charging of the individual hose- or chamber-shaped sieving zones is made out of a general hopper **16** over nozzle-like folded bellows **17** with the possibility of installing the corresponding device in an inclined or vertical position. Figure 3.6i, which is a top view of the device in Figure 3.6h, illustrates an additional principle of the process, which is based on the knowledge that, in case of various goods being treated by sieving, sorting, and filtering techniques, it is advantageous to exert a pressure or suction effect on the material to be sieved, sorted or filtered.

To this end, the process proposes to associate a pair of mats made up of a bottom mat **12‴** and a top mat **13‴** to the zone formed by two adjacent carriers of systems **I, II** and to preshape or pretension the mats **12‴**, **13‴**, if required, in such a way that in the closest distance position of the carriers of systems **I, II** which define a zone, the bottom mat **12‴** assumes a concave sagging, and simultaneously the top mat **13‴** assumes a convex arched position, whereas in the largest distance position of the carriers relative to each other, both mats **12‴** and **13‴** assume a tensioned, momentarily overstretched, parallel position relative to each other.

In this way, a hose-like sorting zone is formed which, as a result of the continuous change in its cross section, exerts a pressure or suction effect on the material to be treated. In the process, the suction effect exerted on the material to be treated can be intensified by producing and maintaining a certain underpressure in the treated material receiving container, as shown in Figures 3.6k and 3.6l. It will be noticed that by means of a drive, e.g., excenter shaft **18**, it is possible to actuate either only one of the systems **I** or **II** or both systems.

Figure 3.6h shows a lateral view of a device according to Figures 3.6h and 3.6i, in which the material to be sieved, sorted or filtered is fed over conveyor band **19** into hopper **16** and from there into the hose-shaped sorting zones, which are actuated by excenter shaft **18**. The treated material is discharged over hopper **20** on the collecting band, whereas the overflow material is conveyed via the overflow hoses **22** to the collecting container **23**.

The hose-shaped sorting zones resulting from the devices shown in Figures 3.6h and 3.6i, are represented again in detail in Figures 3.6k, 3.6l and 3.6o. While according to Figure 3.6k, the top mat **14** is perforated and the bottom mat **13** is an unperforated pressure mat, in Figure 3.6l the perforated mats **14** are mounted below and the unperforated pressure mats **13** above.

While the unperforated mat acts as transport and pressure mat in the arrangement shown in Figure 3.6k, the perforated mat discharges, for example, dust particles of a fraction determined by the size of its sieve holes into an underpressure container, with an excellent interaction of the underpressure maintained in the collecting container and the pressure developed during the tensioning of mats **13, 14**. In the arrangement according to Figure 3.6l, on the other hand, with the perforated mat mounted at the bottom and under the effect of gravity, the material to be treated is transported and subjected to three separate effects, namely, to the pressure exerted by the unperforated mat, to gravity, and to a suction effect due to underpressure in the container **24**.

Figure 3.6o shows two adjacent zones, the perforated mat being mounted below in one zone and above in the other zone. This arrangement is particularly advantageous whenever various materials are to be treated simultaneously in a single device.

Microporous Corrugated Filter Element

D.B. Pall and C.A. Keedwell; U.S. Patent 3,591,010; July 6, 1971; assigned to Pall Corporation describe a process which makes possible the preparation of corrugated microporous filters having ultrafine pores or micropores less than 25 microns in diameter extending from one surface of the material to the other.

The process upgrades coarse fibrous corrugated substrates by application of a microporous layer thereby reducing their pore diameter to the desired ultrafine or microporous dimensions.

The process makes possible a corrugated substrate which can receive a uniform coating and/or impregnation of the particulate material without cracking at the bases of the corrugations or bridging of the corrugations thereby, or shrinkage problems. This requires that the porous substrates be especially prepared for laydown of a microporous layer thereon by flow through of a dispersion of particulate material.

The process employs a substrate in a flat sheet form as the starting material. The preliminary preparation features first that before corrugation the base fold areas of the substrate be treated to reduce their porosity by at least 10%, and preferably from about 15 to 35%. This equalizes flow through during laydown. Next, the corrugations are folded both at their bases and at their peaks to a smooth and substantially uniform radius of curvature, which at their bases is above the prescribed minimum of at least 0.025 inch, and preferably within the range from about 0.04 to about 0.06 inch. The radius of curvature of the peaks or tips of the corrugations is not critical, but is determined by available space and length of corrugation sides. Thereafter, if a generally closed configuration, such as a cylinder, is desired, the ends of the corrugated substrate sheet are joined, preferably in a reverse lapped side seal, utilizing a bonding agent.

The radius of curvature, as given above and below, is measured from the center of curvature of the fold to the upstream surface of the substrate; this radius measurement will include the thickness of the substrate for the tip fold, but will not include it for the base fold.

The thus prepared corrugated porous substrate then is ready for laydown of the microporous layer. This is done by one of the processes described in U.S. Patents 3,158,532; 3,238,056 and 3,246,767. Particulate material is suspended in a liquid, and deposited in one or in a plurality of incremental layers upon the substrate, usually on the surface, but also, if desired, within the pores thereof.

The layer can also be confined wholly or predominantly to impregnation of the pores. The particulate material can all be of the same size, or of two or more sizes, all suspended in the liquid. Gaseous suspensions are unsatisfactory. The desired degree of microporosity of the deposited layer is obtained by varying the type, size, amount and location (on the surface and/or within the pores) of the particulate material deposited, by blending different sizes of material at different points, if desired, and by carefully controlling the state of dispersion of the material.

If impregnation is desired in place of or in addition to coating, the desired reduction in pore diameter of the substrate is obtained by varying the size and amount of the particulate material deposited in the pores from surface to surface of the substrate, blending different sizes at different points, if desired. The particles can be deposited in any amount at any point in or throughout the pores from surface to surface or only at one side or in the center of the substrate, as desired.

Example: A flat sheet of epoxy resin impregnated paper, having an average pore diameter greater than 25 microns and a thickness of 0.036 inch, was subjected to a compression or densification process at the locations of the future bases and tips of the corrugations that were to be formed. During the densification process, the thickness at these points was reduced to 0.026 inch. The corrugations were formed by a standard folding operation that imparted a corrugation base radius of 0.05 inch measured from the center of curvature of the fold on the upstream side of the substrate to the surface only of the substrate, and a tip radius of 0.03 inch measured from the center of curvature of the fold on the interior of the substrate to the upstream surface, including the thickness, of the substrate.

The corrugated sheet was then formed into a substantially cylindrical shape joined in a reverse lapped side seam seal by a strip of epoxy resin bonding agent. The element was 9¾ inches long and had an internal diameter of 1¼ inches and an external diameter of 2½ inches. Twenty-five corrugations gave an outer surface area of 310 square inches. The two ends of the element were then placed in a clamp and sealed by means of silicone rubber gaskets, one of the gaskets having an opening communicating with the interior of the element and an effluent outlet port to permit drainage of fluids contained in the interior of the elements.

The element, so prepared, was then placed in a vessel containing an anchoring layer dispersion. The anchoring layer dispersion was prepared by mixing 20 grams of a liquid bonding agent comprised of a liquid polyepoxide resin (p-aminophenylethyl chlorohydrin) containing 40% by weight of a cyanoethylated primary aliphatic polyamine as a curing agent, 7 grams of potassium titanate fibers averaging 0.5 micron in diameter and 100 microns in length, and dispersing the mixture in 600 cc of water containing 0.4% by weight of hydroxypropyl methylcellulose.

The resultant dispersion had a viscosity in excess of 50 centipoises and had an average droplet size of between 25 and 75 microns. A vacuum equivalent to 15 inches of mercury was then applied through the effluent outlet port to draw the dispersion through and cause the deposition of the particulate material and bonding agent on the external surface of the element.

A microporous layer dispersion was prepared by mixing 28 grams of amosite type amphibole asbestos fibers 0.5 micron in diameter by 300 microns long, 30 grams of the same bonding agent that was used in the anchoring dispersion, 80 grams of butyl carbitol, 28 grams of sodium carbonate, and 1.4 grams of Tamol 850, a water-soluble organic dispersing agent, and dispersing the mixture in 14 liters of water. The resultant dispersion was partly emulsified having an average droplet size of 1.0 to 5.0 microns, and approximately 80% precipitated onto the fiber.

The open ends of the substrate cylinder were closed, and the microporous layer dispersion was then drawn through the sides in four separate applications, each application being equal to approximately 0.25% of the total applied and at an applied vacuum equal to 12 psi. After each application, the element was removed from the pressure vessel and was subjected to an air pressure of 12 psi from the outside to compress the materials thus far applied.

After all of the microporous layer dispersion had been applied, hot air at a temperature of 400°F was passed through for 5 minutes from the outside to the inside to dry the element and cure the resin. The element was then removed from between the clamps and metal and caps were affixed by means of a suitable adhesive.

The filter element thus prepared was found to have a maximum pore diameter of 4 microns, an average pore diameter of 1.5 microns, a water permeability of 1.6 gallons per minute per square foot at an applied pressure differential of 1 psi. The adhesion was sufficient to withstand the application of up to 6 psi in the reverse direction and the voids volume of the microporous layer was about 85%.

Thin Porous Gold Foils

R.R. Russell, H.E. Cline and W. De Sorbo; U.S. Patent 3,594,134; July 20, 1971; assigned to General Electric Company describe a process for preparing a metallic film with substantially parallel and uniform apertures of small cross section and uniform distribution which is useful as a filter. An alloy is cast which in the solid state is comprised of at least two phases. The cast alloy is directionally solidified to produce a body where one of the phases is present as a plurality of substantially parallel rods passing through a matrix comprised of the second or other phases.

The directionally solidified body is etched to remove the rod-like phase to form straight through apertures or, if desired, recesses. A material is placed on an etched area of the etched body to form a negative replica of the recesses or holes. The material is then stripped away and metal is deposited on its negative replica surface. The resulting deposited metallic film is then recovered.

The alloy of the process is an eutectic alloy which is comprised of at least two phases in the solid state. It need only be of a composition which upon being cast and directionally solidified will produce the rod-like phase. Such a composition is the eutectic composition or a composition close thereto. The range that the composition may vary from the eutectic is determinable empirically for the specific alloy. For a majority of these alloys, such range is generally up to about 10% by weight from the eutectic.

There are a number of eutectic alloys which upon being directionally solidified have one phase present in a rod-like form as required by the process. Representative of these alloys are NiAl-Cr, NiW, NiAl-Mo, Al-Al_3Ni, Ta-Ta_2C, CoAl-Co, Cb-Cb_2C, Cb-Th, Ni-$Ni_{17}Th_2$, Ni-Ni_3P, Co-$Co_{17}Y_2$, Fe-FeSb, Cr-C, Ti-B, Ti-Th, V-V_3Si, Ni-Ni_3B, InSb-Sb and Cu-Cr. Typical of alloys having nonmetallic characteristics which are also useful are NaF-LiF, LiF-NaCl, NaF-NaCl and NaF-NaBr.

The negative replica area of the material is used as a substrate on which there is deposited metal to produce the porous metallic film of the process. A wide variety of metals as well as alloys can be used for deposition on the substrate. Representative of these metals is gold, copper, nickel, aluminum, tin, niobium and tantalum. The deposition can be carried out in any conventional manner which does not significantly affect the structure of the replica.

Example 1: In this example, a gold porous film was produced. An alloy comprised of 56.5% by weight nickel, 25.5% by weight aluminum and 19% by weight molybdenum was prepared by melting the components, each of which was about 99.9% pure, under argon in an alumina crucible. The molten alloy was cast under argon in a copper mold to produce a cylindrical ingot about 2.2 cm in diameter and 15 cm in length.

The ingot was placed in an alumina crucible and was directionally solidified at a rate of 6 x 10^{-4} cm per second (0.8 inch per hour). Metallographic examination of both ends and along the length of the ingot showed it to be composed of molybdenum-rich rods in a nickel aluminum-rich matrix where the rod phase was substantially perpendicular to the planes of both ends of the ingot.

The ingot was sliced transversely to the rod growth direction by a cut-off machine. Each slice was about 50 ml thick. The distorted surface layers generated during slicing were mechanically polished to a mirror-smooth surface. Metallographic examination of the resulting polished specimen showed a uniform array of molybdenum-rich rods generally perpendicular to the cut surface.

The specimen was made the anode in a conventional electrolytic cell having an electrolyte of 3% oxalic acid in water. Stainless steel was used as the cathode. A potential of 3 volts DC was applied for 30 minutes. The etched specimen was then rinsed with water and examined metallographically. The molybdenum-rich rod phase had been substantially removed to a depth of 50 microns leaving an etched metallic material wherein the recesses measured approximately 0.3 to 0.5 micron in diameter as determined by an electron beam microscope and an electromicrograph.

The recesses of the etched specimen had a density of about 7 x 10^7 per square centimeter as determined by counting the recesses from a micrograph and knowing the magnification of the micrograph, calculating the number of recesses per square centimeter.

This etched metal specimen was then used as a master to prepare a thin porous gold film by a replication technique. Specifically, a 5 milliliter thick sheet of cellulose acetate was softened in acetone at room temperature and then applied to the etched recess-containing surface of the specimen. After drying, the cellulose acetate film was stripped away. The cellulose acetate surface portion adjacent to the etched surface of the specimen was a negative replica of that surface.

A thin gold film about 1200 A thick was vacuum evaporated onto the cellulose acetate replica surface under a vacuum of below 1 x 10^{-5} mm Hg. The resulting cellulose acetate gold film composite was immersed in acetone at room temperature until the cellulose acetate dissolved leaving the gold film. The gold covering the cellulose acetate projections, i.e., caps, collapsed when the cellulose acetate dissolved to partially close the pores in the gold film.

These caps were removed by immersing the gold film in an etching solution comprised of one part concentrated nitric acid, two parts concentrated hydrochloric acid and three parts water for about 3 seconds at room temperature. The resulting etched porous gold film was examined metallographically. It had straight pores which were substantially cylindrical in form. The pores were also substantially uniform in size which was approximately 0.5 micron in diameter.

Example 2: The etched gold film prepared in Example 1 was treated in this example to produce a controlled reduction of its pores. Specifically, a conventional Watts type nickel plating cell was used with the etched gold film supported on an electron microscope grid as cathode, and nickel as the anode. The electrolyte was maintained at 43°C and was comprised of a mixture of 24 grams of nickel sulfate, 4.5 grams nickel chloride, 3 grams boric acid and 100 cc water. A potential of 1 volt DC was used to carry out the electrodeposition.

The gold film cathode was immersed in the electrolyte for 30 seconds and then was examined metallographically. The diameter of the pores had been reduced substantially uniformly to a diameter of approximately 0.2 micron. The gold film was then reimmersed into the electrolyte for another 30 seconds under the same conditions and again examined metallographically. The holes had been reduced to approximately 300 A. The final thickness of the gold film was about 0.5 micron. The electrodeposited metal greatly strengthened the porous film.

Example 3: The etched metal specimen used as a master in Example 1 was also used in this example to produce a porous gold film. In this example, however, a two step replication method was used.

Acetylcellulose film, about 0.34 mm thick, was cut slightly larger than the etched recess-containing surface of the metallic master. A small amount of methyl acetate solvent was dropped on the surface of the master and spread over it. The acetylcellulose film was then laid over the solvent wet surface from one end to prevent bubble formation underneath. The solvent softened the film sufficiently so that, after drying, when the film was stripped off, it was a negative replica of the master.

A thin gold film, about 5000 A thick, was evaporated onto the negative replica to the surface previously in contact with the master surface. The pressure during evaporation was kept below 1 x 10^{-5} mm Hg.

A layer of molten wax, i.e., paraffin having a melting point of about 40°C, was placed over the gold film and allowed to solidify. The wax-gold film acetylcellulose composite was mounted on a 0.200 stainless steel mesh screen support with the acetylcellulose layer against the screen. The entire resulting assembly was then immersed in methyl acetate at room temperature until the acetylcellulose film dissolved.

The methyl acetate solvent was then heated to approximately 40°C to remove the wax layer which served as a substrate to reinforce the gold film in the subsequent process of separating the acetylcellulose film from it. After the wax had been removed, the gold film supported on the metal screen was rinsed in anhydrous methyl acetate to remove any residual acetylcellulose and/or wax. A light micrograph of the surface of the gold showed the pore size to be 0.6 micron or less with the majority of the pores ranging in diameter from 0.5 to 0.6 micron.

Example 4: In this example, the membrane characteristics of the porous gold film produced in Example 3 were determined. The assembly comprised of the gold film resting on the 200 stainless steel mesh screen support was mounted and epoxied in place between two Lucite washers and allowed to air dry.

This structure was mounted so that the porous gold film covered an opening at one end of a chamber made of Lucite. Through a second opening, water was added to the chamber and pressure could be applied to the water contained in the chamber. A pipette was mounted outside the chamber but adjacent to the screen support to receive all water passed through the porous gold film, i.e., the membrane.

Pressures up to about 0.12 atmosphere were applied to the water contained in the cell. The procedure was repeated using, instead of water, a 50-50 methanol-water solution as well as a 50-50 ethanol-water solution. The pores in the gold film were determined to have a flow permeability directly proportional to the applied pressure and inversely proportional to the solution viscosity as predicted by viscous flow theory, i.e., Poisseuille's Law,

$$\text{Flow Permeability} = \frac{NA\pi R^4(\Delta p)}{8\eta L}$$

For example a pressure of 40 psi or 0.10 atm on the water contained in the chamber resulted in a flow permeability of 0.30 cc/cm^2 sec whereas the same pressure applied to 50-50 volume percent methanol-water solution resulted in a flow permeability of 0.18 cc/cm^2 sec and when applied to 50-50% ethanol-water solution, the flow permeability was 0.11 cc/cm^2 sec.

In plotting flow permeability (cc/sec cm atm) against reciprocal viscosity (centipoise $^{-1}$) the slope of the line using the above equation for flow permeability gives a pore radius R of about 0.3 micron compared with the observed value of less than 0.5 to 0.6 micron. In this respect N, the pore concentration, is approximately 8.5 x 10^7 pores per square centimeter and thickness of the gold film L = 0.5 micron.

Laboratory Filtering Apparatus

A process described by *L.H. Westesson; U.S. Patent 3,515,282; June 2, 1970; assigned to Ingeniorsfirman Nils Weibull, AB, Sweden* relates to a filtering apparatus for laboratory filtering of liquids by sucking up the liquid through a filter. The apparatus has been provided for the purpose of permitting rapid filtering of liquids at routine analyses in connection with continuous production, for instance at the determination of the sugar content in sugar beets delivered to sugar factories.

The filtering apparatus provided by the process is characterized in that the filter is a compressed cellulose cartridge pushed into a suction nozzle which can be optionally connected to vacuum or pressure for sucking liquid through the filter cartridge at filtering or for blowing the filter cartridge out of the nozzle after filterings.

A filtering apparatus of this type can readily be automated inasmuch as the filter cartridge can be exchanged without manual interference with the aid of a program-controlled automatic mechanism for controlling the movement of the nozzle and the application of vacuum or pressure in the nozzle. With the status of the automation techniques this does not involve any difficulties.

In addition, such filtering apparatus is particularly economical in use since the filter cartridge consists of an inexpensive material and the use of glass utensils for carrying out the filtering operation is reduced to a minimum, as is the associated dish washing work.

The filtering apparatus shown in Figure 3.7a comprises a receptacle **10** which at the top is connectible via a valve **11** to a compressed air line **12** and via a valve **13** to a suction line **14**, while it is connectible at the lower end via a valve **15** to one end of a hose **16** and via a valve **17** to a line **18**. At the other end the hose has a funnel-shaped nozzle **19** with a filter cartridge **20**. This filter cartridge consists of cellulose, such as sulfite or sulfate cellulose or a mixture thereof, and can be prepared in the manner described below in connection with Figures 3.7c through 3.7e.

For automation of the use of the filtering apparatus the nozzle can be supported by a member to which is imparted both horizontal and vertical movements indicated by the arrows in Figure 3.7a, in order that the nozzle may be moved between a position above a base **21** and may permit being lowered towards the base, and a position above a cup **22** and may permit being lowered into the cup. Further, the valves may be solenoid valves which are opened and closed by a program mechanism in coordination with the movements of the nozzle.

For the description of the function of the filtering apparatus it is assumed that the nozzle **19** occupies the position shown in Figure 3.7a by full lines, that the cup **22** contains liquid while the receptacle **10** is empty, and that all valves are closed. The nozzle with the filter cartridge **20** is now lowered into the liquid in the cup to the position shown by dash lines **19'**, and valves **13** and **15** are opened.

Liquid is sucked from the cup through the filter cartridge **20** and the filtrate collects in the receptacle **10**. After a predetermined time or at a predetermined level in the receptacle valves **13** and **15** are again closed while valves **11** and **17** are opened instead, so that the filtrate is urged under the action of compressed air supplied to the receptacle into the line **18** to be forwarded to an analyzing or like apparatus which can also be of automatic operation.

After the receptacle has been emptied valve **17** is closed whereupon valve **15** is again opened so that compressed air enters the hose **16** and the nozzle **19** for ejection of the spent filter cartridge **20** from the nozzle. This can take place while the nozzle is over the cup or after the nozzle has been moved to a particular cartridge ejecting station.

FIGURE 3.7: LABORATORY FILTERING APPARATUS

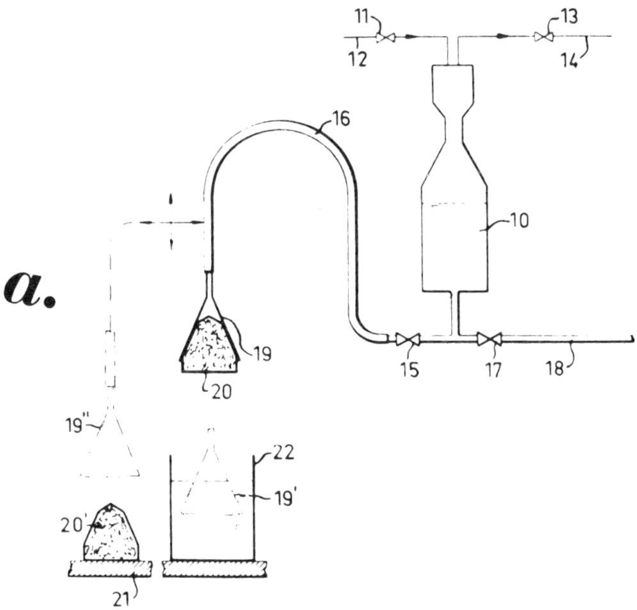

Diagrammatic view of the filtering apparatus.

(continued)

FIGURE 3.7: (continued)

b.

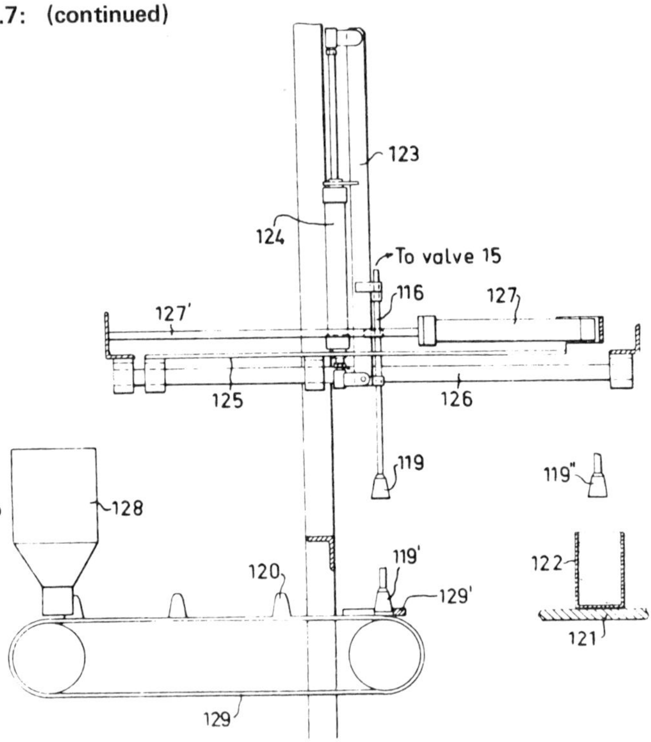

c.

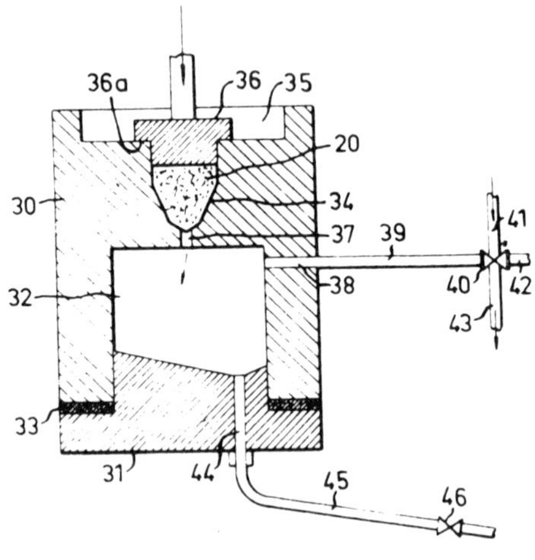

(b) Side view, partly in vertical section, of the construction of the nozzle arrangement.

(c) Diagrammatic view of a mold for manually manufacturing the filter cartridge.

(continued)

FIGURE 3.7: (continued)

d.

e.

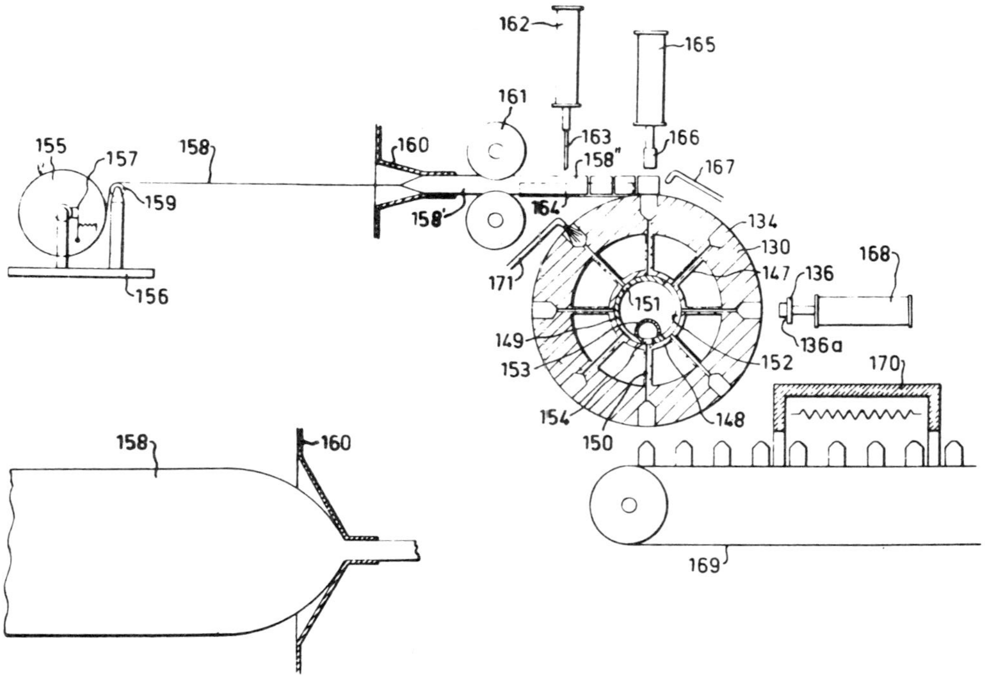

(d) Side view of a machine for the continuous manufacture of filter cartridges.

(e) Partial plan view of the machine.

Source: U.S. Patent 3,515,282

After closure of valves **11** and **15** the empty nozzle is moved to position **19''** over the base **21** and is pressed down over a filter cartridge **20'** placed on the base. When the nozzle is then returned to position **19'**, after the cup, which earlier supplied liquid to the filtering apparatus, has been removed therefrom and exchanged for a new cup containing liquid (all this can be performed by known automatic means), the filter cartridge **20'** is carried along by the nozzle as it is retained therein under slight interference fit, possibly supported by suction in the nozzle in that valves **13** and **15** are kept open while the nozzle is moved from base **21** to position **19'**.

The construction of the filtering apparatus, which is shown in Figure 3.7b, includes a pipe **116** which is vertical and at its lower end has a funnel-shaped widened portion **119**, while it is assumed that the pipe has its upper end connected to valve **15** in Figure 3.7a and via the valve to the system of receptacles otherwise illustrated in Figure 3.7a and which can be incorporated in a polarimeter device for sugar content determination.

Pipe **116** is arranged on and supported by a bar **123** which is vertically reciprocable by means of a double-acting air cylinder **124** which in turn is carried by a sliding carriage **125** horizontally movable on guide means **126**. Travel of carriage **125** is realized by a double-acting air cylinder **127** which is connected with the carriage and the piston rod **127'** which is connected to the guide means.

For each sample sucked up by the filtering apparatus use is to be made of a new filter cartridge, and such cartridges emerge from a supply **128** from which the filter cartridges **120** are fed by means (not shown) one at a time onto a belt conveyor **129**. The filter cartridges travel on the conveyor in upright position to an abutment **129'** where the filter cartridges arrange themselves in a row.

When pipe **116** occupies the Figure 3.7b position and carriage **125** is retracted to the Figure 3.7b position, pipe **116** is vertically lowered by the cylinder **124** so that the funnel-shaped widened portion **119** arrives at position **119'** and is passed over a filter cartridge **120** arrested by the abutment **129'**. The filter cartridge will stick in the funnel-shaped widened portion and as the direction of motion of the cylinder **124** is then reversed the filter cartridge is taken along in the upward movement of pipe **116**.

The carriage is then moved by the cylinder to the right as viewed in Figure 3.7b, so that the funnel-shaped widened portion with the filter cartridge therein will arrive at the Figure 3.7b position **119''**. The filtering apparatus is now in position to permit being lowered into a can **122** located on a base **121**. The movement of the pipe necessary for sucking up filter cartridges can be controlled in a known manner by a program device. After the desired quantity of liquid has been sucked up from the can under filtering thereof, the spent filter cartridge can be ejected into the can in the manner described in connection with Figure 3.7a.

The filter cartridge can be manufactured in a suitable manner by the use of the mold shown in Figure 3.7c, which includes an upper mold part **30** and a lower mold part **31** which is inserted in the first mold part and defines with it a chamber **32**. A packing **33** provides a seal between the mold parts. The mold part **30** forms a mold cavity **34** which conically tapers in a downward direction from an upper cylindrical portion, and a recess **35** concentrically surrounding the mold cavity **34** at the upper open end thereof. A plunger **36** can be introduced into the cylindrical portion of the mold cavity, and the depth of introduction is restricted to the cylindrical part of the mold cavity by a shoulder **36a**. Provided in the mold part **30** is a passage **37** which is in permanent communication with the chamber **32**, and a passage **38** connected to a line **39** which is optionally connectible over a four-way valve **40** to a compressed air line **41**, a line **42** opening into the atmosphere, and a suction line **43**. Arranged in the mold part **31** is a discharge conduit **42** having a shut-off valve **46**.

The filter cartridge is manufactured as follows. Cellulose, such as sulfite or sulfate cellulose or a mixture of both, is introduced into the mold cavity **34** and water is supplied to the recess **35** while the line **39** through valve **40** is kept connected to the suction line **43** and the valve **46** is closed so that vacuum is maintained in the chamber **42** which then functions as a suction box.

Under continued suction of water from the cellulose to chamber **35** the plunger **36** is introduced into the mold cavity in order that the cellulose shall be compressed to the desired density. After finished compression the plunger is reacted, the vacuum in chamber **32** is relieved in that line **39** is connected to atmosphere by adjustment of valve **40**, simultaneously as valve **46** is opened for discharge of the water from chamber **32**, which has been separated from the cellulose. Valve **46** is then closed again simultaneously as valve **40** is connected to the compressed air line **41** so that chamber **32** is placed under pressure for ejection of the resulting filter cartridge **20** from the mold cavity. The filter cartridge is then dried and thus is ready for use in the filtering apparatus described.

The machine, illustrated in Figures 3.7d and 3.7e, for continuous manufacture of filter cartridges comprises a wheel having a wheel rim **130** connected by spokes **147** to a hub **148** which is rotatable on a tubular bearing **149** and connected in a manner not shown to a drive for causing the wheel to rotate. Arranged in the wheel rim is a number of peripherally distributed mold cavities **134** each of which is in communication through a passage **150** located in a spoke with the internal circumferential surface of the bearing **148**.

It is assumed that the tubular bearing **149** is connected to suitable evacuating means and has a short slot **151** and a long slot **152** for attaining communication between each passage **150** and the associated mold cavity **134** and the interior of the bearing **149** during predetermined sections of the path of movement of each mold cavity at the rotation of the wheel.

Besides, there is arranged within the bearing **149** a line **150** which in a manner not shown is connected to a source of compressed air and is connectible via a slot **154** in the bearing **149** to each passage **150** when the latter is in vertical position with the associated mold cavity facing downwardly. The wheel is intermittently rotatable by means (not shown) in increments equal to the angular distances between the mold cavities **134**.

For continuous supply of cellulose to the mold wheel described a roll of cellulose **155** is rotatably mounted in a frame **156**, the rotation of the roll being restrained to a certain extent by a friction brake **157** which is applied against the roll shaft. From the roll a cellulose web **158** passes over a return roller **159** to a nozzle **160** with a successively tapering mouth in which the cellulose web is folded or crinkled into substantially cylindrical cross sectional shape so that the flat cellulose web leaves the nozzle approximately in the form of a rope **158′**.

The rope passes between a pair of pressure rolls **161** which exert traction on the rope for pulling forth the web **158** from the roller **155** and which to this end can be knurled or otherwise made rough on their surfaces contacting the rope **158′**. Following upon the rolls **161** is a knife **163** which is reciprocable by means of a pneumatic cylinder **162** for cutting the rope into short lengths **158″** as the rope is advanced.

On a stationary base **164** of U-shaped cross section these rope lengths are moved into the peripheral surface of the mold wheel by the rope which is advanced intermittently by the rollers **161** in synchronism with the continuous rotation of the mold wheel such that the rope lengths **158″** in order of turn are pushed onto the mold wheel over the mold cavity momentarily facing straight upwards.

Above the mold wheel and opposite the mold cavity is arranged a pneumatic cylinder **165**, and the piston rod of the cylinder has the projecting free end shaped as a plunger **166** for pressing into the mold cavity the rope length which lies on the peripheral surface of the mold wheel and extends over the upwardly facing mold cavity. A pipe **167** which forms a nozzle directed towards the mold cavity and connected to a water conduit is located adjacent the mold wheel to permit pouring water onto the cellulose introduced into the mold cavity.

During the intermittent rotation of the mold wheel new mold cavities are moved in order of turn into position opposite the plunger **166** to permit pressing a supplied rope length of cellulose thereinto and subsequent pouring of water onto the cellulose while the mold cavity which has been filled with cellulose will enter into communication during its continued movement with the interior of the tubular bearing **149** via the slot **152** for sucking water from the cellulose while simultaneously pressing it into the mold cavity under the pressure acting from outside.

After two steps the filled mold cavity has arrived in position opposite a pneumatic cylinder **168** the piston rod of which at the free end has a plunger **136** with a shoulder **136a** thereon. The plunger is introduced into the mold cavity **134** by the pneumatic cylinder **168** for compressing the cellulose therein in the manner described in connection with Figure 3.7c, while water is simultaneously sucked out of the mold cavity through the slot **152**.

The sucking of water out of the mold cavity proceeds also in the next position, but when the mold cavity with the plug of cellulose pressed thereinto arrives in vertically downwardly facing position it is connected to the compressed air lines **153** via the slot **154** for ejection of the resulting plug onto a belt conveyor **169**. This belt conveyor carries the resulting cellulose plugs through a drying oven **170** from which they are delivered in a finished state.

During the intermittent return to vertically upwardly facing position the empty mold cavity moves past a pipe **171** having a nozzle which is directed towards the mold cavity in normal position just ahead of the vertically upwardly facing position in order to spray water into the mold cavity. Simultaneously with the spraying, water can be sucked out in that the mold cavity is in communication via the slot **151** with the interior of the tubular bearing **149** so that the mold cavity and the associated passage **150** are flushed, but one can also have the mold cavity partly filled with water while the communication with the interior of the bearing is kept closed, in order that this water may serve to moisten the cellulose at the later introduction into the mold cavity.

Rigid Open-Cell Plastic Filter Formed from Emulsion

F.W. Markley; U.S. Patent 3,661,267; May 9, 1972; assigned to the U.S. Atomic Energy Commission describes a rigid open-cell plastic foam body which is useful for example as a filter. The plastic foam is prepared by mixing equal parts of a diglycidyl ether of bisphenol A epoxy resin and the reaction product of an excess of a polyfunctional amine and a polyfunctional fatty acid with two parts of water while avoiding the entrapment of air in the mixture to form an emulsion in which water is the continuous phase, casting the emulsion in the desired form and curing the emulsion.

A rigid, open-cell foam filter was prepared by this process and subjected to flow tests to determine the suitability of the material for this purpose. To prepare the filter 50 grams of Epon 828 which is a diglycidyl ether of bisphenol A epoxy resin and 50 grams of Versamid 140 which is the reaction product of an excess of a polyfunctional amine and a polyfunctional fatty acid were mixed gently with 100 grams of water to form an emulsion and the emulsion was cast into a cylindrical form and cured at 75°F for 24 hours.

The cylindrical body thus formed was then machined to remove the impervious skin thereon and to produce a cylinder which was 6.7 cm long, 2 cm in outside diameter and 0.5 cm in inside diameter. Flow tests were then conducted on this filter body to determine the resistivity of the foam to the passage of water.

The flow resistivity of the foam is about 6 min/cm and the material is eminently suited to be employed as a filter. If the material is to be used in a water system, it may be so used without drying; if in any other system, it must be dried first. Examination of a cross section of the material through a microscope following drying shows that the material consists of a large number of balls, about 0.001 inch in diameter, fused together to provide a void space of about 25%. In view of the method of preparation, the void space is continuous.

Cartridge Containing Polymer-Entrapped Aggregate Particles

D.E. Heskett and J.B. Heskett; U.S. Patent 3,538,020; November 3, 1970; assigned to The Kata Manufacturing and Filtering Company describe a generally porous fluid treating device which is characterized by a plurality of fluid treating aggregate particles such as, for example, ion exchange resin, activated charcoal, manganese greens and sawdust and like materials bound together in a closely packed abutting relationship in a matrix formed of a resinous polymeric material. The aggregate particles, though bound, have substantially their entire surface area freely exposed in a manner substantially the same as the distribution of the particles in a loose bed.

An important example of the process specifically concerns a fluid treating cartridge having a porous body of closely packed fluid treating aggregates which are bound together or entrapped in a porous polyurethane matrix. The fluid treating cartridges find practical utility in the removal of particulate matter, odors and other contaminants from both liquids and gases.

Examples of commercially available urethane prepolymer resins which can be used in forming the particle entrapping or aggregate binding porous matrix of the process include a Simoniz Company prepolymer and Nopco Chemical Company prepolymers SX-159D, #H-102-N-T and P-502-T. The Simoniz prepolymer is a prepolymer resin prepared by mixing 444.8 parts by weight of Pluronic L-61, hydroxyl number 50.5 having the general formula

$$HO(C_2H_4O)_a(C_3H_6O)_b(C_2H_4O)_cH$$

with 63.5 parts, by weight, of PPG-400, hydroxyl number 288.8 (a polypropyleneglycol having a molecular weight of about 400), and then combining this polymixture with 235 parts, by weight, of toluene diisocyanate (see U.S. Patent 3,103,408). Nopco SX-159D is a prepolymer resin prepared by reacting polyoxypropylene glycols, having a molecular weight of from 1,000 to 4,000, with toluene diisocyanate, the toluene diisocyanate being present in a sufficient quantity to produce a reaction product having excess isocyanate groups.

Nopco #H-102-N-T is a prepolymer resin prepared by reacting polyoxypropylene glycols having molecular weights of from 100 to 1,000 with toluene diisocyanate, which is present in quantities sufficient to produce a reaction product having excess isocyanate groups. Similarly, Nopco P-502-T is prepared by reacting toluene diisocyanate with castor oil, the toluene diisocyanate being present in quantities sufficient to provide a reaction product having excess isocyanate groups.

The following examples illustrate the process. The percentage weight concentrations referred to in the following examples identify the weight of aggregate with respect to the total weight of aggregate and prepolymer resin.

Example 1: An activated charcoal cartridge containing 95% by weight 20 x 40 mesh activated charcoal was made by first mixing 380 grams of the activated charcoal with 187.5 grams deionized water in a 1 gallon working capacity Baker-Perkins mixer for 3 to 4 minutes.

20 grams of SX-159D were then added thereto with additional mixing for 3 minutes. This mixture was then hand packed in a 2.625 inch ID Pyrex glass tube and cured by passing steam therethrough. After curing, the cartridge was removed from the Pyrex tube and observed to exhibit excellent structural strength.

Example 2: In this example, an activated charcoal cartridge containing approximately 74%, by weight, 30 x 80 mesh carbon was made by first prewetting 350 grams of the carbon with 626.5 grams of water containing 15 grams of Quadrol, a catalyst and crosslinking agent, N,N,N',N'-tetrakis (2-hydroxypropyl) ethylene diamine, and then mixing with 127.8 grams of Simoniz Company prepolymer in a Baker-Perkins mixer for about 1 minute.

The mixture was then hand packed into a 24 inch x 2¾ inch ID Pyrex glass tube and steamed for about 3 minutes. This cartridge was found to reduce the chlorine content of Chicago water from 0.4 ppm to less than 0.1 ppm at a flow rate of approximately 1½ gpm.

Example 3: This example illustrates the making of a fluid treating cartridge which is composed of a composite aggregate, viz, synthetic styrene beads, granulated charcoal and manganese greensand. In this example the term parts refers to parts by volume. To 22 parts of Nopco SX-1595D were added 10 parts water, 25 parts granulated charcoal (average grain size of from 0.025 to 0.05 inch), 25 parts synthetic styrene beads (average particle diameter being approximately 0.025 inch) and 28 parts manganese greensand (average particle diameter being approximately 0.020 inch).

These materials were thoroughly mixed together so that all of the aggregate particles received a uniform coating of the SX-159D urethane polymer. Three drops of tetramethylbutyldiamine were then added and the mixture immediately transferred to an annular mold measuring 9⅞ inches in length and 2¾ inches in outer diameter. The central hollow core measured ¾ of an inch. The volume of the mold was approximately equal to the total volume of the above ingredients in their unreacted condition. The walls of the mold were fixed so as to maintain the aggregate particles in an abutting relationship.

The mold containing these materials was then placed in an oven at a temperature of 175°F for about one-half hour after which time the polyurethane was cured. The filter cartridge was then removed from the mold and, after cooling to room temperature, was tested on drinking water. Performance was satisfactory, thus establishing that the polyurethane matrix can be effectively used in fluid treating cartridges in which a plurality of different aggregates are combined.

FILTRATION AIDS

PARTICULATE FILTERS

Exfoliated and Calcined Vermiculite

I. Preston and J.C. Davidtz; U.S. Patent 3,953,357; April 27, 1976; assigned to Mandoval Vermiculite (Proprietary) Limited, South Africa describe a method for preparing an improved vermiculite product which is especially suitable for use as a filter aid. The process comprises subjecting crude vermiculite to a temperature and residence time sufficient to cause exfoliation without substantial calcination or destruction of its mechanical properties. The exfoliated vermiculite is then pulverized to a relatively fine particle size. The pulverized vermiculite particles are then subjected to temperature and residence time conditions sufficient to cause at least partial calcination. These particles are then recovered and pulverized to a particle size suitable for the contemplated use.

The resultant product can then be used as a filter aid in competition with other filter aids manufactured from such materials as perlite and diatomaceous earth. In addition, the filter aid properties possessed by this process are dramatically superior to those possessed by vermiculite which has been merely exfoliated.

The first step of this process, the exfoliation step, is conducted under conditions whereby the resultant exfoliated particles are substantially unchanged with respect to structural strength. This may be accomplished, for example, by heating the crude vermiculite under conditions whereby its particles are exfoliated at furnace residence times which are longer in duration than those presently practiced in the art and furnace temperatures which are lower than those in established exfoliation procedures.

In general, the temperature of the oven or kiln in which the particles are exfoliated is in the range of 150° to 750°C. It has been found practical to use a countercurrent rotary kiln having a flame temperature of about 1100°C. The kiln temperature is controlled in the range of 550° to 750°C. Crude vermiculite having a particle size of about 500 microns is introduced into the kiln for a residence time of about one-half to three minutes, depending on kiln temperature. It is only necessary that the expansion step be conducted so as to substantially preclude calcination of the vermiculite particles. Hence temperatures of 600°C and particles which pass a 200 micron sieve are suitable for the process.

As the exfoliated particles are removed from the kiln, they are permitted to cool and are further pulverized and classified through use of any known particle size classification device, such as screens. This intermediate crushing and classification step should be such so as to produce particles which substantially pass through a 200 micron sieve. Preferably, about 55% of the particles should pass a 20 micron sieve.

In the calcination step the pulverized exfoliated particles are heated in a furnace or kiln at a temperature of about 850° to 1200°C. The residence time utilized varies from about 20 to 40 minutes. Preferably the particles are calcined at a temperature of about 850°C for 30 minutes. When higher temperatures are utilized the residence time should be lower. For example, it has been found that suitable particles result from calcining at 1200°C for about 20 minutes. In general temperatures in the range of 880° to 1150°C, with a residence time of about 24 minutes, produce excellent products. It is necessary to carry out the heating step under such conditions that the temperature of the particles themselves is at least 600°C so as to induce at least partial calcination.

The following examples illustrate the process.

Example 1: Crude vermiculite particles ranging in size from -2 millimeters to +710 microns were heated in a furnace with a temperature of about 500°C at a residence time of about two minutes, to provide exfoliated vermiculite particles. The particles were then pulverized and classified to an intermediate particle size in the range of -200 microns. These particles were then partially calcined in a furnace for about 24 minutes, at a temperature of about 1000°C. The particles leaving the furnace were then cooled.

Since partial agglomeration takes place during the calcination step, the resultant vermiculite was fed to a roller crusher to de-lump the material. They were then air-classified to provide selected particle size ranges.

Example 2: Crude vermiculite, which was crushed and sized to -2 millimeters to +710 microns (80% on grade), was heat-treated in a first heating step to produce exfoliated vermiculite of approximately the same particle size in a kiln at temperatures of 650°C in the exhaust flues and 780°C in the combustion chamber, the residence time in the kiln being approximately 2½ minutes.

The exfoliated vermiculite was air-cooled and pulverized to produce a product having a particle size analysis of about 94% -45 microns and 50% -20 microns. The pulverized exfoliated vermiculite was then fed into the kiln when the temperature of the exhaust flue was 880°C and the combustion chamber was at a temperature of 1150°C. The material was retained in the kiln for a residence time of about 24 minutes.

This heat-treated product was allowed to cool. It was then de-lumped in a roller crusher and classified to give an end product having the following product size analysis: 97%, -88 microns; 90%, -45 microns; 47%, -20 microns.

Example 3: The materials of the above examples were experimented with as filter aids. The same volume of solution utilizing equal weights of different filtering aid materials under exactly the same conditions were used. The following results were obtained: diatomaceous earth, 75 seconds filtration time; exfoliated vermiculite prepared in accordance with this process, 74 seconds filtration time; conventionally exfoliated vermiculite pulverized to obtain a particle size equivalent to that of this process, 178 filtration time.

As can be seen from these results when conventionally exfoliated vermiculite, which had not been prepared in accordance with this process, was utilized as a filter aid, it was found that the filtration time was more than double the time required with either the vermiculite of the process or diatomaceous earth. Great care was taken to assure standardization of the experimental results. Accordingly, equal weights of each material were utilized as well as equal uniformity of particle size. The vermiculite prepared in accordance with the process demonstrated outstanding properties as a filter aid.

Example 4: The following experiments were conducted to demonstrate the effect of the partial calcination heat treatment step on the efficacy of vermiculite treated in accordance with the process as a filter aid.

Crude vermiculite was exfoliated, crushed and classified, partially calcined and recrushed and classified. The calcination was conducted in a furnace at a temperature of 1000°C over a range of furnace residence times. Samples of equal weight of vermiculite were then utilized as a filter aid. The effect of residence time on filtration time is depicted in the table below:

Residence Time (minutes)	Filtration Time (seconds)
5	134
10	101
15	94
20	83
25	67
30	62
35	60
40	55

It can be seen from the above table that the partial calcination step yields dramatic changes in the efficacy of the vermiculite product as a filter aid.

Beneficiation of Diatomaceous Earth

A.P. Kouloheris; U.S. Patent 3,572,500; June 18, 1968; assigned to Cities Service Company describes a process whereby diatomaceous earth can be purified of sand and clay and reduced to a smaller particle size by wet attritioning raw diatomaceous earth to reduce the diatomite particle size, hydraulically classifying the wet attritioned material to remove sand and quartz, dispersing clay and centrifugally classifying to remove said clay.

Wet attritioning can be done in the presence of a clay dispersant and centrifugal classification can be done before or after the quartz and sand are removed.

In the process, low-grade diatomaceous earth deposits are beneficiated by liberating the diatomite, clay and quartz constituents thereof and separating the constituents by specific classification techniques. With low-grade deposits containing appreciable quantities of both sand and quartz, a wet attritioning action is first employed to break down agglomerates of the raw material. The diatomites, quartz and clay are thus liberated from each other. The attritioning action employed is also sufficient to reduce the particle size of coarse diatomite particles to less than about 44 microns.

In degree and time of agitation, the wet attritioning action is moderated so as to avoid reducing the diatomite content of the raw material below about 5 microns in particle size. Before separation of the clay from the diatomite is accomplished, the dispersion of the clay constituent into its natural -2 micron environment should be assured. A clay dispersant is mixed with the material being treated for this purpose.

Separation of the clay is accomplished by a centrifugal classification, as in a solid bowl centrifuge or a battery of microcyclones suitable for a 2 micron cut. The quartz constituent of the raw material can be separated from the diatomites by a hydraulic classification operation. For this purpose, a +325 mesh fraction containing the quartz is separated from a -325 mesh fraction, which contains a large proportion of the diatomite content of the raw material.

The clay dispersant may be added to the raw material during the wet attritioning action to facilitate the dispersion of the clay into its -2 micron environment. If the hydraulic classification for quartz removal precedes the clay removal operation, additional clay dispersant may be added to the clay-diatomite slurry following quartz removal to assure the complete dispersion of the clay in its -2 micron environment. Alternately, all of the clay dispersant may be utilized either in the initial wet attritioning step or in the separate dispersion step following separation of the quartz from the raw material.

The total amount of clay dispersant employed is generally within the range of from about 1 to 14 pounds per ton of solids in the raw diatomaceous earth material being treated. When a clay dispersant is mixed with the raw material both during the initial wet attritioning step and following quartz removal, it is generally desirable to add a major portion (for example, about 75%) of the clay dispersant in the dispersion step following quartz removal.

When a raw diatomaceous earth material containing from about 5% to about 10% clay is treated, the beneficiated diatomite fraction constitutes an improved feed for the production of high-grade filter-aid material by a conventional calcination operation. If the clay content ranges from about 10% to about 40% by weight, however, the beneficiated diatomite fraction comprises an improved feed for processes intended for the production of high-grade filter-aid products from low-grade raw diatomaceous earth material. The beneficiated diatomite fraction in this instance is also useful in that it may be sold for secondary diatomaceous earth uses as a filler, insulating material, abrasive, and the like.

When a portion of the quartz is present as silt having a particle size distribution of from about 30 to about 40 microns, a modified process must be employed in order to separate the silt from the diatomite fraction. Techniques such as quartz flotation and elutriation may be employed for this purpose.

Irrespective of the particular case employed, the beneficiated diatomite fraction suitable for calcination may, in accordance with conventional practice, be passed through a dryer prior to the calcination operation. To reduce the necessary drying capacity, the beneficiated diatomite fraction may be concentrated prior to drying. This concentration may be accomplished by filtering, as in a vacuum filter. The resulting filter cake to be fed to the drying operation would thus have a lower water content, as for example on the order of 30 to 35% by weight as opposed to a water content of 50% or more in the beneficiated diatomite fraction prior to filtration. In order to reduce the load on the filter, an intermediate slurry thickening operation may be included, with a conventional flocculant added to increase the rate of settling of the solids in the thickener.

The process, therefore, represents a significant means for beneficiating low-grade diatomaceous earth deposits. A high proportion of the diatomite content of the raw material may be recovered in the beneficiated diatomite fraction. The thus-beneficiated fraction may be employed as an improved feed for the production of high-grade filter-aid material. The process thereby permits the use of low-grade deposits conventionally discarded or rejected. The method also enables the processing of low-grade deposits so as to avoid the necessity for employing costly selective mining techniques.

Foaming Volcanic Ash

T. Kuroiwa, Y. Isayama, A. Yokota, T. Kamio, K. Jinnai, E. Abe, K. Kimura, M. Hioki and H. Noguchi; U.S. Patent 3,847,806; November 12, 1974; assigned to Director-General of the Agency of Industrial Science and Technology, Japan have found that in the dehydration treatment of sludge from a city sewage system by filtration, the addition of an adequate amount of foaming granular volcanic ash deposits serves to render the suspended solid particles easily filterable and to promote the filtration treatment and that incineration of a filter cake obtained by such filtration treatment gives a granular lightweight foamed material useful as filler.

The term "foaming volcanic ash deposits" is used here to mean volcanic ash deposits capable of foaming by heat treatment and forming a porous substance or hollow spherical particles. Examples of such foaming volcanic ash deposits include Kansas volcanic ash deposits and Shirasu.

The above-mentioned Shirasu is one of the volcanic eruptives having the property of an acidic pumice and has been retained as such except for minimum use as natural lightweight construction material and road-base material. The use of Shirasu is especially preferable because Shirasu itself is widely distributed in Southern Kyushu, Kanto, Tohoku and Hokkaido districts of Japan and is easily available at a low price.

According to the process, granular particles of the foaming volcanic ash deposits are added to sewage sludge in such a ratio that 3 to 10 parts by weight of the volcanic ash deposits are added for 1 part by weight of the total evaporation residue of the sludge and the mixture is thoroughly mixed and then filtered. Thus, the sludge can be dehydrated effectively without using any coagulating agent. Preferred coagulating agents utilizable jointly with the foaming volcanic ash deposits include, for example, Kurifloc LC-599, Hymorock M-166 and Primafloc C-7.

Example 1: Sewage sludge is extracted from a sludge digestion tank in a Kinko sewage treatment plant belonging to Kagoshima Waterworks Bureau. A given amount of Shirasu and/or a given amount of a coagulating agent is added to the sludge, the given amount being based on 100 parts by weight of a total evaporation residue of the sludge. The mixture is dehydrated by filtration under reduced pressure of 500 mm Hg, using a vacuum filtering machine. Table 1 shows the amount of sludge treated at a given area for a given period of time and the moisture content of the resultant filter cake.

Shirasu used in this example is extracted at Furue, Kanoya City, Kagoshima prefecture and has the chemical and mineral compositions shown below.

Results of Chemical Analysis		Mineral Content	
SiO_2	70.36%	Volcanic vitreous material	66.31%
Al_2O_3	13.16%		
FeO	1.41%	Feldspar-quartz	31.11%
Fe_2O_3	0.92%		
CaO	2.58%	Hypersthene, Magnetite, Other	2.58%
MgO	0.62%		
Na_2O	3.20%		
K_2O	2.95%		
TiO_2	0.32%		
MnO	0.06%		
P_2O_5	0.02%		
H_2O^-	1.04%		
H_2O^+	3.49%		

For the purpose of comparison, the amount of sludge treated in the case of adding nothing to the same sludge and the moisture content of the filter cake obtained are also shown in Table 1 below.

TABLE 1

	Experiment No.*							
Additive	1	2	3	4	5	6	7	8
Shirasu	0	0	500	500	1,000	0	300	300
Ferric chloride	6	7	1	0	0	0	0	0
Slaked lime	10	50	10	0	0	0	0	0
Organic coagulant	0	0	0	0	0	0.8**	0.4**	0.3***
Results								
Amount of sludge treated ($kg/m^2/hr$)	5	5.5	11	5.5	8.5	4.8	13	9.0
Filter cake treated ($kg/m^2/hr$)	5.5	8	67	33	94	4.8	52.5	36
Moisture content (wt %)	80	78	70	60	50	80	40	45

*Parts by weight, dry basis.
**Primafloc C-7.
***Hymotock (SIC) M-166.

The filter cakes obtained in Experiments 4 and 5 of Example 1 are dried by heating at 100°C until moisture is evaporated. These filter cakes are then pulverized, charged into a rotary kiln having a diameter of 150 mm and a length of 2,000 mm and then incinerated at 950°C for about two minutes. A gray incinerated material is thus obtained which mainly contained lightweight foamed particles. The properties of the products are shown in Table 2.

TABLE 2

Experiment No.	Content Shirasu (parts by weight)	Weight per Unit Volume (kg/l)
4	500	0.51
5	1,000	0.42

Particle Size of Incinerated Material (wt % through sieve)

Experiment No.	0.59*	0.297*	0.149*	0.074*	0.044*
4	96.2	76.3	53.2	31.1	21.3
5	97.5	80.7	61.0	39.1	24.4

*Sieve size, mm

Example 3: The filter cakes obtained in Experiments 3, 7 and 8 of Example 1 are incinerated for about two minutes at 980°C in a manner similar to that described in Example 2. The properties of the products thus obtained are shown in Table 3.

TABLE 3

Experiment No.	Content Shirasu (parts by weight)	Coagulating Agent	Parts by Weight	Weight per Unit Volume (kg/l)
3	500	Ferric chloride	1	0.55
7	300	Primafloc C-7	0.4	0.40
8	300	M-166	0.3	0.35

Particle Size of Incinerated Material (wt. % through sieve)

Experiment No.	0.59*	0.297*	0.149*	0.074*	0.044*
3	95.8	76.7	42.8	30.0	22.5
7	97.5	80.0	59.5	38.5	24.6
8	99.1	82.5	64.4	45.6	31.0

*Sieve size, mm

Example 4: The lightweight foamed products obtained in Examples 2 and 3 are admixed with ordinary Portland cement. The mixtures are converted into cement mortar and tested to determine the flow, the compression and bending strength after hardening for 28 days, the specific gravity after air drying, the water absorption rate after 24 hours, and thermoconductivity. For the purpose of comparison, a similar cement mortar is made by using a known filler and similarly tested. The results of these tests are shown in Table 4 below.

TABLE 4

Type Particles	Mixing Ratio by Volume (cement/particles)	Cement Ratio (%)	Flow (mm)	Compression Strength (kg/cm^2)
Experiment No. 4	1:3	60	150	110.0
Experiment No. 3	1:3	62	150	101.5
Experiment No. 7	1:3	65	150	98.7
Mitsui pearlite	1:3	110	180	65.1

Type Particles	Bend Strength (kg/cm^2)	Specific Gravity	H_2O Absorption	Thermal Conductivity*
Experiment No. 4	40.0	0.95	30.2	0.28
Experiment No. 3	38.5	0.82	31.1	0.25
Experiment No. 7	30.5	0.79	32.0	0.23
Mitsui pearlite	17.3	0.80	31.6	0.13

*Measured in $kcal/m^2/hr/°C$.

Protein Colloid and Silica Filter Aid

D.W. Davis, J.M. Baloga and B.C. Olmstead, Jr.; U.S. Patent 3,658,184; April 25, 1972; assigned to Johns-Manville Corporation have found that the technique of filter aid filtration may be improved by using a protein colloid in conjunction with conventional filter-aid-based materials. More specifically, it has been determined that certain protein colloids (that is, gelatin and glue) having specific gel strength, may be combined either as a coating upon or as a physical mixture with the filter aid in the liquid to be treated, and preferably used under proper pH conditions to maintain the protein colloid below its isolectric point, generally about a pH of 8, to remove efficiently and effectively insoluble impurities contained in the water by standard filter aid filtration techniques.

Specific advantages have involved improved head loss rates to clarity ratio as contrasted with the filter aid per se, improved clarity at the initial portion of the filtration cycle, improved ability to handle a high influent turbidity while maintaining consistent and constant effluent turbidity, and improved ability to hold low effluent turbidity despite influent fluctuations without immediate filter blockage and shortened filtration cycles.

It has been determined that the addition of the protein colloid in an amount between 0.02 and 20.0 parts per million (ppm) based on the liquid to be treated, when added in combination with the filter aid, improves the filter liquid clarity. With regard to the treatment of water, 0.2 to 2.0 ppm is generally sufficient. As to industrial treatment, the amount of addition is dependent upon the degree of contamination and the desired purity. The amount of filter aid used varies, dependent upon the liquid being treated, but generally varies between 2 and 200 ppm.

Before discussing the specific examples, it is believed desirable to appreciate fully the filtration effects of certain commercially available filter aids. A class of diatomaceous silica filter aids (Celite) has been produced having a variety of flow rate and clarifying capacities. Generally speaking, the higher the clarifying capacity of the filter aid, the lower the flow rate. The relative flow rates of these filter aids are shown in Figure 4.1a.

The following equipment and test procedure were used to evaluate the process. Two single-element test filters were used with direct feed and pressure supplied by a self-priming centrifugal pump. Slurried body feed and other ingredients were pumped into feed lines against raw water pressure. A detention tube was used in one of the filter lines to give a three-minute contact time for body feed or additives ahead of the filter.

FIGURE 4.1: PROTEIN COLLOID FILTER AID

a.

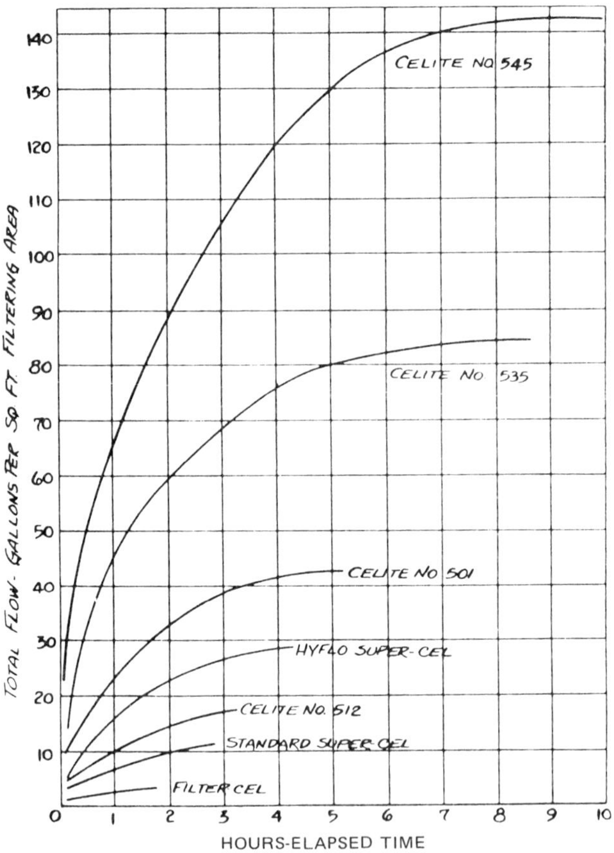

Relative Rates of Flow Obtained with Celite Filter Aids in an Average Industrial Clarification

(continued)

FIGURE 4.1: (continued)

b.

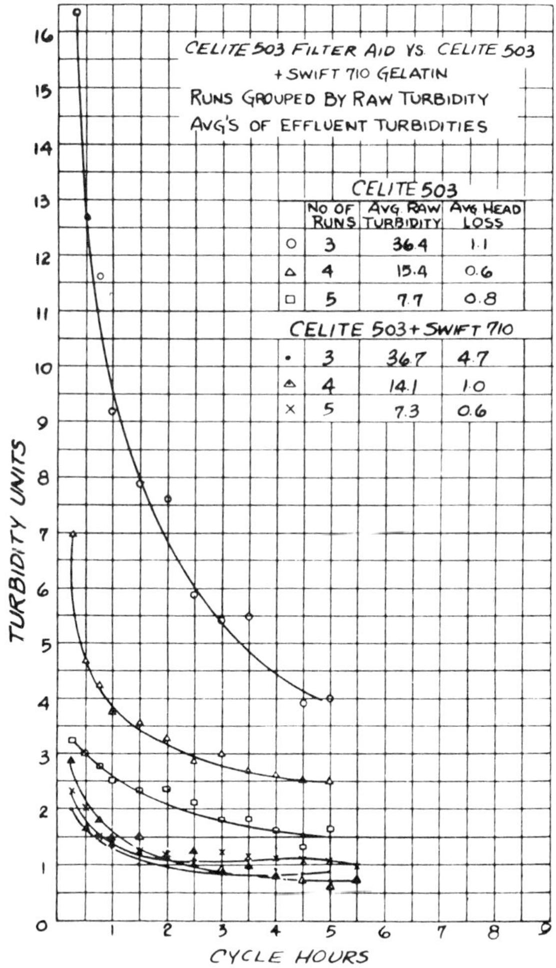

Effect of Swift 710 Gelatin on Filter Aid Performance

Source: U.S. Patent 3,658,184

The filter medium per se was precoated in a conventional fashion at one-tenth of a pound per square foot of filter area. Turbidity calculations were made by taking samples from the common raw water header and from each filter effluent line. The body feed additions were made up as a slurry in a 50-gallon tank using the raw water for the carrier. The slurry was fed at about 150 milliliters per minute to the raw water line ahead of the filters using the diaphragm pump. In some runs the regular filter aid was slurried up in the tank and then sufficient protein colloid solution was introduced to obtain the desired addition.

When separate additions of protein colloid were desired, a separately stirred solution tank and diaphragm pump were installed with a separate pumping line to the raw water line. Thus the filter aid was pumped into the raw water stream separately from the protein colloid solution.

In the following table the amount of protein colloids used is stated as mg/g as well as ppm. The mg/g is the milligrams of additive used per gram of body feed and is the term best applicable to coated filter aid. The term "head loss" is the average increase in the pressure per hour required to maintain one gallon per square foot flow rate. It is stated in pounds per hour with the average pressure filter having 35-pound pressure rise available for a cycle. Filtration cycles were generally terminated by time without reaching terminal head losses. It is important to note that in this type of filtration, and generally in diatomite filtration, the cycle is terminated by pressure increase rather than by breakthrough. Most cycles were run 5 to 6 hr before termination to make direct comparison more valid.

Examples 1 through 8: Gelatin (Swift 710) additions of between 0.02 and 0.08 ppm were employed, varying according to the above procedure. The coated examples were made by coating Swift 710 gelatin on Celite 503 filter aid at three levels of gelatin addition. The Swift 710 gelatin is a food-grade gelatin of about 300 Bloom gel strength and coatings were made at levels of 1.9, 2.9 and 5.8 mg of gelatin per gram of filter aid. A 1% solution of gelatin was sprayed into a ribbon mixer containing the Celite 503 filter aid. The batches were dried below 220°F, above which temperature the gelatin would degrade. The data from these runs are summarized below, showing significant differences in head loss and effluent turbidity when low concentrations of gelatin were used and effectiveness of coated filter aid vs separate addition of filter aid and the gelatin material. Coated precoat filter aid was used.

Swift 710 Coated Celite 503* Filter Aid at Various Coating Concentrations

Run Number	Swift 710 Concentration mg/g	ppm	BF/Turbidity Ratio	Average Turbidity Raw	Effluent	Head Loss (lb/hr)
1	1.9	0.03	1.5/1	11.3	1.4	2.6
2	2.9	0.05	1.4/1	11.3	1.35	2.7
3**	5.8	0.08	1.6/1	8.2	1.1	5.6
4	2.9	0.03	1.9/1	6.3	1.6	0.36
5	2.9	0.03	1.8/1	6.6	1.3	0.64
6**	2.9	0.03	1.8/1	6.6	1.2	0.77
7**	2.9	0.05	2.9/1	6.4	1.1	0.29

*All runs using coated filter aid for precoat and body feed.
**Three-minute detention ahead of filtration.

Coated Filter Aid vs Separate Additions of Celite 503 Filter Aid and Swift 710

Run Number	Addition Method	Swift 710 Concentration mg/m	ppm	BF/Turbidity Ratio	Average Turbidity Raw	Effluent	Head Loss (lb/hr)
8	Separate	1.45	0.03	1.9/1	11.8	1.15	1.1
9	Coated	1.9	0.07	2.25/1	16.3	1.0	1.1
10	Separate	1.6	0.06	2.4/1	16.3	1.5	0.7
5	Coated	2.9	0.03	1.8/1	6.6	1.3	0.6
11	Separate	2.0	0.04	2.8/1	7.4	1.5	0.7
12	Coated	5.8	0.56	2.2/1	43.1	4.9	0.9
13*	Separate	5.3	0.48	2.1/1	43.1	1.0	6.8

*Three-minute detention ahead of filtration.

The improved clarity achieved by the low concentration treatment is represented graphically in Figure 4.1b. Each line on the graph represents the average filtered water turbidity for a group of runs made under similar conditions. Three groups with raw water turbidity ranges of 5 to 10, 10 to 20 and 30 to 50 are shown. Similar groups are also shown using body feed of Celite 503 filter aid, a high flow rate filter aid, and using Celite 503 filter aid with Swift 710 gelatin.

The improved average clarity of the runs with the gelatin additions is shown in a comparison of the lines of equal raw water turbidity. The excellent clarity at the beginning of the run is also illustrated in this figure. All average turbidities for the run with the gelatin addition were below 3 turbidity units after 15 minutes of the cycle, while without gelatin addition the average turbidity ranged from 3.25 to 16.4, dependent upon the raw water turbidity.

Excellent removal was achieved at low turbidity level. Filtered water with turbidities below the maximum acceptable limit is produced from raw water when turbidity ranges from 7 to 36 units. The lines for filtered water when using gelatin for all three turbidity groups are in the same band showing a constantly good filtered water turbidity.

Substituted Oxazolines as Filter Aid Conditioners

A process described by *B.J. Davis; U.S. Patent 3,648,847; March 14, 1972; assigned to Reichhold Chemicals, Inc.* relates to a composition and method for improving the properties of filter aids in nonaqueous systems so that these modified filter aids can easily form strong, free-flowing filtration beds to remove suspended solids from polymers or essentially nonaqueous polymer solutions in relatively nonpolar solvents, lubricants, vegetable oils and similar systems wherein difficulty is met in properly wetting and depositing a free and rapid filtration bed composed of a good precoat on the filters of the filtration system for removal of solids from polymer solutions or oils.

Thus, it has been found that certain materials having marked surface activity occlude the hydrogen bonding sites on the filter aid, actively trap water present in the system in this occluded layer, leaving a lyophilic layer exposed on the surface of each filter aid particle so no adverse forces are exerted to reduce the rate of filtration over the cohesive strength of the precoat to itself or to the filter cloth.

The resultant system is essentially insensitive to water being present at levels as high as 500 parts per million and actually serves to remove this water as an entrapped water-in-oil layer during filtration. Furthermore, it has been found that these modifiers also exert a suspending effect on the filter aid to prevent packing and agglomeration yet allow the filter aid to form a fixed cohesive, free-flowing bed.

It has been discovered that the most effective filter aid treating agents are members of a family of substituted oxazolines having the configuration shown on the following page where R is a straight chain paraffin or olefin with a C_8 to C_{22} chain length. The C_{12} to C_{16} chains were most effective. Oxazolines with no hydroxylmethyl groups were ineffective. The dihydroxylmethyl oxazoline configuration gave better results in the presence of higher water levels than the monohydroxymethyl derivatives, which were found to be more effective than oxazolines containing no hydroxymethyl groups.

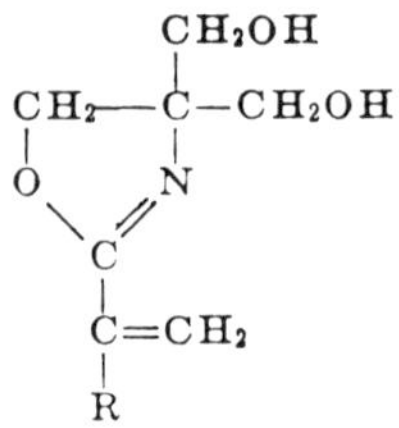

or

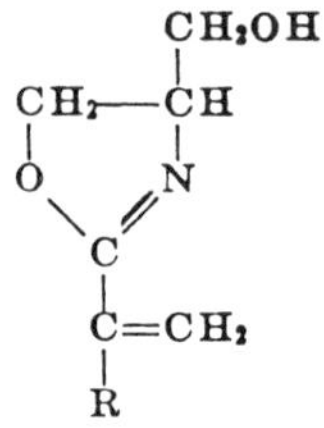

Examples where the above-described filter aid conditioner is applicable include olefinic polymers, such as polymers from isoprene, piperylene and the like, or mixed olefins with appended aromatic groups such as styrene, methylstyrene and the like. They also include terpene resins as derived, for example, from dipentene, phellandrene, camphene and the like. They include further aliphatic and aromatic hydrocarbon polymers such as mineral oils, coumarone-indene resin, polymeric oils such as vegetable oils and lubricants.

Evaluation tests were run in plant conditions where the above-specified compounds were filtered in the absence of a solvent and also dissolved in a nonpolar solvent. The technique involved is typified by the following example.

Example: Piperylene was polymerized at 40% solids in xylol using 2% anhydrous aluminum chloride as a catalyst. Polymerization was run at 20°C. When polymerization was completed 5% hydrated lime was added for neutralization of the catalyst, and the system heated to 100°C. At this point, the system was filtered through a Sparker filter press, wherein the precoat was prepared with xylol and diatomaceous earth containing a small percentage of asbestos fiber. The efficiency of filtration was measured by the number of gallons of filtrate processed before the bed had to be changed.

The filter aid modifiers were added to the prepared filter aid slurry before making the precoat bed on the filter by simply stirring it into the solvent-filter aid system. The amount of modifier can vary between about 0.025 and about 1.0% by weight on the weight of the filter aid with little effect on its efficiency. Over 1.0% was deemed wasteful.

Initial laboratory tests demonstrated that the dihydroxymethyl oxazoline derivative was most effective in systems containing up to 500 parts per million water and that of this type of oxazoline those having hydrophobic chains of C_{12} to C_{16} were most effective. The oxazoline derivative employed in examples 1 through 6 had the following structure.

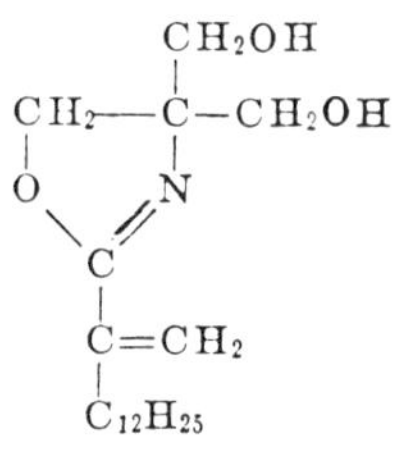

Such compounds are known as Deltazoline, with the above called Deltazoline A-14.

Examples 1 through 6: Runs were made employing the piperylene resin solution described on the preceding page.

TABLE

Percent Deltazoline A-14 in Filter Aid	Filtration Capacity of 558 ft^2 of Filter Area
None	Precoat not acceptable due to agglomeration
1.0	35,000 gallons/cycle
0.5	34,000 gallons/cycle
0.1	35,000 gallons/cycle
0.05	33,000 gallons/cycle
0.025	32,000 gallons/cycle
0.01	10,000 gallons/cycle

It can be seen above that the marked efficiency of the oxazoline derivative is evidenced in concentration of 0.025% by weight of the filter aid or over.

Improved Dispersability of Diatomite in Hydrocarbon Liquids

Filter aids such as diatomite and perlite tend to flocculate when added to hydrocarbon liquids. This flocculation frustrates the formation of a satisfactory filter cake or precoat of the filter aid for filtration purposes. To overcome this problem, according to a process described by *G. Coombs, E.R. Butterworth and G.R. Bell; U.S. Patent 3,905,910; September 16, 1975; assigned to Johns-Manville Corporation*, a concentrate is made by intensively mixing a small amount of filter aid with more surfactant than is required to disperse the filter aid in a hydrocarbon liquid. This concentrate is then gently blended in one or more steps with additional filter aid to produce a final mixture having the desired amount of surfactant absorbed by the particles of the filter aid. This treated filter aid product will rapidly disperse in hydrocarbon liquids.

Any type of mixer that provides intensive mixing action without producing excessive shearing action can be used to produce the free-flowing first concentrated mixture. It is preferred to use a mixer having conventional high-speed choppers. A twin-shell blender equipped with conventional agglomerate-breaking devices can also be used.

In making the first concentrated mixture many conventional dispersants or surfactants can be used. The primary requirement of the surfactant is to disperse rapidly the treated filter aid particles in at least one hydrocarbon liquid, and preferably in a wide variety of hydrocarbon liquids. Some surfactants that have been found useful in the process include Gafac RS 610, Arquad 2HT75, Centrolex I and Actiflow 68UB, Deltazolines A-8 through A-22 and Zelec NE. The optimum concentration of these surfactants in the final mixture is usually below 0.5%, and and in many cases is about 0.2 to 0.3 weight percent.

After the first concentrated mixture is obtained, it is then gently mixed with sufficient filter aid material to produce an intimately blended final mixture having the desired concentration of surfactant. This can be accomplished by one or more gentle blending steps. For example, the surfactant concentration of the first concentrated mixture can be reduced in one step to the final desired concentration by adding all of the remaining filter aid and then gently blending to form an intimate mixture. It has been found advantageous to convert the first concentrated mixture to an intimately mixed second concentrated mixture whose surfactant concentration is lower than that of the first concentrated mixture, but higher than the final desired concentration and then to convert this second concentrated mixture to an intimate mixture having the final desired surfactant concentration. This latter technique can more readily produce a more intimate mixture.

Many types of blending mixers can be used so long as their action on the material being mixed is gentle. The preferred type of mixer for the gentle blending step or steps is a ribbon-type mixer, but tumbling-type mixers, double-bladed mixers and mixing-type screw conveyors would also be suitable.

The final mixture having the desired concentration of surfactant absorbed by the surfaces of the filter aid particles has the same appearance and flow characteristics as untreated filter aid and, therefore, can be handled, packaged, shipped, stored and used in the same manner as conventional filter aid material. This is particularly advantageous to the user who, with the product of the process, is not burdened with the problems inherent in storing and metering a correct amount of a surfactant into a hydrocarbon liquid either prior to or after addition of an untreated filter aid as suggested by the prior art.

Example 1: One-hundred milliliters of water were placed in a first 100 milliliter graduate and 100 milliliters of toluene, a conventional hydrocarbon liquid, were placed in a second 100 milliliter graduate. Five grams of Hyflo Super-Cel diatomite filter aid were added to each graduate and the two graduates were shaken under the same conditions to disperse the diatomite in the two liquids. The diatomite dispersed rapidly and thoroughly in the graduate containing water, but appeared to form flocs in the graduate containing toluene.

After standing overnight, the diatomite had settled in the bottom of the graduate containing water to a volume of about 13 to 15 milliliters, while in the graduate containing toluene the diatomite had only settled to a volume of about 35 to 37 milliliters. The flocs or clumps of diatomite particles formed in the toluene had prevented the diatomite from settling to an extent necessary to form a satisfactory precoat or cake for a filter.

Example 2: One-hundred milliliters of toluene were added to a 100 milliliter graduate and thereafter 0.3%, based on the weight of the diatomite to be added, of Gafac RS 610 surfactant was added to the toluene and thoroughly mixed. Five grams of diatomite were then added to the toluene surfactant mixture and shaken in the same manner used in Example 1. After setting overnight, the diatomite had only settled to a volume of 18 to 20 milliliters. Although this is an improvement over the use of diatomite in straight toluene, the diatomite still did not disperse as well or settle to the same degree as it did in water.

Example 3: One-hundred milliliters of toluene were added to a 100 milliliter graduate and thereafter 5 grams of diatomite were added and the graduate and its contents were shaken as in Examples 1 and 2. Next, 0.015 gram of Gafac RS 610, which is 0.3% of the diatomite added, was added to the toluene-diatomite slurry and the graduate and its contents were again shaken as before. After standing overnight, the diatomite had only settled to a volume of 30 to 32 milliliters. This manner of adding the surfactant was essentially ineffective.

Example 4: A concentrate was made by intensively mixing 0.015 gram of Gafac RS 610 with 0.5 gram of diatomite. One-hundred milliliters of toluene were added to a 100 milliliter graduate. The concentrate and 4.5 grams of untreated diatomite were separately added to the toluene and the graduate and its contents were shaken as before. After standing overnight, the diatomite had only settled to a volume of 18 to 20 milliliters. Although this represents an improvement, the diatomite was still not dispersed as well nor did it settle to the same degree as it did in water.

Example 5: A 3% concentrate was made by intensively mixing 0.015 gram of Gafac RS 610 with 0.5 gram of diatomite. This concentrate was then gently blended with the remaining 4.5 grams of diatomite to form 5.015 grams of a final mixture containing very close to 0.3% of the surfactant, based on the weight of the diatomite present. This 5.015-gram mixture was added to 100 milliliters of toluene in a 100 milliliter graduate and shaken as in the previous examples. After standing overnight the diatomite had settled to a volume of only 13 to 15 milliliters, indicating a degree of dispersion and settling equal to that displayed by diatomite in water.

In all of the previous examples, the type of diatomite described in Example 1 was used. In these examples only one type of surfactant was used in order to change only one variable from test to test. The other surfactants mentioned earlier in the specification display the same general type of results as Gafac RS 610, although the magnitudes of change from test to test vary somewhat. Table 1 is provided for convenience in comparing the results of Examples 1 through 5.

TABLE 1

	Example No.					
	1		2	3	4	5
	Water	Toluene				
Condition of filter aid immediately after shaking	well dispersed	severe flocculation	some flocculation	severe flocculation	some flocculation	well dispersed
Height of filter aid in 100 ml graduate after standing overnight (ml)	13-15	35-37	18-20	30-32	18-20	13-15

The previous examples illustrate the effectiveness of the technique and the resultant product of the process on laboratory-scale batches. The following example verifies the advantages of the process on a production-sized batch.

Example 6: Twenty-five pounds of Celite 503 diatomite filter aid were placed in a conventional mixer having a plow speed set for about 110 rpm and a chopper speed of about 3,600 rpm. The plows and choppers were started and immediately about 10.7 pounds of Gafac RS 610 surfactant were gradually fed into the mixer at such a rate that all of the surfactant was added to the mixer about 90 seconds from the time the plows and chopper were started. After all of the surfactant was added, the mixer was allowed to continue running another 15 seconds. The dump door in the mixer was then opened and the choppers and plows were allowed to operate for 30 seconds to discharge the free-flowing first concentrated mixture having a surfactant concentration of about 30%, based on the total weight of the mixture.

The previously described process is repeated one or more times to produce a desired supply of first concentrated mixture. For hundred fifty pounds of diatomite were added to a clean ribbon mixer. The ribbon mixer was started and 50 pounds of the 30% concentrate mixture were added and mixed for ten minutes to produce a second concentrated mixture containing about 3% surfactant. Four hundred fifty pounds of diatomite were added to a clean ribbon mixer. The mixer was started and 50 pounds of the second concentrated mixture were added and allowed to mix for 10 minutes, after which the mixer was dumped. The resultant final mixture, having a surfactant concentration of about 0.3%, was packaged in a conventional manner.

Ten samples each representing successive fifty-pound bags of the resultant product were then tested for dispersion as previously described. Five-gram portions of each were dispersed in 100 milliliters of toluene. All produced settled volumes within plus or minus 1.5 milliliters of the 15 milliliters settled volume of the same weight of untreated filter aid dispersed in water thus showing a high degree of uniformity throughout the finished product.

Example 7: Five-gram samples of the final mixture produced by Example 6 were added to 100 milliliters of seven different hydrocarbon liquids, each in 100 milliliter graduates, and shaken and allowed to set overnight in the same manner used in Examples 1 through 5. The results of these tests are reported in Table 2 and show the effectiveness of the method and product of the process.

TABLE 2

Hydrocarbon Liquid	Height of Filter Aid in 100 ml Graduate After Standing Overnight (ml)
Toluene	12
Naphtha	15.5
Perchloroethylene	11
Pentane	12
Hexane	20
Stoddard Solvent	17

Although, in Example 6 separate ribbon blenders were used for each of the gentle blending steps, a single blender could have been used if properly sized. The mixing times will vary somewhat depending on batch size, but over-mixing should be avoided as excessive mixing will degrade the filter aid structure and decrease the flow rate of the precoat.

OTHER PROCESSES

Ground Polyacrylonitrile or Polyamide Fibers

A process described by *H.R. Brost; U.S. Patent 3,880,754; April 29, 1975;*

assigned to Bayer Aktiengesellschaft, Germany relates to a filtration aid consisting of ground polyacrylonitrile or polyamide fibers which still show a fibrous structure. Ground polyacrylonitrile or polyamide fibers with a denier range of from 1 to 50 dtex are preferably used as the filtration aid. The use of these ground polyacrylonitrile or polyamide fibers as a filtration aid in settling filtration affords significant advantages over conventional filtration aids by virtue of their outstanding thermal and chemical stability.

In addition, it has been found that mechanical separation is better by comparison with conventional filtration aids. This applies both as regards the absorption capacity of the settling layer in regard to the suspended matter and as regards the residual content of such suspended matter in the treated water.

It is only necessary to use half the quantity by comparison with conventional filtration aids to obtain comparable filtration effects. Normally, the filtration aid is used in a quantity of 1 kg/m^2 of the filter surface. Where the filtration aids according to the process are used, it is possible by employing a quantity of 0.5 kg/m^2 of filter surface to obtain a settling layer having a thickness which, where conventional filtration aids are used, could only be obtained by employing a quantity of 1 kg/m^2 of filter surface. The process is illustrated by the following examples.

Example 1: Comparison Example – A condensate-treating installation was used in in which the suspended matter was retained by a settling filter with a filter surface of 40 m^2. A suspension of regenerated cellulose fiber powder was used in a total quantity of 40 kg in order to form a settling layer. Following homogenization in the settling vessel, settling took place in a matter of 6 to 7 minutes.

The filter was then switched into the condensate stream cooled to 30°C. A pressure loss of 0.3 kg/cm^2 was obtained at a metric filtration rate of 9.2 m/hr. The iron content of the effluent fluctuated between 5 and 15 μg of Fe/l. Filtration was stopped at a pressure loss of 1.6 kg/cm^2 by which time the filter had been in use for 146.5 hours and the throughput of water amounted to 53,912 m^3. Analysis of the settling layer showed that 11 g of suspended matter had been absorbed per square meter of filter surface.

Example 2: The procedure was as in Example 1 except that 20 kg of ground polyacrylonitrile fibers were used as a filtration aid. The pressure loss amounted to 0.2 kg/cm^2 at the corresponding metric rate of 9.2 m/hr. The effluent had an iron content of around 5 μg of Fe/l. Filtration was stopped when the pressure loss amounted to 1.6 kg/cm^2. At this point, the filter had been in use for a period of 376 hours and the throughput of water amounted to 135,360 m^3. Analysis of the settling layer showed that 89 g of suspended matter had been absorbed per square meter of filter surface.

Example 3: Following homogenization, 20 kg of ground polyacrylonitrile fibers were deposited in the installation described in Example 1. The condensate to be treated had an average temperature of 89°C. A pressure loss of 0.28 kg/cm^2 was measured at a filtration rate of 9.35 m/hr. After a period in service of 445 hours (= 166,430 m^3), a pressure loss of 165 kg/cm^2 was obtained. During this period, the iron content of the effluent was on average 8 μg/l. Analysis of the settling layer showed that 108 g of suspended matter had been absorbed per square meter of filter surface.

Example 4: Three hundred fifty grams of ground polyamide-6 fibers were deposited in a test installation with a filter surface of 0.7 m^2. The condensate to be treated had an average temperature of 91°C. The hourly throughput amounted to approximately 650 liters corresponding to a metric filtration rate of 9.3 m/hr. The pressure loss settled at 0.3 kg/cm^2 and rose over 326 hours to 1.6 kg/cm^2. The iron content of the effluent was always less than 10 $\mu g/l$. Analysis of the settling layer showed that 82 g of suspended matter had been absorbed per square meter of the filter surface.

Example 5: The following tests were carried out to determine to what extent ions and organic substances are given off to the treated water. Twenty grams of the different filtration aids were suspended in 250 milliliters of demineralized water. After a running time of 72 hours at 95°C, the following results were obtained:

Filtration Aids	Oxidizability, mg $KMnO_4$/l of extract	Conductivity, $\mu S/cm$ (20°C)
Demineralized water	2.2	0.5
Regenerated cellulose powder	184	85
Ground polyamide-6	3.2	5.2
Ground polyacrylonitrile	3.0	5.8

Polyvinyl Bromide-Monoalkylaniline Products

I. Shinohara and J. Aoyagi; U.S. Patent 3,969,239; July 13, 1976; assigned to B.R. Chemical Co., Ltd., Japan describe a method for using a macroamine polymer as a flocculating agent, the polymer having a structural unit shown by the following formula:

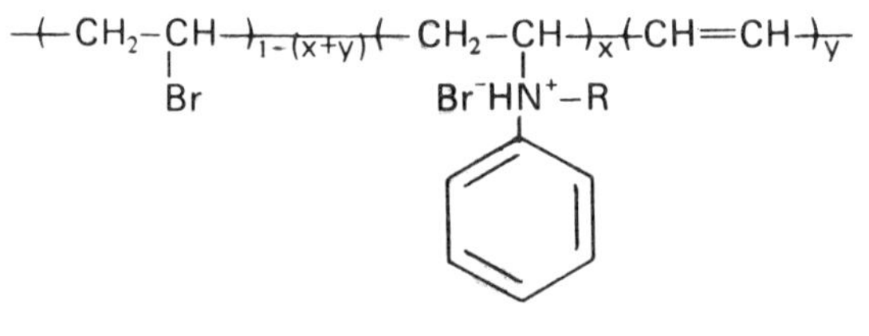

where R represents an alkyl group containing 1 to 4 carbon atoms, x is a mol decimal fraction between about 0.65 and 0.95, and y is a mol decimal fraction between about 0.23 and 0.04. The polymer can be obtained by reacting polyvinyl bromide with an N-monoalkylaniline having 1 to 4 carbon atoms in the alkyl group.

The polymer of the process is especially useful for the disposal of various types of waste water or drainage, such as those drained from iron works, aluminum processing works, dyeing works, pulp and paper-making works, salt electrolyzing works and other industrial waste water; for the purification of city water, sewage, etc. and for the recovery of metals from catalysts, electrolytes, etc.

On the following page are several specific examples of the process.

Example 1: A four-necked flask having a capacity of one liter fitted with a stirrer, thermometer, reflux condenser and nitrogen feed pipe was charged with 200.0 grams N-methylaniline and heated under an atmosphere of nitrogen. When the liquid temperature reached 150°C there was introduced 10 grams of polyvinyl bromide (produced by emulsion polymerization process, having numerical average molecular weight, measured by viscosity method, of 134,000) to the flask and reaction was conducted under stirring at that temperature for 8 hours.

After the completion of the reaction, an excess of ethyl acetate was added to thereby precipitate the polymer which was then separated by filtration. The polymer so obtained was washed with water and dried in vacuum at room temperature until the weight became constant.

The aminated polymer obtained in this way was analyzed by means of elementary analysis, halogen analysis and infrared absorption spectrum methods. The proportion of N-methylaniline introduced was 82.00% and the degree of double bond formation was 8.12%. The polymer so formed was found to be completely soluble in water.

Example 2: The procedure of Example 1 was repeated except that the reaction was carried out for a period of 5 hours. The proportion of the introduction of N-methylaniline in the water-soluble polymer so obtained was 70.12% and the degree of the formation of double bond was 22.21%.

Example 3: The procedure of Example 1 was repeated except that the reaction was carried out for a period of 12 hours. The introduction of N-methylaniline in the resulting water-soluble polymer was 91.00% and the degree of double bond formation was 5.12%.

Example 4: The procedure of Example 1 was repeated except that 100.00 grams of N-methylaniline was used instead of 200.0 grams of N-methylaniline. The introduction of N-methylaniline in the resulting water-soluble polymer was 81.0% and the degree of double bond formation after the reaction was 9.02%.

Example 5: A four-necked flask, same as that used in Example 1, was charged with 170.0 grams of N-ethylaniline and heated under an atmosphere of nitrogen gas. When the temperature of the reaction liquid reached 150°C, 10.0 grams of polyvinyl bromide (having an average molecular weight of 134,000) was continuously added and the reaction carried out at that temperature for 10 hours.

After completion of the reaction, the reaction mixture was treated in a manner similar to that employed in Example 1. Analysis of the resulting water-soluble polymer showed an N-ethylaniline introduction of 75.00% and the degree of double bond formation to be 10.12%.

Example 6: The procedure of Example 5 was repeated, except 252.66 grams of N-isopropylaniline was used instead of N-ethylaniline. The introduction of N-isopropylaniline in the resulting water-soluble polymer was 73.15% and the degree of double bond formation was 15.03%.

Example 7: The procedure of Example 5 was repeated except that 278.83 grams of N-n-butylaniline was used instead of N-ethylaniline. The introduction of N-n-butylaniline was 79.2% and the degree of double bond formation was 21.8%.

Example 8: To 300 ml of aqueous bentonite suspension at the concentration of 100 ppm each high macroamine polymer obtained in Examples 1, 2, 3 and 5 was added in a given proportion. Each suspension was subjected to stirring at a rotation speed of 130 rpm for the first 5 minutes then at a rotation speed of 30 rpm for a further 5 minutes. After the completion of stirring, each solution was allowed to stand still for 5 minutes and the supernatant liquid was carefully collected. The turbidity of the supernatant was determined with the use of a turbidimeter (or nephelometer) to measure remaining turbidity. The results obtained are shown in the following table.

	. . . Amount of Macroamine Polymer Added, ppm. . . .				
	0.5	**1.0**	**2.0**	**4.0**	**8.0**
Polymer of	. . . . .Remaining Turbidity, degrees				
Example 1	0.6	0.3	0.2	0.2	0.2
Example 2	22.8	18.4	15.1	8.4	5.3
Example 3	0.3	0.2	0.1	0.1	0.1
Example 5	2.5	1.6	0.9	0.7	0.6

Amine Additives for Cement Slurries

According to a process described by *H. Stettler; U.S. Patent 3,753,902; Aug. 21, 1973; assigned to Sandoz Ltd, Switzerland* water-suspended inorganic masses, especially cement slurries, of improved filtrability are obtained by adding thereto cation-active polyamines and/or polyamidamines, guanidine derivatives and optionally $Ca(OH)_2$. Compounds which are especially suitable for the process are those of formula

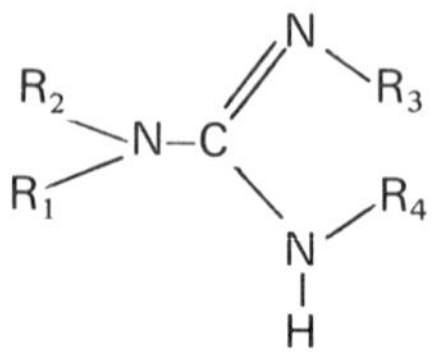

where R_1 stands for a hydrogen atom,

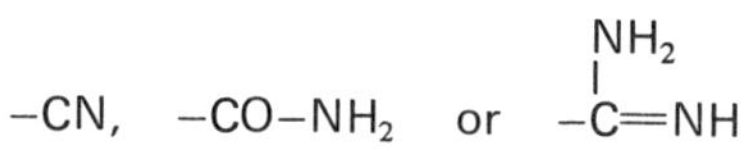

R_2 or R_4 for a hydrogen atom, a lower alkyl or hydroxyalkyl or a mononuclear aryl group; R_3 for a hydrogen atom or a substituted or unsubstituted amino group which is bound through a lower alkylene group and may form part of a heterocyclic ring; and R_4 or R_2 for a substituted or unsubstituted amino group which is bound through a lower alkylene group and may form part of a heterocyclic ring.

Examples of compounds of the formula are the products obtained by condensation of formaldehyde with compounds of the formula

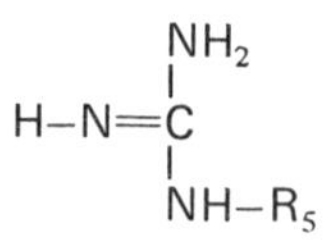

In the formula on the preceding page, R_5 represents hydrogen, $-CN$ or $-CO-NH_2$ optionally in the presence of ammonium salts of strong inorganic acids.

Examples of cation-active polyamide amines suitable for the stated purpose are the polycondensation products of the esters of acrylic, phenylacrylic, methacrylic and crotonic acids with alkylenediamines. The polymers which are produced by condensation of methyl acrylate with ethylenediamine, diethylenetriamine, di-N-(β-aminoethyl)-methylamine and di-N-(γ-aminopropyl)-methylamine respectively and quaternated with dimethyl sulfate, diethyl sulfate or epichlorohydrin may be instanced as examples, of which preferably the polymerization product of methyl acrylate and diethylenetriamine, partially quaternated with diethyl sulfate or the partially quaternated polymerization product of ethylenediamine, methyl acrylate and epichlorohydrin are used.

Polyalkylene imines of high molecular weight which may, if desired, be hydroxylated (for example, polyethylene imine) may be named to exemplify the suitable polyamines.

If the polyamines and polyamide amines are used singly, their action is manifest only if the solid content of the suspension is not in excess of 25%. The cationic polyamines and polyamide amines, however, in conjunction with guanidine derivatives, show effective action in suspensions and slurries with a solid content of up to 65%. In some circumstances they are effective in smaller amounts than the guanidine derivatives alone. Given a suspension for which 0.01 to 0.1% of a guanidine derivative relative to the solid content is required, an amount of 0.001 to 0.03% of one of the polymers will be found sufficient, although the addition can be increased to 0.5% with advantage. Thus the optimum quantitative relation of the polyamine or polyamide amine to the guanidine derivative is 1-500 to 10.

Example 1: The filtration aids are stirred into 80 grams of a crude cement slurry containing 50% solids, of which 20% is clay and 80% chalk (percentages on the dry weight). The slurry further contains the normal cement additives. The slurry is filtered through a press fitted with a synthetic-fiber filter of 3 centimeter diameter at 20°C and 24 atmospheres excess pressure.

The filtration aids used are calcium hydroxide, a guanidine solution containing 0.032 gram of dry substance, and a high-molecular-weight cationic polyethylene imine. Prior to use the last is prepared as an aqueous solution containing 20% active substance with a specific gravity of 1.07 at 20°C and a viscosity of 500 to 1,000 centipoises measured at 20°C and 20 rpm with a Brookfield viscometer. The results are shown in the following table.

Filtration Aid	Amount*	Filtration Time (minutes)	Residual Water Content of Filter Cake (percent)
-	-	180	22
$Ca(OH)_2$	0.3	20	23
Guanidine derivative Polyethylene imine $Ca(OH)_2$	0.01 0.03 0.3	15	22

*Amount of filtration aid in relation to solid content of slurry.

Example 2: If the crude cement slurry specified in Example 1 is filtered in the presence of 0.1% of the above guanidine derivative and 0.375% of the compound described in Example 3.2 of Belgian Patent 721,332, a filter cake of 22% residual water content is obtained at a filtration time of 16 minutes.

Example 3: A chalk suspension made up of 32 grams of a chalk containing a high proportion of montmorillonite relative to the total solid content and 48 grams of water are used in this example. A commercially available polymerization product (Product A, probably obtained by reacting 1 mol of methyl acrylate with a slight excess of 1 mol of ethylenediamine and alkylating with a slight excess of 0.5 mol of diethyl sulfate used in the form of an aqueous solution containing 0.0016 gram of dry substance and a guanidine derivative solution according to Example 1 of 0.032 gram dry substance were added to the chalk suspension with vigorous stirring.

The suspension was filtered through a filter cloth of synthetic fiber in a suction funnel of 3 centimeter diameter at 20°C and 24 atmospheres pressure. The filtrate was collected in a graduated measuring flask and the filtration time measured in minutes at short intervals, starting from the moment when the pressure reached 24 atmospheres. The volume of filtrate at each interval of measurement was noted in milliliters.

Parallel filtration tests were conducted, the first with 80 grams of the chalk suspension without a filtration aid and every successive one with 80 grams of the suspension and an addition of Product A. The test results are shown in Table 1.

TABLE 1

Filtration Aid	Filtration Aid * (percent)	Filtration Time (minutes)	Residual Water Content of Filter Residue (percent)
–	–	42	18
Product A	0.005	26	18
Product A Guanidine derivative	0.005 0.100	8	18

*Amount of filtration aid in relation to solid content.

Table 2 below gives the amounts of filtrate in milliliters which are obtained in the stated times, employing the procedure of the example, with the addition of another commercial polymerization product (Product B, probably obtained by reacting 1 mol methyl acrylate with a slight excess of 1 mol of diethylenetriamine and a slight excess of 1 mol of epichlorohydrin) alone and in combination with the guanidine derivative mentioned in Example 1 and again with the addition of the guanidine derivative alone.

The percentages are relative to the dry content of the solutions and the suspensions. The column headings give the weight percentages of the filtration aids relative to the solid content.

TABLE 2

Filtration Time (minutes)	0.005% Product B	0.005% Product B 0.1% Guanidine Derivative	0.1% Guanidine Derivative
3	13	25	24
5	16	32	32.5
8	21	39	37.5
10	-	41	-
15	29	-	-
20	31.5	-	-
25	36	-	-
30	41	-	-

Ground Waste Newsprint

S. Boorujy; U.S. Patents 3,677,413; July 18, 1972; 3,574,098; April 6, 1971; both assigned to Environmental Sciences, Inc., has found that suitably prepared waste newsprint may be utilized to remove suspended particulate material having a high biological oxygen demand from sewer water which has undergone primary and secondary purification. It has been found that by first powdering such waste newsprint in a hammer mill, then scorching the powdered newsprint in a rotary kiln until it is a brown or cocoa color, then washing the scorched newsprint in perchloroethylene or other solvent to remove the residual oils retained in the dried ink of the newsprint, by further grinding the cleaned and scorched newsprint to a particle size that will pass through a 400 mesh screen and by then introducing this processed newsprint into the sewer water at the rate of approximately 100 to 200 pounds to one-million gallons of domestic sewage, containing approximately 250 parts per million biological oxygen demand particles (ppm BOD), that 95% of such BOD particles will be removed at a rate sufficient to meet the practical demands of a functioning municipal sewage disposal system.

Referring to Figure 4.3, there is shown a tertiary filtering unit constructed according to the process. Water which has undergone primary and secondary treatment (that is which has been passed through a settling tank and a stone filter bed) is pumped into closed tank **6** through pipe **4**. While the dimensions of the tank may be varied to suit particular installations, it is envisioned that a 3-foot diameter tank, approximately 5-feet tall would contain sufficient volume to process approximately 1,200 gallons per hour or 28,800 gallons per day of domestic sewage.

Processed waste newsprint is introduced into pipe **4** via a feed line **5** and the mixture of water to be treated and a slurry of processed newsprint under the force imparted by the pump **7** enters the tank. The correct proportion of newsprint to sewer water is established by regulating the rotational speed of a proportionating wheel of proportionator **3** whose buckets scoop up the newsprint from the hopper **3a**.

It has been found that a proportion of 100 to 200 pounds of processed newsprint per million gallons of influent has been satisfactory in accomplishing the desired purification of domestic sewage. A correspondingly greater amount of processed newsprint is needed to purify industrial sewage in which the BOD content is higher.

FIGURE 4.3: FILTRATION UNIT EMPLOYING GROUND NEWSPRINT

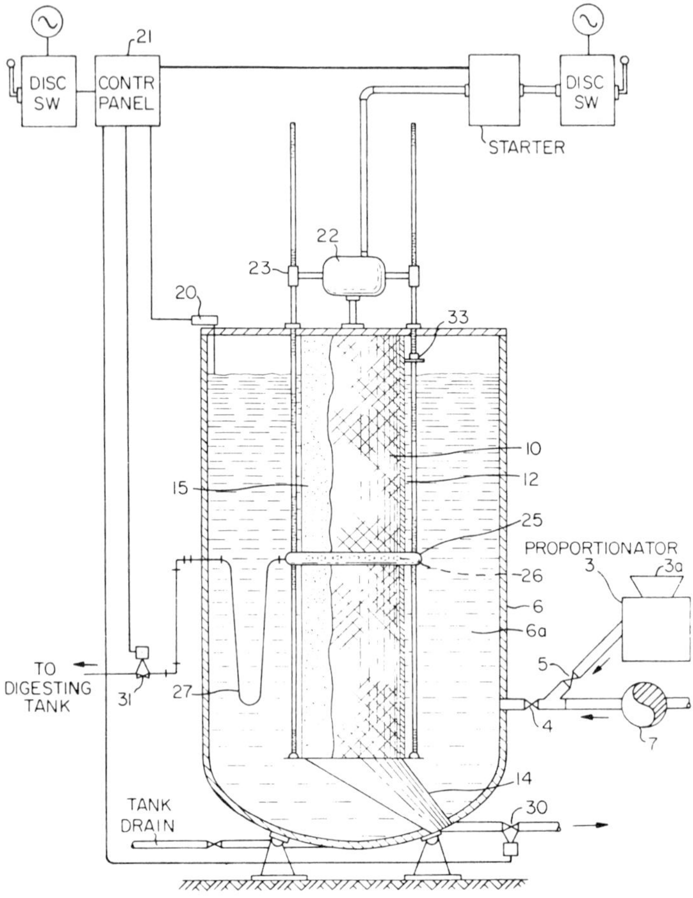

Source: U.S. Patent 3,677,413

Within the tank there is a central core **10** which may be in the form of a wire screen or perforated cylinder of suitable corrosion-resistant material, such as stainless steel. The central core or screen, in one illustrative example, may have one-quarter-inch square or round apertures. Positioned externally to the central core is a filter medium **12** which advantageously may be formed from a roll of felted filter material having a thickness of approximately one-quarter inch and having a porosity which is adapted to pass approximately 25 cubic feet per minute of air per square foot at a pressure of 8 ounces per square inch or, what is the same thing, 50 gallons of water per hour per square foot. Material having these properties may be of wool, Orlon or Teflon felt and is available from many manufacturers. The pressure of the water in the tank causes penetration of the filter medium and central core, and the water enters discharge funnel **14**, from which the effluent may be discharged into a river or stream. The action of the processed powdered newsprint is such that it will remove suspended solids and organic materials from the influent, such materials depositing themselves on the surface **15** of filter media **12**. The build-up of the deposit on the surface increases the pressure differential between the liquid in the outer portion **6a** of tank **6** and that within the discharge funnel **14**. Pressure sensing device **20**, which advantageously may be set to respond when the pressure differential reaches 10 pounds per square inch, provides a signal to control panel **21**. The control panel, responding to the signal from pressure probe **20**, shuts valve **30**, preventing the discharge of effluent from funnel **14**, opens valve **31** in ventable cleaning line **27**, and sets drive motor **22** into operation.

The drive motor, via drive train **23**, causes cleaning scraper ring **25** to commence travelling along the outside of filter media **12** to clean the surface **15** of the deposits, causing the detected pressure differential. Cleaning ring **25** is advantageously fabricated to be in the form of a hollow toroid with apertures or slots **26** on its inner circumference. The deposits enter the apertures of the cleaning ring and are carried away through cleaning line **27** to a digesting tank.

The cleaning action of the ring will thus be observed to be unique in that it cleans the external surface of the filter media under the back pressure of the already filtered water which is inside the central core. This cleaner water in the central core, which is under approximately 10 pounds pressure, finds a vent to the atmosphere through open valve **31** over a path which is travelled in the direction from the inside of the central core through the filter media, the apertures **26** of the cleaning ring **25**, and the cleaning line **27** to the digesting tank which is at atmospheric pressure.

When surface **15** has been cleaned, valve **31** is closed, valve **30** is opened, and drive motor **22** is deenergized by control panel **21**. The cleaning operation may be terminated after the ring has been caused to pass over the entire length of the filter media **12**, either once or as many times as is found desirable under given operating conditions. By suitable connection to central panel **21**, a conventional limit switch **33** may advantageously be employed to reverse the direction in which cleaning ring **25** is driven by drive motor **22** when the ring has traversed the length of the surface **15**.

The powdered newsprint, which has been found efficacious in cleaning sewer water having biological oxygen demand, is advantageously manufactured by the following process.

The waste newsprint is first powdered in a hammer mill of conventional design to a fineness that will pass through a 75 to 80 mesh screen. The material which passes through this screen is then subjected to a perchloroethylene bath to remove the residual oils which are customarily employed in manufacturing the printer's ink which is on the newspaper. In this regard it is important to appreciate that this process, in one of its aspects, requires waste newsprint which has actually been printed upon. This is because it is believed that certain of the chemicals, albeit in small proportion, that are found in the printer's ink, and also these chemicals which are found in the type of paper customarily used as newsprint, are efficacious in chemically attacking the suspended solids. For example, in bleaching the newsprint so that it is sufficiently white to be acceptable for newspaper purposes, there is caused to remain approximately 5 parts per million of chlorine in the paper. In the printers ink there also may be of the order of 3 parts per million of iron chloride which also is believed to play a part in the oxidation and precipitation of the suspended solids in the sewer water. There is also a certain amount of residual oil in the printer's ink, but it is absolutely essential that this oil be removed as otherwise the powdered newsprint will form a colloid with the BOD particles and prevent their precipitation on the surface of the filter medium. Accordingly, the powdered newsprint passing through the 75 to 80 mesh screen is washed with perchloroethylene to remove this trace of oil from the printer's ink.

The powdered, de-oiled newsprint is then scorched to a brown or cocoa color, advantageously, in a rotary kiln of the type similar to that employed in calcining. The scorching opens the fibers of the powdered newsprint and provides a degree of activated carbonization. However, the carbonization cannot be allowed to progress too far or the iron chloride and chlorine traces mentioned previously will be destroyed. Scorching is also essential so that the next grinding process can proceed to the desired degree of fineness. The second grinding process which is now carried on is also performed in a hammer mill and is continued until the newsprint will pass through a 400 mesh screen. The material passing through the 400 mesh screen is then suitable for mixing with the sewer influent as described above.

FILTRATION OF MOLTEN ALUMINUM AND HEAVY METAL SEPARATION

MOLTEN ALUMINUM

Demountable Cartridge

T.A. Nowak; U.S. Patent 3,747,765; July 24, 1973; assigned to Kaiser Aluminum & Chemical Corporation describes a demountable cartridge for filtering molten aluminum which comprises in combination a top plate, a bottom plate, at least one and advantageously a plurality of rigid filter tubes or elements disposed between such plates, and means for fastening the assembly together and for maintaining the filter media and top and bottom plates in sealed relation with each other. Suitable rigid filter elements are described in U.S. Patent 3,524,548.

As there described, the rigid filter medium is a fired, shaped article which includes a granular refractory material compatible with molten aluminum, such as alumina, fused alumina, high purity bauxite or diaspore, boron nitride or other suitable refractory material which is known to resist attack by molten aluminum. The granular material is bonded together with a vitreous material containing less than 10% silica and particularly as containing from 15 to 80% boron oxide, from 5 to 50% calcium and from 2 to 60% aluminum oxide and not more than 10% silica. From 4 to 20% of the bonding material is advantageously used in making the rigid filter media. This suitable bonding material is described in U.S. Patent 2,997,402.

The rigid filter media made as described in U.S. Patent 3,524,548 are excellent filter means for molten aluminum metal but are rather fragile and care must be taken in the handling and use of the filters or cartridges in order to avoid breaking or shattering, with consequent increase in costs of operation.

The device of this process enables more efficient handling of the rigid media filter and cartridge and reduces breakage loss significantly. The plates and tubes are quickly assembled, and this can be done by labor of minimum skill or training. The device can be picked up and is ready for placement because it is not necessary to wait for mortar joints to set. The cartridge can be readily disassembled when necessary or when the casting operation has been completed, by loosening the means holding the plates, tubes and tie rod together, lifting off the top plate and removing the used filter tubes.

A particular advantage of the filter cartridge of this process is that such a cartridge has been found to have several times the usable working life of a comparable but all ceramic cartridge. Furthermore, when breakage or exhaustion of the filter element occurs, the cast iron parts of the device of the process can be reused with replacement of several sets of the ceramic filter tubes.

A most important advantage of this process is the ability to pick up the assembled cartridge with a simple hook by lifting on the tie rod or bar. Previous all ceramic assemblies were picked up by complicated three point tongs which were difficult to operate at high temperatures. The all ceramic assemblies were usually picked up by the top plate and if the mortar joints failed, the entire cartridge was dropped and destroyed.

Figure 5.1a shows a filter cartridge device in a perspective view; Figure 5.1b is a cross section of the assembly of Figure 5.1a taken on line 2–2. Figure 5.1c is a perspective view toward the lower surface of the top plate of the assembly shown in Figure 5.1a; Figure 5.1d is a perspective view toward the top surface of the bottom plate of the device of Figure 5.1a; Figure 5.1e is a perspective view of a tie rod of the assembly of Figure 5.1a. Figure 5.1f is a vertical cross-sectional view of the assembly, showing it in place in a filtering zone or unit, and ready for use; Figure 5.1g is a vertical cross-sectional view showing another tie rod; and Figure 5.1h is a device for grasping, lifting and transporting a filter cartridge.

Referring to Figure 5.1a through 5.1e, the demountable assembly **10** includes a top plate **11** having in it a plurality of holes or apertures **12** and a bottom plate **13**, both plates being of cast iron. Disposed between the plates and sealed respectively to the lower surface **22** of plate **11** and the upper surface of plate **13** are a number of filter tubes **15** of the rigid fired refractory material. That is to say, these tubes consist essentially of a shaped, fired mixture of alumina particles having a maximum diameter of between 1.165 mm and 2.8 mm, bonded with from 4 to 20% of a prefused vitreous bonding material containing from 15 to 80% boron oxide, from 5 to 50% calcium oxide and from 2 to 60% aluminum oxide, and not more than 10% silica.

As shown in Figure 5.1b, plates **11** and **13** and tubes **15** are held firmly together by a tie rod **16** of the type shown in Figure 5.1e. Plates **11** and **13** are each provided with a central aperture at **17** and **18** respectively. In this case, the walls of **18** are threaded and the lower end **19** of the tie rod is also threaded so that it can be threadedly connected tightly to plate **13** within aperture **18**, the tie rod being circular in horizontal cross section.

The tie rod is also provided with an annular shoulder **20** at its upper end, to fit into annular recess **14** in the lower surface **22** of top plate **11**; and a lower annular shoulder **21** fits into recess **34** in upper surface **23** of lower plate **13**. At its top end the tie rod is provided with an extension **24** of reduced cross section which extends through aperture **17** and upwardly from top plate **11**.

The plates and tie rod are made of heat resistant cast iron. Rigid filter tubes **15** are circular in horizontal cross section, although the cross-sectional configuration is not critical and other shapes can be selected if desired. However, the circular cross-sectional tubes are generally most suitable to make and use.

Each such tube is provided with an internal passageway **25** which communicates with an aperture **12** in upper plate **11**, the walls of passageway **25** and of aperture **12** being in alignment or registration so that a smooth flow of metal results. An annular recess **26** is provided in the wall of each aperture **12** to accommodate or receive the upper end of tube **15** in such relationship. Interposed between the surface of plate **11** and the end of each tube **15** is a gasket **27** of Fiberfrax which is an alumina silica fibrous sheet material useful at temperatures above 2000°F. Disposed in upper surface **23** of plate **13** are a number of circular recesses **28** which register with the apertures **12** in upper plate **11** and which receive the lower end of tubes **15**.

FIGURE 5.1: DEMOUNTABLE CARTRIDGE

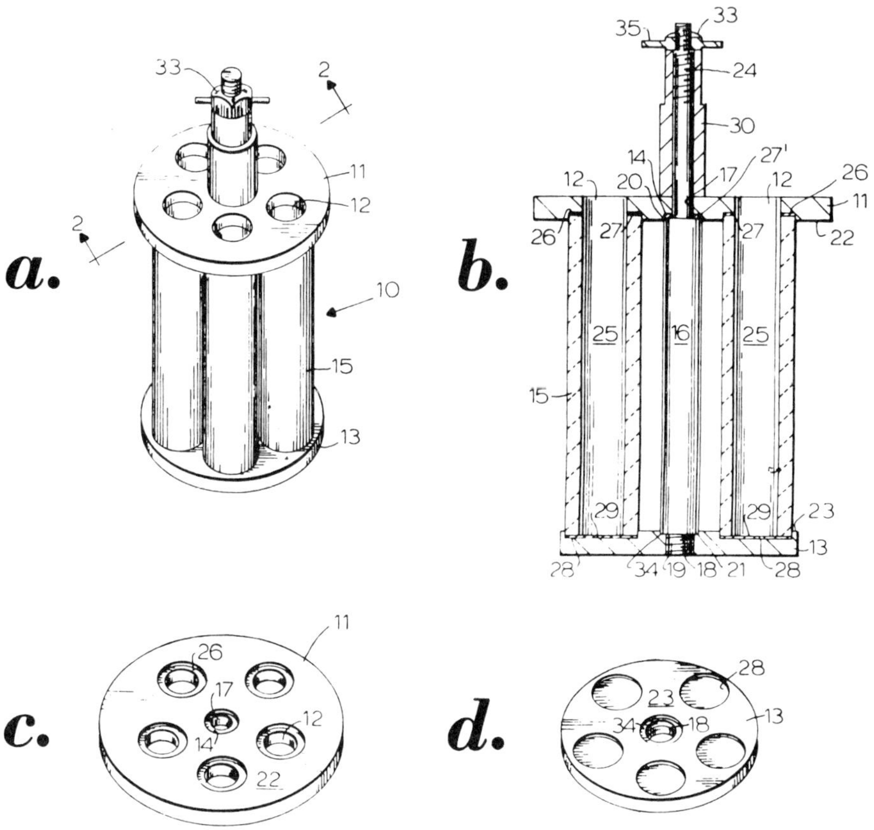

(continued)

FIGURE 5.1: (continued)

e.

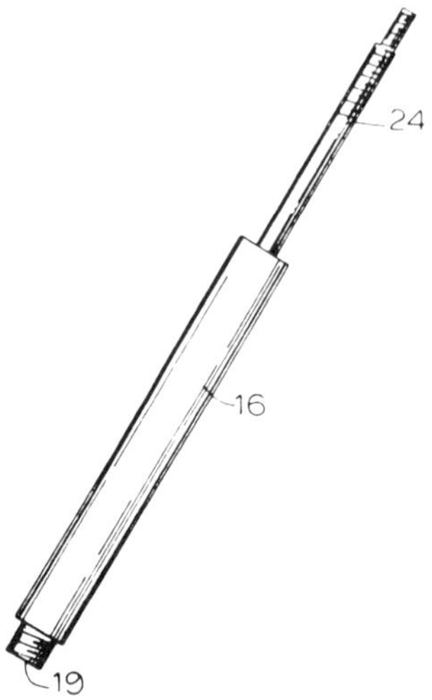

f.

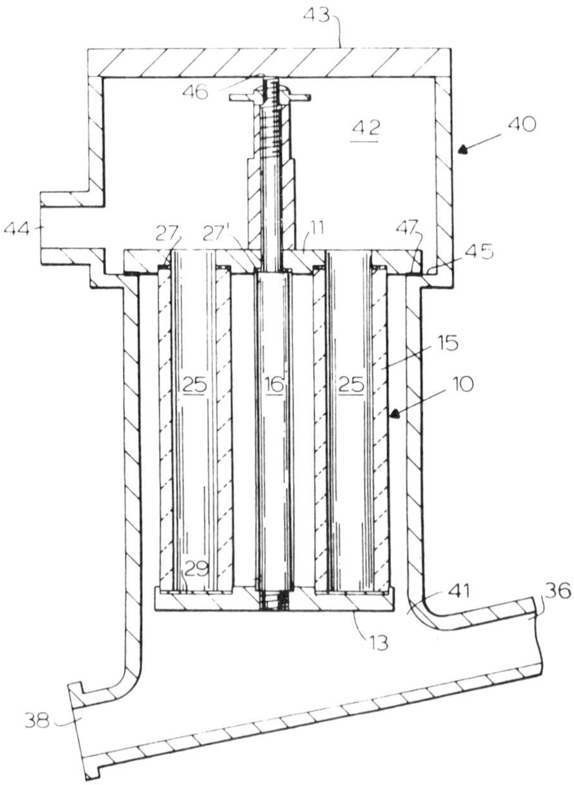

(continued)

FIGURE 5.1: (continued)

g.

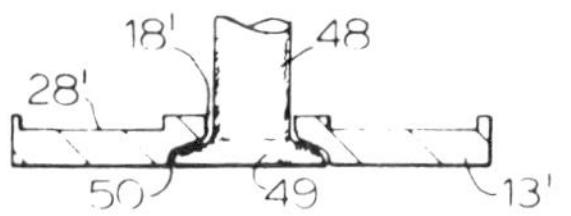

h.

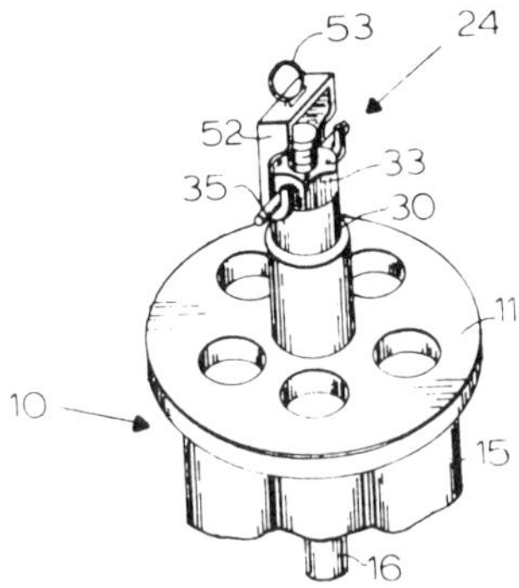

Source: U.S. Patent 3,747,765

A sheet or disc **29** of a like sealant material, such as Fiberfrax, is interposed also between the surfaces of recesses **28** in plate **13** and the ends of tubes **15**. A thicker layer **37'** of Fiberfrax is also placed between the surface of upper plate **11** and the upper end of rod **16**, because as the rod expands upon heating, this layer will compress accordingly and accommodate for such expansion. The upper portion, or extension **24** of rod **16** is at least partially threaded.

In assemblying this device, tie rod **16** is threadedly connected to lower plate **13** and the required number of tubes **15** are set in their respective recesses **28**, after which top plate **11** is placed over the tops of the tubes and tie rod **16**, with suitable implacement of the Fiberfrax discs and gaskets, and the whole fastened together by suitable means. A sleeve **30** is placed over the upper extension **24** and a retaining nut **33** is screwed down over the upper portion of the extension to tighten the sleeve and the extension against plate **11** and to hold the entire assembly in secure attachment. Instead of the tightening nut, any other suitable quick disconnect device can be used, for example, a cam arrangement as will be described below.

When the cartridge is heated to the temperature of operation, the cast iron tie rod expands lengthwise more than the ceramic tubes so that the sleeve and top plate must be tightened against the tops of the filter tubes. Conversely, upon cooling after a drop or casting operation is finished, the fastening device is quickly loosened to permit adjusting again the position of the sleeve and plate. A pair of arms **35** extend outwardly from retaining nut **33** and assist in lifting and transporting the entire assembly.

In Figure 5.1f is shown the implacement of the device in the filtering zone or housing **40** having an entry pipe **36** for introduction of hot molten metal from a melting zone or the like and an outlet **38** provided with a suitable cap, to assist in draining and cleaning the device. The filter cartridge or assembly **10** is placed in filtering chamber **40** which includes lower or inlet portion **41** for introduction of hot metal coming through inlet **36**, and upper or outlet chamber **42** which receives filtered metal from filter tubes **15** and conducts it to outlet **44**, whence it flows to casting stations.

Chamber **42** is provided with a lid or cover **43** so that the hot filtered metal is kept out of contact with the air. If desired, and advantageously, an inert gas can be introduced into chamber **42** above the level of the metal by conventional means to assist in maintaining nonoxidizing conditions. The top plate **11** of filter cartridge **10** rests at its outer edge on an annular shoulder **45** provided at the upper end of lower portion **41**. Bottom plate **12** is supported by tie rod **16** and is of smaller diameter than the top plate and also of smaller diameter than the interior of lower portion **41** so that passageway for molten metal is provided around the circumference of the lower plate.

Other means of conducting molten metal into contact with filter tube **15**, such as holes in plate **13** could be used, alternatively. The upper end **46** of tie rod **16** is at such distance from the upper surface of plate **11** that cover **43** rests on surface **46** and thereby plate **11** is held down in sealed connection with shoulder **45** against the force of metal being filtered through tube **15** when the device is in operation. Advantageously, compressible sealant gasket **47** is placed between shoulder **45** and the outer edge of plate **11** to ensure that no leakage will occur. Instead of a lid or cover, a bar across the top of outlet chamber **42** can serve to hold down the rod and cartridge.

In Figure 5.1g there is shown an alternative lower portion of a tie rod useful herein. In this arrangement, tie rod **48** is provided at its lower end with flange or disc-shaped end piece **49**. Plate **13'** is provided with the central aperture **18'** and with an annular recess **50** at the lower end thereof so that the disc-shaped end **49** of tie rod **48** fits snugly into recess **50**, the central stem of tie rod **48** extending upwardly through aperture **18'** in plate **13'**.

In Figure 5.1h there is shown one means of grasping, lifting and transporting the filter cartridge of this process, all but the top portion of the device being cut away for simplicity. In this process, sleeve **30** is in place resting on plate **11** and held tightly against that plate by means of nut **33** which is screwed tightly against the top of the sleeve. Extending arms **35** are provided on nut **33**. A double hooked grasping means **52** is slipped over arms **35** , and is provided with a top eye or grasping means **53** by which the device can be lifted, either by hand or by suitable mechanical means. The grasping means **52** can be quickly disengaged when the device **10** is in place.

Example: A filter cartridge was assembled as shown in Figure 5.1a through 5.1e and was preheated to 1300°F in an electric oven at the rate of 350°F per hour. The assembly was retightened by screwing down the nut **33** to take up the thermal expansion of the tie rod **16** before the cartridge was removed from the preheater.

The preheated cartridge was then lifted out of the preheater and placed in the filter chamber **40** with the proper gaskets. Molten 7075 aluminum alloy was filtered through the assembly at a temperature of 1300°F and at a rate of 300 pounds per minute while casting two 15 x 49 inch sheet ingots. After an ingot length of 114 inches had been reached the metal flow was stopped. The filter cartridge was allowed to remain in the filter chamber while the ingots were removed from the casting station.

Casting was again started and molten 7075 aluminum alloy at 1300°F was filtered through the cartridge at the rate of 300 pounds per minute. The process was repeated as described above until filter cartridge was no longer usable or after approximately 100,000 pounds of molten aluminum had been filtered. Filter pressure buildup was good, indicating no leakage for the cast, and no filter tubes broke in the handling. The quality of the metal cast was also good, with good removal of solid particles of impurities.

In earlier work *H.A. McDonald and R.C. Snyder; U.S. Patent 3,524,548; August 18, 1970; assigned to Kaiser Aluminum & Chemical Corporation* describe a rigid filter which is a fired, shaped article consisting essentially of a granular or particulate refractory material, which is compatible with molten aluminum bonded with a vitreous or glassy material. The granular material is refractory composition or batch in an average particle size from 0.165 mm to 2.8 mm, advantageously from 0.25 mm to 2.5 mm maximum diameter. The vitreous or glassy material is present in an amount of from 4 to 20%, based on total weight of the fired composition.

The granular material is any substance capable of resisting molten aluminum and which will maintain its solid form and dimensional stability at the temperatures of molten aluminum. An especially advantageous granular aggregate for use in the filters of this process is alumina, especially fused alumina, or tabular alumina, which can be high purity commercial grade or made from naturally occurring alumina such as bauxite or diaspore; or there can be used fused mullite, boron nitride, titanium diboride or other suitable refractory material known to resist attack by molten aluminum.

Ceramic Filter Medium

H. Uchiyama, K. Kakimoto, Y. Fujiwara, S. Yamagishi and J. Hayakawa; U.S. Patent 3,939,079; February 17, 1976; assigned to Mitsui Mining & Smelting Co., Ltd. and Tokyo Kokyu Rozai Co., Ltd., Japan describe a filter medium for molten aluminum which permits an increased amount of filtration. The medium is prepared by blending 100 parts by weight of aggregate particles, 5 to 18 parts by weight of an inorganic bonding material having 10 to 50% by weight SiO_2 and about 5 to 20% by weight B_2O_3 and containing at least 70% by weight of a nonvitreous substance such as talc or boric acid, and 3 to 15 parts by weight of a combustible substance such as wheat flour.

The mixture is kneaded with a caking material and water, the kneaded mixture is molded and dried and then the dried mixture is calcined at a temperature not lower than about 1100°C. The size and number of the air permeable pores in the filter medium can be controlled by adjusting the amounts of the bonding material and combustible substance.

The following examples illustrate the process. The inorganic bonding materials used in the examples were formed from individual components having the following chemical analysis values (by weight). Not all components are used in each example, and due to analytical error the figures may not always total 100%.

	Al_2O_3	SiO_2	CaO	MgO	B_2O_3	Fe_2O_3	Ignition* Loss
Talc	1.30	59.85	0.86	28.85	-	3.10	5.21
Calcium carbonate	-	-	54.89	0.93	-	-	43.72
Gairome-Clay	32.19	48.83	0.80	0.06	-	1.39	16.40
Calcined alumina	99.20	0.02	-	-	-	0.03	-
Boric acid	-	-	-	-	56.40	-	43.52
Wollastonite	-	51.32	48.01	-	-	-	-
Glaze**	8.08	43.45	3.89	18.74	25.81	-	-

*Merely for analytical purposes, not an important criterion of usefulness.
**Commercially available glaze for porcelain.

Example 1: 23 parts by weight of talc, 23 parts by weight of calcined alumina, 20 parts by weight of Gairome-Clay, 12 parts by weight of boric acid, 7 parts by weight of calcium carbonate, and 15 parts by weight of commercially available glaze for porcelain were mixed and pulverized for 48 hours in a ball mill to adjust the particle size of the resulting mixture such as to pass a 200 mesh Tyler screen. The resulting mixture had a chemical composition, by weight, of 32.15% of Al_2O_3, 30.05% of SiO_2, 4.78% of CaO, 9.52% of MgO, and 10.63% of B_2O_3. Its ignition loss was 12.27%.

9 parts by weight of the resulting mixed pulverized product and 6 parts by weight of wheat flour were blended with 100 parts by weight of commercially available fused alumina of a particle size of No. 24 (as stipulated in JIS R6001), and the blend was further kneaded with 2.5 parts by weight of dextrin and 4 parts by weight of water. The kneaded mixture was molded into a tube which was open at both ends and which had an inner diameter of 60 mm, an outer diameter of 100 mm, and a length of 900 mm, removed, and, after drying, the molded product was calcined for 8 hours at 1350°C in a heavy oil combustion flame.

The resulting product had a bulk density of 2.08, an apparent porosity of 45.0, a compression strength of 230 kg/cm^2 and a grade number of 160. When molten aluminum was filtered through this tubular molded article, the amount of aluminum which could be filtered was about 1.45 times as large as that with a conventional filter medium. When this calcined product was immersed for 20 days in a bath of aluminum alloy 63S at 770°C, no change was observed.

Example 2: Twenty parts by weight of calcined alumina, 20 parts by weight of Gairome-Clay, 17 parts by weight of wollastonite, 15 parts by weight of talc and 8 parts by weight of boric acid were blended with 20 parts by weight of the same glaze as used in Example 1, and the mixture was pulverized for 48 hours in a ball mill to provide a pulverized mixture with a particle size such as to pass a 200 mesh Tyler screen. The resulting mixture had the chemical composition of 28.09% of Al_2O_3, 36.17% of SiO_2, 9.24% of CaO, 8.09% of MgO, and 9.67% of B_2O_3, and had an ignition loss of 7.54%. 13 parts by weight of the resulting pulverized mixture and 10 parts by weight of rice bran were blended with 100 parts by weight of commercially available fused alumina of a particle size of No. 30 (stipulated in JIS K6001), and the blend was thoroughly kneaded with 5 parts by weight of sulfite pulp waste liquor as a caking material (available from any paper mill) and 2 parts by weight of water. The kneaded mixture was molded into a tube open at both ends and having the same size as in Example 1. After drying, the molded tube was calcined for 10 hours at 1300°C in a heavy oil combustion flame.

The product had a bulk density of 1.98, an apparent porosity of 47.7%, a compression strength of 195 kg/cm^2 and a grade number of 180. When molten aluminum was filtered through the resulting filter medium, the amount of molten aluminum filtered was about 1.7 times as large as that with conventional filter media. When the calcined product was immersed for 20 days in a bath of aluminum alloy 63S at 770°C, no change was observed.

Example 3: 100 parts by weight of fused alumina having a particle size of No. 36 (as stipulated in JIS R6001) was mixed with 16 parts of the same pulverized mixture as was obtained in Example 1 and 13.5 parts by weight of wheat flour. The mixture was thoroughly kneaded with 3 parts by weight of dextrin and 6 parts by weight of water, and the mixture was molded into a tube open at both ends having the same size as in Example 1. After drying, the molded tube was calcined for 10 hours at 1350°C. The calcined product had a bulk density of 1.89, an apparent porosity of 50.0%, a compression strength of 152 kg/cm^2 and a grade number of 240. The amount of molten aluminum filtered was about 1.9 times as large as that with conventional filter media.

The most important characteristics of the filter medium are as follows: it can be formed to have a grade number of 100 to 300, most preferably 150 to 180; a bulk density of 1.70 to 2.27, most preferably 2.00 to 2.25; and an apparent porosity of 39.0 to 51.0%, most preferably 42.0 to 46.0%. The grade number is the number of cubic feet of air per minute that will pass through one square foot of the filter medium at a pressure of 2 inches water column.

Ceramic Foam Filter

J.C. Yarwood, J.E. Dore and R.K. Preuss; U.S. Patent 3,962,081; June 8, 1976; M.J. Pryor and T.J. Gray; U.S. Patent 3,947,363; March 30, 1976; M.J. Pryor and T.J. Gray; U.S. Patent 3,893,917; July 8, 1975; all assigned to Swiss Aluminium Ltd., Switzerland describe a highly efficient, ceramic foam material for use in filtering molten metal, especially molten aluminum.

The ceramic foam material is characterized by having an open cell structure with a plurality of interconnected voids surrounded by a web of the ceramic material. The filter is further characterized by a combination of critical characteristics. The filter has an air permeability in the range of from 400 to 8,000 x 10^{-7} cm^2. In addition, the filter has a porosity or void fraction of 0.80 to 0.95. Still further, the filter is characterized by from 5 to 45 pores per linear inch and an effective range of filter thickness of from ½ to 4 inches. It has been found that the foregoing filter is particularly useful in the filtration of molten metal, especially molten aluminum.

In the normal situation one use a relatively fine filter having an air permeability of from 400 to 2,500 x 10^{-7} cm^2, a porosity or void fraction of 0.80 to 0.95

and from 20 to 45 pores per linear inch (ppi), especially when filtering an aluminum alloy of the 5,000 series. However, if the input metal is particularly dirty, one should preliminarily filter the metal through a relatively coarse ceramic foam filter having a pore size of between 5 and 20 ppi, air permeabilities of 2,500 to 8,000 x 10^{-7} cm^2 and porosities or void fractions of between 0.90 and 0.95. This may be accomplished by providing a single ceramic filter with a gradation of properties or by using a series of filters of differing porosity. In addition to the foregoing, therefore, the process provides a method of filtering molten metal through a ceramic foam filter utilizing a molten metal flow rate through the filter of from 5 to 50 cubic inches per square inch of filter area per minute.

The ceramic foam is prepared from an open cell, flexible foam material having a plurality of interconnected voids surrounded by a web of the flexible foam material. Typical materials which may be used include the polymeric foams such as polyurethane foams, and the cellulosic foams. Generally, one may utilize any combustible organic plastic foam material which has resilience and ability to recover its original shape.

The foam must burn out or volatilize at below the firing temperature of the ceramic material which is employed. Naturally, the characteristics or dimensions of the polymeric foam should roughly correspond to the desirable dimensions of the resultant ceramic foam material. Generally, one utilizes a foam material having a thickness of from ½ to 4 inches since this is the desired thickness of the ceramic foam material. The following examples illustrate the process.

Example 1: A polyurethane foam was provided having a thickness of 2 inches. An aqueous ceramic slurry was provided having the following composition: 47% Al_2O_3, 13% Cr_2O_3, 3.5% kaolin, 1% bentonite and 14.5% colloidal aluminum orthophosphate added as an aqueous solution with an equal amount of water. The slurry contained 82% solids and 18% water.

The foam material was immersed in the slurry and kneaded to remove air and substantially fill voids with the slurry and also to coat the fibrous webs of the foam with the slurry. The resultant impregnated foam was removed from the slurry and subjected to compression through preset rolls to squeeze approximately 80% of the slurry out of the foam by passing the impregnated foam through the preset rollers. The foam material sprung back to its original dimension after passing through the rollers and had the fibrous polyurethane filaments coated with a substantially uniform residue of the ceramic slurry.

The material was oven dried at 125°C for 1 hour, heated slowly at a heat up rate of 0.5°C per minute to 500°C to drive off the water and allow the polyurethane fibers to volatilize and/or burn out without collapsing the ceramic and without destroying the filamentary ceramic configuration. The foam was held at 500°C for 1 hour and was subsequently heated to 1350°C at a rate of 1°C per minute, held at 1350°C for 5 hours to permit the ceramic to sinter together and thereby provide an open cell ceramic foam material having a configuration of the original polyurethane foam material. The characteristics of the resultant foam were as follows.

Permeability	1,425 x 10^{-7} cm^2
Porosity	0.87
Pore size	30 pores per linear inch
Thickness	2"
Structural uniformity	Excellent

Example 2: A 50,000 lb. charge of aluminum alloy 5252 containing 2.3% magnesium, 0.04% silicon, 0.05% iron and 0.06% copper was melted in a gas fired open hearth furnace and fluxed with chlorine gas according to conventional practice. This unfiltered metal was then cast into three 20 x 53 inch cross section rolling ingots weighing 10,000 lb each.

A second charge of the same composition was melted and prepared for casting in accordance with the same practice except, however, the metal was passed through the ceramic foam filter prepared in Example 1 installed in the pouring trough before the metal was cast into rolling ingots. A metal head of only 6 inches was required to prime the filter and during the casting the running head loss built up from ⅛ inch to 1 inch after 30,000 lb of metal had been filtered therethrough.

Pressure filter tests were run on metal taken from upstream and downstream of the filter during the casting of the filtered metal and from the pouring trough during the casting of the unfiltered metal. The filtration effect of the ceramic foam filter of the foam of the process proved to be excellent.

Cross sections of the pressure filter disc from the unfiltered metal were compared with cross sections of the pressure filter disc from the filtered metal. These cross sections clearly showed that little or no residue was present in the metal filtered with ceramic foam filter of the process, whereas, considerable amounts of residue existed in the cross section of the pressure filter disc from the unfiltered metal. Similarly, a pressure filter disc was obtained upstream of the ceramic foam filter and resulted in the presence of considerable residue. This clearly shows the effectiveness of the ceramic foam filter of the process.

The pressure filter test described above is a method of concentrating and examining the nonmetallic particulate in a 20 to 25 lb sample of molten aluminum. The molten aluminum in this test is carefully ladled into a preheated 25 lb clay graphite crucible into the base of which is set a 30 mm diameter, 3 mm thick porous silica disc plug. 90% of the metal is forced through the disc by application of air pressure and the remaining metal solidified in situ. The disc and adjacent metal are then sectioned, polished and examined by normal metallographic techniques to reveal the quantity of nonmetallics filtered out.

Example 3: The ingots prepared in Example 2 were all rolled down to 0.030 inch gage and samples of the sheet from coil locations corresponding to the head and butt sections of the ingots were tested to reveal the incidence of linear defects per unit area of sheet. The sheet manufactured from filtered metal was found to contain about 20 times fewer linear defects than did the sheet manufactured from unfiltered metal as indicated in the table below.

In addition to the foregoing, the spent ceramic foam filter utilized in Example 2 was examined metallographically.

Considerable oxide stringers and fine nonmetallic particulates were found to be captured in the web of the filter, showing further evidence of the filter's filtration power. Good mechanical properties were obtained and no metal contamination was found to exist in the resultant product due to the use of the ceramic filter.

Example 4: A further 50,000 lb melt of aluminum alloy 5252 was prepared as in Example 2. In this case, the metal was filtered through a tabular alumina bed in accordance with conventional practice prior to casting into rolling ingots for comparative purposes. The ingot thus cast was rolled down to 0.030 inch gage sheet and samples were removed at locations in the resultant coil corresponding to the ingot butt, a location 20 inches from the ingot butt and the ingot head.

The samples were then tested to reveal the incidence of linear defects. The results of this inspection are shown in the table below together with similar data carried out on unfiltered metal and ceramic foam filter metal from Example 2. The results in this table are given as percentages using the unfiltered metal as basis for comparison.

Comparative Linear Defect Counts

Type of Metal	Butt	20" from Butt	Head
Unfiltered metal	100%	100%	100%
Bed filtered metal	150%	25%	10%
Ceramic foam filter metal	10%	10%	10%

These results clearly show the reduced quality of the butt region of the ingot produced by filtration with the commercial bed filter as compared with that produced by the filter of the process.

Halide Compounds

K. Ogiso; U.S. Patent 3,907,962; September 23, 1975 describes a method of preparing a filter medium for the filtration of molten aluminum or a molten aluminum alloy. More than two halide compounds are used, of which at least one halide is selected from the group consisting of the compounds sodium fluoride, potassium chloride, silver chloride, potassium fluoride and sodium nitrate, and of which the second halide is selected from the group of compounds having a melting point of more than 1000°C consisting of calcium fluoride, magnesium fluoride, aluminum fluoride, barium chloride, cryolite, and cerium fluoride and mixtures thereof. The compounds are heated until they are molten and then mixed thoroughly with some agitation to obtain a homogeneous melt and the thus obtained melt is subjected to a molding step.

Examples 1 through 4: The composition of the molding materials for the filter medium of the process are as follows:

Ingredients	Content (%) Based on the Mixture Weight
Composition of Example 1	
Calcium fluoride	10-40
Borax	5-15
Magnesium fluoride	30-60

(continued)

Ingredients	Content (%) Based on the Mixture Weight
Man-made cryolite	5 - 15
Lithium carbonate	5 - 15
Sodium chloride	5 - 15
Composition of Example 2	
Barium fluoride	30 - 60
Potassium chloride	10 - 30
Sodium chloride	5 - 20
Potassium fluoride	5 - 20
Lithium chloride	5 - 20
Composition of Example 3	
Potassium fluoride	20 - 40
Sodium chloride	20 - 40
Alumina	10 - 30
Composition Example 4	
Calcium fluoride	30 - 50
Lithium chloride	10 - 20
Sodium nitrate	10 - 20
Titanium metal powder	5 - 10

The raw materials of each example are heated to about 1000° to 1400°C, that is, to their melting points with thorough stirring after the substances become liquid. Each composition thus obtained is cast into a mold of graphite with a rapid cooling treatment so that such disadvantageous effects, such as segregation phenomenon, or some phase separation phenomenon between the ingredients does not take place.

The composition of Example 1 comprises metal halides and a carbon compound. The composition of Example 2 comprises metal halides only. The composition of Example 3 comprises metal halides and a ceramic substance. The composition of Example 4 comprises metal halides, a nitrogen compound and metal powder. The filter medium made of each composition in Examples 1 to 4 has a striking permanency and endures against wearing out in severe circumstances of high temperatures of the molten aluminum phase.

Each filter medium made of the composition in Examples 1, 3 and 4 has a specific gravity of value range of from about 2.4 to 2.6, and these mediums can be conveniently used for the metal filtering process by putting them in a piled mass form between two porous filter walls in the molten aluminum metal. On the other hand, the filter medium made of the mixture of Example 2 has a specific gravity of about 3.0 to 3.2, and can be conveniently used for the process by putting the medium in a piled mass form on a porous filter plate wall at any place in the molten aluminum metal, or on the same kind of plate wall positioned at the bottom side of the molten metal liquid to be filtered.

HEAVY METALS

Isocyanate Resin Adsorbent

N. Oda, Y. Horie, M. Idohara and T. Hukui; U.S. Patent 3,959,149; May 25, 1976; assigned to Nippon Soda Company Limited, Japan describe an economical method to remove or capture free chlorine, heavy metals or heavy metallic compounds from liquids and gases, for example, service water, river water, drainage or sewage from a factory, laboratory or hospital, air, exhaust gas from cars or factory, which are sometimes contaminated with free chlorine, heavy metals and heavy metallic compounds.

It has been discovered that isocyanate resins, which are prepared by reacting polyisocyanates with one or two compounds selected from the group consisting of polyol compounds and a functional monomer, selectively adsorbs free chlorine, and heavy metals, such as Cr, Mn, Cd, Co, Ni, Cu, Hg, Pb, metallic ions and heavy metallic compounds. A functional monomer can be defined as a compound having at least 2 active hydrogen atoms to react with the isocyanate group and at least 1 functional group selected from the group consisting of:

$$-\underset{\underset{S}{\|}}{C}-,\quad -\underset{\underset{S}{\|}}{C}-\underset{\underset{S}{\|}}{C}-,\quad -S-,\quad -S-S-,\quad -SH,\quad -\underset{\underset{O}{\|}}{C}-,\quad >NH \text{ and } =N-.$$

Active hydrogen atom may be combined in the functional monomer as a radical of H_2N-, $HS-$,

$$\underset{\underset{R}{|}}{HN}-$$

(wherein R is an alkyl radical), HO_3S-, $HOOC-$, $HON=$, H_2NCO-,

$$-\underset{\underset{H}{|}}{N}CO-$$

H_2NO_2S- and $HO-$ and some preferable examples of functional monomers are thiourea, rubeanic acid, tetramethylene-bis-dithiocarbamate, diaminothiophenyl, diaminodiphenyldisulfide, diaminobiphenyl and diphenylthiourea.

Silver, mercury, silver compounds and mercuric compounds are especially adsorbed by the resin particularly when the functional monomer has the radical

$$-\underset{\underset{S}{\|}}{C}-,\ -\underset{\underset{S}{\|}}{C}-\underset{\underset{S}{\|}}{C}-,\ -S-,\ -S-S-,\ -\underset{\underset{H}{|}}{N}- \text{ or } -\underset{\underset{O}{\|}}{C}-,$$

for example, thiourea, dithiooxamide, diphenylthiourea, tetramethylene-bis-dithiocarbamate, dithioglycolic acid, 2,3'-diaminodiphenylsulfide, 3,3'-diaminodiphenyldisulfide, urea and diphenylcarbazide. The following examples illustrate the process.

Example 1: 38 parts of powdered thiourea were gradually added to 87 parts of TDI-80 at 100°C with vigorous stirring. The mixture became creamy, and gradually turned yellow, and when the temperature rose to 110°C, a solidified resinous material was obtained. A content of —NCO group in the obtained material was about 10%.

Example 2: Toluylenediamine and phosgene were reacted with a solvent of o-dichlorobenzene and N_2 gas was bubbled into the reaction mixture in order to remove phosgene and hydrochloric acid and TDI tar containing 25% of —NCO groups was obtained as a residue by distillation of TDI under a reduced pressure of 10 mm Hg.

Twenty-five parts of the obtained TDI tar was ground, melted at 160°C, and reacted with vigorous stirring with 12.7 parts of thiourea which was the equivalent weight for the —NCO group content in the TDI tar. The mixture gradually turned to a solidified resinous material and 8 parts of o-dichlorobenzene were added to enable stirring. The reaction was continued for 1 hour and after 2 hours of digestion of the reaction in a drying chamber at 160°C, dark (black-brown) resinous foam material containing free —NCO was obtained.

Examples 3 through 21: Example 1 or 2 was repeated for the following isocyanates and functional monomer. Isocyanates, functional monomer, reaction condition etc. were listed in the following table.

Ex.	Isocyanates	Parts	Functional Monomers	Parts	Reaction Conditions: Temp. (°C)	Reaction Conditions: Corresponding to Example	Appearance of Resinous Material
3	TDIT	103	Thiourea	38	100-120	1	yellow
4	TDIT	103	Rubeanic acid	60	140-160	1	red-brown
5	TDI-80	174	Rubeanic acid	60	130-150	1	red-brown
6	TDI-80	174	Dithiodiglycolic acid	91	100-120	1	yellow
7	TDI-80	174	Ammonium dithiocarbamate	120	70-100	1	yellow brown
8	TDI-80	87	Tetramethylene-bis-dithiocarbamate	120	80-100	1	yellow brown
9	TDI tar	87	Diphenylthiourea	114	140-160	2	black (foamy)
10	TDI 80	174	Urea	30	80-100	1	pale yellow (foamy)
11	TDI tar	100	Urea	18	140-160	2	black (foamy)
12	MDI	125	Thiourea	38	80-100	1	pale yellow (foamy)
13	MDI (crude)	140	Thiourea	38	100-120	2	black (foamy)
14	TDI dimer	87	Thiourea	19	120-140	2	brown
15	TDI trimer	87	Thiourea	13	130-150	2	brown
16	NDI	105	Thiourea	38	120-140	2	brown (foamy)
17	TODI	132	Thiourea	38	100-120	2	brown (foamy)
18	PDI	80	Thiourea	38	70-90	2	yellow (foamy)
19	HMDI	84	Thiourea	38	100-120	2	pale yellow
20	LDI	106	Thioglycolic acid	91	100-120	2	pale yellow
21	TDI-80	87	Chloranilic acid	52	100-120	2	black-violet

TDI, tolylenediisocyanate; MDI, diphenylmethanediisocyanate; TODI, toluidinediisocyanate; NDI, naphthalenediisocyanate; HMDI, hexamethylenediisocyanate; PDI, phenylenediisocyanate; LDI, lysine diisocyanate; MDI (crude), undistilled MDI.

Example 22: 8.7 parts of TDI-80 and 1.6 parts of thiourea were gradually mixed and 10.5 parts of polyethyleneglycol-400 was added dropwise under vigorous stirring. The reaction mixture generated heat, became viscous and the temperature rose slowly to 80°C. Then 2 parts of water was added as a foaming agent under vigorous stirring. When the mixture became sticky and generated foams, stirring was stopped and the mixture was digested in a drying chamber at 80° to 100°C for 3 hours.

Example 23: Seventy-six parts of TG-3000 (polyethertriols), 20 parts of thiourea, SH-190 (silicon compound; foam controller), 0.3 part of T-9 (organotin compound; catalyst for foaming agents) were mixed at 20°C for 1 minute and then 100 parts of TDI-80 was added and mixed with vigorous stirring at 20°C for 7 to 10 seconds.

The mixture generated foam and the foamed substance was allowed to stand for 24 hours at room temperature and further cured at 120°C for 3 hours.

Example 24: Five hundred sixty parts of TDI-80 and 100 parts of TG-300 were mixed at 100°C for 20 minutes and prepolymer containing 30% of —NCO groups was obtained. On the other hand, 100 parts of TG-300, 2.0 parts of SH-190, 0.3 parts of T-12, 0.1 parts of Dabco, 3.4 parts of water and 63 parts of thiourea were mixed for 1 minute with vigorous stir at 30°C and then 400 parts of above mentioned prepolymer was mixed at 20°C and the mixture was stirred for 2 to 5 minutes and generated foam. The foaming material was allowed to stand at room temperature for 24 hours and then cured at 130°C for 3 hours.

Example 25: Two hundred thirty parts of TDI tar, which was obtained in Example 2, 50 parts of thiourea and 1,800 parts of methylethylketone were mixed well and the mixture was gradually heated to dissolve completely. One thousand four hundred parts of diatomaceous earth 1 to 2 mm in diameter were added to the mixture and then methylethylketone was slowly distilled off under reduced pressure. The obtained diatomaceous earth containing TDI and thiourea were heated at 140° to 160°C for 2 hours. The finished diatomaceous earth contained 14% of resinous material.

Example 26: Thirty grams of resinous adsorbent obtained in Example 22 were ground to 14 to 32 mesh (Tyler), dipped in 1N HCl aqueous solution for 24 hours and rinsed with water to the extent of pH 6 of rinsing water. Obtained resinous adsorbent was packed in a glass tube having 15 mm diameter. Ten liters of aqueous solution containing $HgCl_2$ at a concentration of 10 ppm of Hg were passed at pH 6 through the glass tube at a space velocity (SV) rate of 5 and the content of $HgCl_2$ in the treated aqueous solution was measured and less than 0.01 ppm of $HgCl_2$ content was found.

This example was repeated except that an aqueous solution containing Hg, Pb, Cd, Ni, Mn, Cr, B, Co, Cu, As and Al respectively at a concentration of 20 ppm of each metal was passed through the tube at pH 4. Contents of these metals in the treated aqueous solution were measured by emission spectroanalysis and it was found that content of Hg under detection sensitivity (0.001 ppm) and contents of other metallic compounds of Pb, Cd, Ni, Mn, Cr, B, Co, Cu, As and Al were 17 to 20 ppm and consequently the resinous adsorbent had a selectivity for mercury compounds.

Example 27: The resinous adsorbent obtained in Example 23 was ground to 50 to 100 mesh (Tyler) and 0.1 g of adsorbent was added to an Erlenmeyer flask with 100 ml of aqueous solution containing $HgCl_2$ at a concentration of 10 ppm Hg. The flask was shaken for 20 hours at room temperature and filtrate obtained by filtration of the resinous adsorbent was measured for content of Hg and it was found that mercury adsorption equilibrium concentration was 0.08 ppm and adsorption equilibrium amount was 10 mg/g.

Example 28: Nine grams of foamed substance obtained in Example 23 were packed in a glass tube having an inside diameter of 15 mm. The thickness of the layer was 30 cm. Water containing 100 ppm of free chlorine was passed through the glass tube at a rate of SV = 8. After the treatment, the content of free chlorine was measured and identified as less than 0.1 ppm by colorimetric analysis with the ortho-tolidine method. The treatment was repeated for service water containing 0.4 ppm of chlorine except the rate of SV was 15. The content of chlorine in the treated water was less than 0.1 ppm.

Formaldehyde-Modified Nut Waste

According to a process described by *J.M. Randall, E. Hautala, A.C. Waiss, Jr., and J.A. Kuhnle; U.S. Patent 3,944,415; March 16, 1976; assigned to the U.S. Secretary of Agriculture* heavy metal ions are removed from water by contact with formaldehyde-modified nut waste or tree bark. As a first step in the process, the starting material—peanut skins or other nut waste or tree bark—is modified by reaction with formaldehyde. The reaction is conducted in an aqueous system under acidic conditions.

Typically, peanut skins (1 part) are mixed with about 10 parts of water containing about 0.5 to 1% of an acid. Acids which are suitable for the purpose are sulfuric, hydrochloric or phosphoric. After the peanut skins have been mixed with the dilute acid, 15 to 20% (based on the weight of peanut skins) of formaldehyde is added.

The mixture is stirred and then allowed to stand for 2 to 3 hours at 50° to 70°C. Following this standing period the liquid is separated from the modified skins, which are washed and dried. The residual acidic liquid may be recharged with formaldehyde and used again as above. The dry modified peanut skins are ready for use as is or they may be ground prior to use. Alternatively, the peanut skins may be ground prior to treatment with formaldehyde.

The formaldehyde-modified peanut skins may be formed into a bed and the water containing heavy metal ions can be allowed to percolate therethrough. Alternately, the modified skins may be placed in a column and contaminated water can be pumped therethrough. The critical point is that contact be effected and maintained for a period long enough for the heavy metal impurities to be absorbed by the modified peanut skins. Generally, the gravity flow rate of the liquid through a column of modified peanut skins provides a sufficient contact time. The gravity flow rate of water through modified peanut skins ground 1 mm in diameter is approximately 2 gallons per square foot per minute.

A particular advantage of the process lies in the fact that it can be adapted easily to present water purification systems whether industrial or municipal. Since most water purification systems involve a filtration procedure, the addition of an extra filter containing formaldehyde-modified nut waste or tree bark would be routine. The following examples illustrate the process.

Example 1: Preparation of Formaldehyde-Modified Peanut Skins – Peanut skins were obtained from a local nut packer. One kilogram of unwashed peanut skins was mixed with 10 l of 0.2 N H_2SO_4 and 500 grams of 35% aqueous formaldehyde was added. The mixture was stirred well and then allowed to stand for 2 hours at about 50°C. The modified peanut skins were separated from the reaction liquid, washed with water, and then air dried.

Example 2: Removal of Mercuric Ion from Dilute Solution Using Formalde-

hyde-Modified Peanut Skins — A column 30 cm in height and 12 mm in inside diameter was packed with 5 grams of formaldehyde-modified peanut skins (prepared as in Example 1). The column was loaded with 495 ml of an aqueous solution of mercuric acetate containing 1 mg of mercuric ion per ml. The column was then washed with three 100 ml portions of distilled water. These washings were analyzed for mercuric ion content by atomic absorption, which is sensitive to 1 ppm of mercuric ion. The results are summarized in the table below. To provide a point of reference a similar experiment was conducted with unmodified peanut skins.

Absorbent	Mercuric Ion Applied (mg)	Mercuric Ion Concentration, ppm 1st Wash	2nd Wash	3rd Wash	Total Mercuric Ion Absorbed (mg)
Formaldehyde-modified peanut skins	495	<1	<1	<1	495
Unmodified peanut skins	495	<1	<1	<1	495

The effluent that passed through the column containing the formaldehyde-modified peanut skins was colorless whereas the effluent that passed through the unmodified skins had an intense reddish brown color.

Example 3: Removal of Nickel and Silver Ions from Dilute Solution by Contact with Formaldehyde-Modified Peanut Skins — A column 24 cm in height and 25 mm in inside diameter was packed with 16 grams of formaldehyde-modified peanut skins (prepared as in Example 1). The column was loaded with 3.5 liters of an aqueous solution of nickel chloride containing 20 mg of nickel ion per liter or a total of 70 mg of nickel ion. After the effluent had passed through the column, the column was washed with three 1 liter portions of distilled water. These washings and the effluent were analyzed for nickel ion content by atomic absorption.

A similar experiment was carried out with an aqueous solution of silver nitrate containing 197 mg of silver ion per liter. A total of 3 l of solution (537 mg of silver ion) was passed through the column which was 35 cm in height and 20 mm in inside diameter and was packed with 20 grams of formaldehyde-modified peanut skins. The column was washed with distilled water as described above. Analysis was again made by atomic absorption, which is sensitive to 0.1 ppm nickel ion and 0.01 ppm silver ion. The results of the above experiments are summarized in the table below.

Ion Applied	Amount (mg)	Concentration (ppm) Effluent	Wash	Total Ion Absorbed (mg)
Nickel	70	<0.1	<0.1	70
Silver	537	<0.01	<0.01	537

In earlier work *A.C. Waiss, Jr., J.A. Kuhnle and M.E. Wiley; U.S. Patent 3,719,473; March 6, 1973; assigned to the U.S. Secretary of Agriculture* described a process by which mercury is effectively removed from water by contact with a nut waste such as peanut skins, walnut expeller meal and peanut hulls.

Waste Brine from Mercury Cells

J.H. Entwisle and R.W. Griffiths; U.S. Patent 3,718,457; February 27, 1973; assigned to Imperial Chemical Industries Limited, England describe a process for recovering mercury from waste brine from mercury cells. The process comprises adding sulfide or hydrosulfide ions to the brine to precipitate mercury sulfide, trapping the mercury sulfide on a mixed media filter in which the average size of particles in successive superimposed layers decreases and the specific gravity of the particles increases from the top of the filter downwards, removing the mercury sulfide from the filter by backwashing with an aqueous liquor, separating the mercury sulfide from the backwash liquor and treating the sulfide to produce metallic mercury.

The materials which can be used to make up the filter may be any suitable material as long as in at least two directly superimposed layers the average size of the particles in the successive layers decreases and the specific gravity increases from the top of the filter downwards. Examples of suitable materials include anthracite coal, activated carbon, graphite, quartz, flint, silica, magnesia, alumina, garnet and sand.

It is preferable that the waste brine be treated to remove insoluble matter, for example, graphite, before addition of the sulfide or hydrosulfide. This may be carried out by any convenient means, for example, by filtration of the brine, for example, with a filter of the type used to remove the mercuric sulfide precipitate or a precoat filter, such as a precoat charcoal filter. Other additional or alternative means for the insoluble matter removal include using a hydrocyclone. This treatment may be carried out before or after any air-blowing to remove chlorine. The process is illustrated in the following examples in which ppm stands for parts per million, weight/volume.

Example 1: 100 liters of effluent brine from a mercury cell which had been air blown to reduce the available chlorine to 70 ppm and which contained 10 ppm Hg was passed through a reaction vessel at a rate of 20 l/hr and treated continuously with sodium sulfide solution to maintain the oxidation potential measured with a platinum electrode and a saturated calomel electrode at +0.1 v.

Mercuric sulfide was precipitated as an extremely fine, black precipitate and was filtered by passing the brine through a 5 cm diameter mixed media filter comprising a 2.5 cm deep top layer of 7 to 10 BS mesh (particle size of 1.7 to 2.4 millimeters) anthracite having a specific gravity of 1.4, a 2.5 cm deep intermediate layer of 18 to 25 BS mesh (particle size of 0.6 to 0.85 mm) flint having a specific gravity of 2.65 and a 5 cm deep layer of 30 to 36 BS mesh (particle size of 0.4 to 0.5 mm) alumina having a specific gravity of 3.9. This bed was supported by a 2.5 cm layer of 6 to 8 BS mesh (particle size 2.0 to 2.8 mm) alumina and a 3.8 cm layer of 0.6 to 1.3 cm diameter pebbles on a perforated titanium plate. The effluent brine after filtration contained less than 0.1 ppm Hg and no free chlorine.

The precipitate was removed from the bed by passing air upwards through the bed and then passing water upwards through the bed to fluidize it and then remove the precipitate from it. The backwash liquor was collected in a separating funnel and the mercuric sulfide precipitate allowed to settle. The precipitate, which was coarse and dense, settled in a few minutes and was easily separated from the supernatant liquor. The precipitate recovered from the filter weighed 1.04 grams representing 90% mercury recovery.

Example 2: Effluent brine from a mercury cell, which had been air blown to reduce the available chlorine to 70 ppm, was passed through two filters in parallel of the type described in Example 1 except that they each had a diameter of 15 cm. The total flow rate through these filters was 1 m^3/hr. This filtering reduced the level of insoluble graphite in the brine. The brine was then treated as described in Example 1 to precipitate mercuric sulfide. The brine was then filtered on a filter of the type described in Example 1 except that it had a diameter of 10 cm and each layer of material had a depth of 23 cm. The filtered brine was analyzed for mercury content and hence removal efficiency.

The precipitate was removed from the bed by passing air at a pressure of 0.3 to 0.4 kg/cm^2 gauge upwards through the bed for 2 to 3 minutes. Water at the rate of 8 to 10 l/min was then passed upwards to remove the precipitate. The backwash liquor was then collected and analyzed for mercury content. The precipitate could easily be removed in a separating funnel. The results of a number of runs are shown in the following table.

Run No.	Initial Concentration of Hg in Brine (ppm)	Flow Rate Through Filter (m^3/hr)	Volume of Brine Filtered (m^3)	Weight of Solids from Backwash (g)	Estimated Mercury Recovery (percent)	Removal Efficiency Half Way Through Run (percent)
1	1.8	0.25	19	104	93	96
2	1.8	0.52	11.5	143	96	92
3	1.8	0.25	17.5	138	77	98
4	2.0	0.37				>95

Recovery of Particulate Copper Solids

W.N. Schjerven, Jr.; U.S. Patent 3,696,928; October 10, 1972; assigned to Western Electric Company, Incorporated describes a method of recovering copper solid particulate material from a rinse water. In this process, a filter aid in the form of a relatively thin layer of the solid particulate material is initially formed on a portion of an indexable filter medium. This is accomplished by combining solid particulate material of the type which is to be recovered, with water, to produce a relatively thick slurry.

This slurry is passed through the filter medium and the material which filters out is subsequently leveled to a predetermined uniform depth to form the filter aid layer or bed. The rinse water then is passed through the filter aid and the underlying portion of the filter medium, which cooperate to remove solid particulate material from the rinse water.

Subsequently, as the filter medium is indexed, a filter aid layer is formed on succeeding portions thereof and the filtered solid particulate material and the solid particulate material forming its associated filter aid layer are removed. After removal of the solid particulate material from the filter media, the media may be incinerated to recover any remaining particulate material. The process is of particular value in the recovery of copper particulate material or solids from water which has been used to rinse pickling solution from rolled copper rod prior to the copper rod being drawn into electrical conductor wire. In this connection, the copper rod, after being dipped in a pickling solution, which includes sulfuric acid, certain other process enhancing agents and water, is dipped successively in a series of rinse tanks before being subjected to a wire drawing operation. Periodically, as the water rinse tanks become contaminated with the copper solids, alkaline water is added to the rinse water to neutralize its acidity, preferably to a neutral to slightly alkaline condition, to reduce any copper in solution (in the form of copper sulfate) to a concentration of less than that of the copper ppm.

The tanks then are drained and the copper solids are removed from the neutralized water so that it can be reused or discharged into a sewer, and so that the copper solids can be reused in manufacturing operations or used for other purposes. This process is based on the principle of utilizing the copper solids to be recovered from the rinse water as their own filtering material, in cooperation with a supporting filter medium.

Manganese Oxide Coated Acrylic Fibers for Radioisotope Recovery

A process described by *W.S. Moore; U.S. Patent 3,965,283; June 22, 1976* involves coating acrylic fibers with manganese oxides so strongly and heavily and without serious damage to the fiber that the resulting materials can continuously remove metals and radioisotopes even from water in which the impurity concentration may be as low as 1 ppm.

The mechanism by which the treated fibers of this process purify water is by a combination of surface adsorption and ion exchange. The fibers act as a matrix into which manganese oxides are incorporated and through which liquids such as water may be circulated. Important to the success of this process in removing trace constituents from water is the great strength of the bond between the manganese oxides and the acrylic fiber. This strength is due to a chemical bonding between the coating and fiber.

A surprising aspect of this process is the heavy loading of acrylic fibers with manganese oxide without any appreciable damage to the fiber. Rayon and cellulose were similarly treated, but these fibers became brittle and broken. Hence, such materials are not too practical for purifying water by themselves. An additional material would be required to hold the weakened and damaged fibers together. Any type of acrylic fiber may be used. The characteristics sought in a fiber are strength and cohesiveness.

Example 1: 500 grams of acrylic fiber (Acrilan, 3.0 denier, type B-16) were covered with 2 liters of 6 N NaOH solution at 80° to 90°C; when the fiber began to develop an orange-red color, the reaction was stopped by squeezing the excess NaOH from the fiber. The treated fiber was immersed in 1 liter of 5 N $MnCl_2$ at 30° to 40°C. The $MnCl_2$ solution was maintained at pH 1 to 2 with HCl to prevent precipitation of the manganese. The soaked fiber was drained and immersed in 1 liter of 6 N NaOH at 25°C to precipitate $Mn(OH)_2$. The fiber was then separated and left exposed to the air to oxidize the $Mn(OH)_2$ to a mixture of hydrated MnO, MnO_2 and MnOOH. When the fiber had completely blackened, it was packed in polyethylene bags for later use.

Example 2: 500 grams of acrylic fiber (Acrilan, 3.0 denier, type B-16) were immersed into a 0.5 N $KMnO_4$ solution at 70° to 80°C for 10 min. After removal from the $KMnO_4$ solution, the fiber was washed with water and drained of excess moisture by a wringer. To remove the collected material from the fiber, any standard retrieval process would suffice; e.g., to remove radium from the Mn-treated fiber, the sample (5 to 200 g) is covered by 6 to 8 N HCl, and boiled for 1 hr. This procedure converts MnO and MnO_2 to $MnCl_4$, which decomposes to $MnCl_2$ and Cl_2, leaving the solution clear and the fiber literally bleached white. About 1 liter of acid is required for 100 g of fiber. A better method and one which allows the fiber to be reused at least 5 times comprises placing the used fiber in a volume of 1 to 2 N HNO_3 at least equal to four times the volume of fiber.

To demonstrate the strength of the bond between the oxide coating and the acrylic fiber, a quantity of fiber coated by the first method was towed for 8 hr behind a ship traveling at a speed of 18 km/hr with an oxide loss of only 30%. This strong bond permits the use of fibrous material to be used in very cold and high pressure environments such as those found 3 to 4 km below the surface of the ocean. In an experiment described in Moore, W.S., et al, "Extraction of Radium from Natural Waters Using Manganese-Impregnated Acrylic Fibers" in *J. Geophy. Res.* 78, p. 8880-6, Dec. 20, 1973, the depth profile of ^{226}Ra between 3,200 and 3,740 meters below the surface of the ocean was determined within a 5% margin of error by using a coated acrylic fiber prepared by Method 1.

The experimental results and conditions summarized in Table 1 were obtained with the coated acrylic fiber of Example 1 except for Sample 6. That sample was a $Fe(OH)_3$-acrylic fiber prepared according to the method disclosed in Krishnaswami, S., et al, "Silicon, Radium, Thorium and Lead in Seawater: Insite Extraction by Synthetic Fiber" in *Earth Planet Sci. Letter,* 16, pp. 84 to 90, 1972. Only 11% of the radium in a duplicated sample was removed. The results are summarized in Table 1. The experiment comprised passing samples of seawater through 8-cm columns containing a quantity of filtering fiber.

TABLE 1

Sample	Fiber Coating	Conditions	^{226}Ra % Removal
1	5 g Mn	Single pass, 20 liters, 100 ml/min	95±5
2	5 g Mn	Single pass, 11 liters, 100 ml/min	93±5
3	10 g Mn	Two passes, 20 liters, 100 ml/min	95±5
4	10 g Mn	Two passes, 20 liters, 100 ml/min	95±5
5	80 g Mn	Continuous circulation of 700 liters at 10 l/min for 300 min	90±5
6	80 g Fe	Continuous circulation of 700 liters at 10 l/min for 300 min	11±1
7	40 g Mn	First pass of 170 liters at 10 l/min	56±4
8	40 g Mn	Second pass of 170 liters at 10 liters per minute	23±2
9	40 g Mn	Third pass of 170 liters at 10 l/min	12±1

As can be seen from the above result, the fibers of this process have an exceptional filtering proficiency and represent a radical improvement over the $Fe(OH)_3$-coated acrylic fiber which is a highly regarded adsorption material. To demonstrate the filtering proficiency of the fibrous materials with a heavily contaminated liquid, the above experiment was repeated with a 20 liter seawater sample spiked with 50 dpm ^{226}Ra. The flow rate was 100 ml/min. More than 95% of the radium was retained by one pass, and an additional 3% was removed by a second pass through fresh fiber. The spiked seawater had less than 0.5 dpm ^{226}Ra after two filtrations through the fiber.

The proficiency of the manganese oxide fibers of this process in extracting trace elements from seawater surfaces is demonstrated by the following experiment which compared samples prepared in Example 1 with a $Fe(OH)_3$-acrylic fiber prepared as before and with an untreated acrylic fiber. The experiment is reported in greater detail in Moore, W.S., et al, "Extraction of Radium from Natural Waters Using Manganese-Impregnated Acrylic Fibers, in *J. of Geophy. Res.* 78(36): p 8880-8886, Dec. 20, 1973.

Large volumes of surface waters were sampled by towing the fiber. Between 100 and 500 grams of fiber (on a dry weight, untreated basis) were fluffed and packed into a PVC cylinder 23 cm long and 22 cm in diameter. The ends were closed with PVC sheets, each having 85 holes 1 cm in diameter. A sheet of 18 mesh fiber glass screen was sealed between the covers and the cylinder to prevent washout of the fiber.

The sampler was towed at approximately 18 km/hr 20 meters behind the ship. In calm seas it sampled from about 500 cm below the surface; however, in rough seas and especially in billowing seas it often sank to several meters and broke surface. Even then the sampler has little tendency to skim over the surface. The results are summarized in Table 2. Again the results show that the fibers of this process provide highly efficient filters and represent a marked improvement over the $Fe(OH)_3$-acrylic fiber.

TABLE 2

Sample	Coating Type	Tow Time (min)	^{226}Ra dpm/100 g Fiber
1	Mn	300	220
2	$Fe(OH)_3$	300	5.6
3	Mn	85	156
4	Mn	295	305
5	Mn	320	250
6	Mn	85	103
7	Mn	510	137
8	Mn	300	181
9	Mn	310	152
10	Untreated	285	0.64
11	Mn	0	0.12

BACTERIAL AND MEDICAL APPLICATIONS

BACTERIAL

Microporous Submicron Filter Media

According to a process described by *B.S. Goldberg; U.S. Patent 3,862,030; January 21, 1975; assigned to Amerace Esna Corporation* microporous fluid-permeable filter media are provided which are especially adapted for removing microscopic or ultrafine particles of submicron size from the medium in which they are suspended. The microporous fluid-permeable submicron filter media comprise a polymeric resinous matrix having an inorganic filler dispersed throughout the resinous matrix, and a network of microvoids or pores formed in the resinous matrix between the particles of dispersed filler and the resinous matrix, and between the neighboring particles of the filler. The size distribution of such pores is relatively nonuniform.

The filter media may be fabricated from poromeric or microporous plastic sheets made from a mixture compounded from the following essential ingredients: (1) a thermoplastic resin binder, for example, resinous polyvinyl chloride; (2) a solvent, for example, cyclohexanone, which serves to plasticize or solubilize the resin binder as well as help form some of the pores; (3) an inorganic filler such as silica hydrogel or precipitated hydrated silica, or any other carrier or substrate for volatile matter which serves to capture a controlled amount of volatile matter whereupon the volatile matter may later be removed under controlled conditions to leave behind the desired system of micropores, as will be more fully explained below; and (4) a volatile nonsolvent pore former such as water which is capable of being absorbed by the inorganic filler.

It is preferable to employ a vinyl chloride resin binder of the "EP" or "easy processing" type. An EP resin is characterized by resin particles which are porous and highly absorbent, as compared to commercial resin particles having a hard glossy beaded appearance. A good example of a suitable thermoplastic resin binder is a nonplasticized gamma vinyl chloride homopolymer resin such as Geon 103 EP. The vinyl chloride resin binder may also be a copolymer of vinyl chloride and a small amount (for example, up to about 15%) of a monoethylenic monomer, i.e., vinyl acetate, vinylidene chloride, propylene, or ethylene. Exemplary copolymers of the latter type suitable for use with the process may be a propylene modified vinyl chloride resin such as that commercially available from the Air Products Company under the trademark Airco 401, or a vinyl acetate modified vinyl chloride resin also commercially available from Air Products Company under the trademark Escambia 6240. The following examples illustrate the process.

Example 1: Submicron filter media were fabricated utilizing the following procedure. A microporous sheet was prepared by dry blending 35.0 lb of Hi Sil 233, a precipitated hydrated silica, and 25.0 lb of Geon 103 EP F10 polyvinyl chloride resin having a particle size of about 20% through a 140 mesh screen in a low shear liquids-solids blender for approximately 3 minutes. Thereafter, and during continued agitation, 50.5 lb of solvent (cyclohexanone) were added over a 20 minute period by means of a pump.

Water in an amount of 48.2 lb was then added to the mix in the agitating blender over a subsequent 20 minute period to form a damp, free-flowing powder. The powder was then introduced into a screw extruder having a barrel temperature of approximately 120°F and passed between the rolls of a calender to obtain a substantially flat, uniformly dimensioned sheet. The latter was then passed through an extraction bath of water at 170°F and subsequently dried in a hot air oven at 225°F for 6 minutes.

From the resulting substantially flattened semirigid sheet, a filter element 0.018 inch thick and 47 mm in diameter was cut and heat sterilized by immersion in a steam bath for about 60 minutes. The sterilized filter element was then mounted in a filter housing, and 100 cc of water containing 800 *Escherichia coli* bacteria, average size 0.5 micron in breadth and 2.5 microns in length, was passed through the above filter. The filtrate was analyzed for bacteria by count technique as well as growth via inoculation in EMB, BBL, and trypticase soy broth. In all cases no bacteria count or broth growth was observed. The throughput or flow rate of the filter element was measured by conventional methods at 0.4 gal/min/sq ft under a pressure gradient of 10 psig.

Example 2: A fresh filter element was prepared following the procedures of Example 1. A suspension of 800 *Staphylococcus aureus* bacteria (0.75 micron in average size) in 100 cc of water was then passed through the filter. The resulting filtrate was analyzed by count technique as well as inoculated into sheet enterococcus broth, trypticase soy broth and staph agar 110 broth. In all cases no bacterial count nor any growth was observed.

Example 3: A fresh filter element was prepared in accordance with Example 1 and was used to filter 800 *Enterobacteria bacilli* bacteria (average size 0.5 micron in breadth and 2 microns in length) in 100 cc of water. The resulting filtrate was analyzed by bacteria count and also inoculated into EMB, BBL, and trypticase soy broths. No bacterial count nor any bacterial growth was observed in the broths.

Example 4: A fresh filter element was prepared in accordance with Example 1 and was used to filter 800 Enterococci bacteria (0.75 micron in average size) in 100 cc of water. The resulting filtrate was counted as well as inoculated into select enterococcus, trypticase, soy broth, and staph agar 110. In all cases, no bacterial count nor any growth was observed.

Example 5: A fresh filter element was prepared in accordance with Example 1 and mounted in a filter housing without the benefit of sterilization. 100 cc of water having a 0.1% suspension of powdered carbon black was passed through the filter specimen under a pressure gradient of 10 psig. The resulting filtrate was viewed under a microscope and no black particles were observed.

It has been found that increased flow rate can be obtained without increasing the size limit of the smallest particles capable of being filtered and therefore without decreasing the retentive capability of the filter. One preferred method for increasing flow rate without sacrificing retentivity is to prepare the microporous submicron filter media using a previously prepared quantity of reground microporous material in lieu of a portion of the combined amount of the binder and filler ingredients during fabrication of the microporous sheet material from which the filter media may be obtained.

Generally, the amount of reground microporous material which is to be added should replace from about 15 to 30% of the original total weight of the dry ingredients, that is, of the binder and the filler. The particle size of the reground microporous material may vary within convenient limits and is not especially critical. For example, a particle size of about 90% through a 70 mesh screen has been found to be satisfactory. The following example illustrates the foregoing.

Example 6: The same procedures employed in Example 1 were adhered to using the following ingredients in the following amounts given in pounds:

Resin (Geon 103 EP F 10)	20.0
Silica (Hi Sil 233)	28.0
Solvent (cyclohexanone)	50.5
Water	48.2

However, 12.0 lb of the microporous sheet material prepared in accordance with Example 1 and reground to a 70 mesh particle size were added to the mix in the Patterson-Kelly low shear blender, and before the damp powder mix was extruded, calendered into sheet form, extracted and dried. Several filter specimens 0.018 inch thick were then prepared and used respectively to filter the bacterial suspensions of Examples 1 through 5. In all cases no bacterial count nor any growth were observed in the resulting filtrates. The flow or throughput rate of the filter specimens made in part from regrinds of Example 1 was measured at 0.8 gpm/sq ft under a pressure gradient of 10 psig or 100% greater than that of the filter media of Example 1.

In Examples 1 and 6, the ratio of water to solvent was 0.955/1. It has been found that an increase in the amount of water relative to the amount of solvent will also increase the flow rate of the filter media of this process.

Bactericidal Membrane in Filter Unit

A process described by *A. Nishino, Y. Iura, M. Kubo and M. Tateisi; U.S. Patent 3,872,013; March 18, 1975; assigned to Matsushita Electric Industrial Company, Japan* relates particularly to a water purifier for drinking water having one or more bactericidal membranes which inhibit the propagation of bacteria in the purifier after removal of chlorine or its derivatives from the supplied city water.

A water purifier according to the process has one or more bactericidal membranes, comprising a porous substratum sheet made of water and chemicals-resistant material such as a synthetic fiber, and porous coatings of a sparingly water-soluble bactericidal substance selected from copper, silver, gold, and their chlorides, bromides and iodides cohered on the surfaces of the substratum sheet, placed between the filter and water outlet. The purifier may further comprise granules of an alkaline substance such as calcium carbonate to enhance the bactericidal efficiency of a heavy metal ion.

The water purifier features many desirable properties such as excellent and long-lasting bactericidal effect, high resistance and durability of the bactericidal membrane against water pressure changes, low restriction to water flow, and no need for enlargement of existing purifier or complicated operation.

Referring to Figure 6.1a, a water purifier according to the process is assembled in a case **10** which is a combination of a cylindrical body **11** and a cap member **12**. The body has a water inlet **13** at its bottom to be connected to a city water faucet, and the cap member has an outlet pipe **14** and an air relief opening **15** covered with a plug **16**. In the case **10** is installed a cartridge **17** in a manner to provide a space **18** between the cartridge and the interior surface of the case.

The cartridge has water inlet openings **19** at the bottom which may be sealed before use by seal members **20** formed of a water-soluble material, and an air relief opening **21** at the ceiling covered with a filter member **22**, allowing only air and water to pass through it. The cartridge contains finely powdered activated carbon **23** and a filter unit **24** to which the outlet pipe **14** is connected through an opening **25** at the ceiling of the cartridge. The filter unit **24**, which is preferably folded many times as shown in Figure 6.1b to have as large a surface area as possible in a limited space in the cartridge, comprises, as seen from Figure 6.1c, one or more porous supporting members **26**, for example a large mesh net or lattice, filter cloths **27** and at least one porous bactericidal membrane **30** in a sandwiched construction. A portion of the outlet pipe **14** inserted into the filter unit **24** has many tiny openings **28** to let in filtered water.

It is a feature of a water purifier that the bactericidal membranes **30** are placed in the filter unit so that the membranes are soaked in the dechlorinated and filtered water. A bactericidal membrane shown in Figure 6.1d comprises a porous substratum sheet **31** and coatings of a bactericidal substance **32** formed porously on the surfaces of the sheet.

It is known that traces of the ions of heavy metals such as mercury, silver, copper, gold and lead act as bactericides. Therefore, these metals can be used as bactericides for water either in the form of pure metal or in the form of sparingly soluble compounds, except harmful ones. Considering the solubility in water and the effect on tastes and odors of the treated water, examples of selected bactericidal substances for the membrane **30** are silver, copper, gold, and their halides, oxides and sulfides, the most preferable among them being silver chloride.

A substance selected from these examples may be introduced into a water purifier by mixing it with the activated carbon **23**, but the result is considered unpromising. The reason is that the mixed substance separates from the activated carbon and settles down at the bottom of the cartridge **17** due to its markedly higher specific gravity compared with that of activated carbon, and that chlorine ions absorbed in the activated carbon strongly hinder the dissolution of the mixed substance, when it is a preferable chloride, by the common-ion effect.

Therefore it is contemplated to use porous sheets of water and chemicals-resistant materials as the substratum for the bactericidal substance in place of activated carbon. The sheet **31** must be harmless and should have sufficient physical strength besides the mentioned chemical property. Some metals such as silver, gold and platinum may be used, but they are all too costly for ordinary use. Various synthetic resins and fibers, cotton, glass, asbestos and unglazed pottery are more inexpensive examples of the sheet material.

Considering porosity, formability, flexibility, and adaptability to the coatings **32**, the suitable form of the sheet **31** is cloth, nonwoven cloth or paper, and the most preferred is a nonwoven cloth of a synthetic fiber of nylon, rayon, polyester fibers, acrylonitrile fibers, acetate fibers, polypropylene fibers, partially acetal-formed polyvinyl alcohol fibers, or blends of these fibers. Ordinary cloths, canvas and Japanese paper may also be used.

When one of the above mentioned pure metals is employed as the bactericidal substance of the membrane **30**, the metal can work as the substratum **31** and as the coatings **32** simultaneously if it is formed with many openings like a net.

The width and length of the sheet **31** are scarcely limited because the finished membrane is cut into pieces of desired shape and size, but the thickness should be determined depending on the kind of the sheet and design of a water purifier. The preferable thickness is generally within a range of about 0.01 to 5 mm. Although a larger thickness may be employed, a larger space in the cartridge **17** is required, and the resistance of the membrane **30** to water flow is increased, while a sheet **31** thinner than 0.01 mm cannot have the necessary strength to withstand the water pressure changes and various shocks taking place during long usage. As for the coatings **32**, the preferred thickness ranges from 0.4 to 50 microns.

The coatings of a sparingly water-soluble substance such as silver chloride are formed with uniform thickness and adequate porosity by an ingenious and efficient method as described below.

The sheet is at first immersed in an about 10% aqueous solution of silver nitrate to coat and impregnate it with the solution, then immersed in an about 5% aqueous solution of sodium chloride to convert the coated and impregnated silver nitrate into silver chloride. After removal from the solution, the coated sheet is washed with water and air dried. Similar silver chloride coatings may alternatively be formed by immersing the sheet in an about 5% aqueous ammonia solution containing about 4% silver chloride as dissolved complex salts, and drying the coated and impregnated sheet to evaporate ammonia and water.

It is preferable to place the membranes **30** so as to face the entire interior surfaces of the filter cloth **27** from the viewpoint of bactericidal effect, but some

FIGURE 6.1: BACTERICIDAL WATER PURIFIER

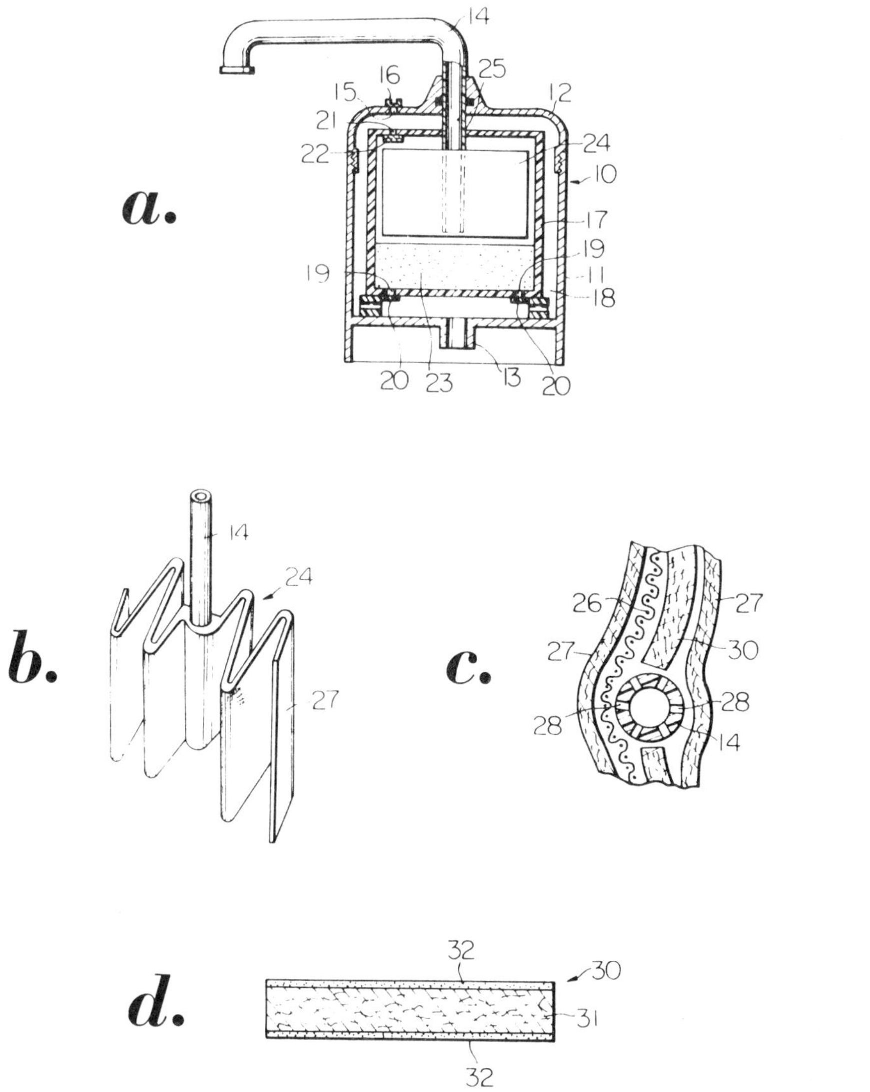

(a) Longitudinal sectional view of water purifier.

(b) Schematic perspective view of the filter cloth unit of the purifier shown in Figure 6.1a.

(c) Schematic cross-sectional view of the filter cloth unit of Figure 6.1b.

(d) Schematic sectional view of a bactericidal membrane.

Source: U.S. Patent 3,872,013

portions of the interior space of the filter unit **24** except for the portions near the outlet pipe **14** may be left vacant of the membrane.

When city water is supplied into the purifier through the inlet **12**, the water flows into the cartridge through the openings **19** dissolving the seal members **20**. By the turbulent and upward water flow, the activated carbon **23** is stirred and carried to the outer surface of the filter cloth **27**, forming a layer on the surface. The chloride ions, hypochlorous acid, and other impurity matters in the supplied water are absorbed in the thus formed layer of activated carbon, and the resulting purified water, which has been deprived of bactericidal power, comes into the inner space of the filter unit **24** through the filter cloth.

The purified water flows along and through the membranes **30**, dissolving a trace of the coated heavy metal compound **32**. As a result, the water recovers an effective bactericidal power by the action of the dissolved heavy metal ion, so that bacteria contamination of the purified water, the cartridge **17** and the outlet pipe **14** is effectively prevented.

A measurement of Ag^+ ion concentration in the purified water obtained from the water purifier of Figure 6.1a after 8 hours from the start of operation gave the value of 23 ppb. Then, specially prepared water containing 30,000 bacteria per ml was put into the purifier and left there for 6 hours. A determination after 6 hours revealed that the bacteria in the water had been completely destroyed and that the Ag^+ concentration of 23 ppb of the water purifier had the expected bactericidal effect.

The water purifier maintains its excellent bactericidal power substantially unchanged for months even when used under severe water pressure changes to the extent of a range from 0.3 to 20 kg/cm^2. The long-lasting effectiveness mainly depends on the strong cohesion of the coatings **32** to the sheet **31** and a large effective surface area of the coatings **32**.

Stable Triiodide-Containing Resin

J.L. Lambert and L.R. Fina; U.S. Patent 3,817,860; June 18, 1974; and U.S. Patent 3,923,665; December 2, 1975; both assigned to The Kansas State University Research Foundation describe a method of treating bacterially contaminated water for direct production of disinfected water ready for use, where the water is rendered bacterially sterile by passing it through a strongly basic anion exchange resin containing combined triiodide (I_3^-) without forming a detectable concentration of iodine in the disinfected water.

The triiodide is combined in a stable form with the basic groups of the resin. Although the dissociation of the iodine of the resin-insolubilized triiodide is extremely low, the triiodide resins are highly bactericidal, the triiodide groups reacting on demand with bacteria in water suspensions to kill the bacteria.

In the process, a strong base anion exchange resin is reacted with triiodide ions to form a stable compound of extremely low dissociation in water. The association of the triiodide ions with the basic exchange sites (cationic groups) of the resin is such as to block the sites from normal ion exchange with ions in the solution. The amount of iodine, triiodide, or other oxidizing form of iodine liberated in the water is negligible, being below the ordinary limits of detection.

Nevertheless, the resin compound is a potent bactericide. By some mechanism not yet fully understood, the insolubilized triiodide groups of the resin interact with bacteria without forming a detectable concentration of iodine (or any oxidizing form thereof) in the water. The killed bacteria do not remain attached to the resin, but will readily pass on through a bed of the resin. Complete bacterial sterilization of highly contaminated water can thus be achieved without introducing objectionable amounts of iodine into the water.

The amount of iodine liberated, if any, by the interaction of the bacteria with the insolubilized triiodide is below the limits of detection by the highly sensitive cadmium iodide-linear starch test. The disinfected water is thereby ready for immediate use as drinking water, or for other applications where bacterially sterile iodine-free water is desirable. No further treatment will ordinarily be required. The following examples illustrate the process.

Example 1: A typical batch of quaternary ammonium (strong base) ion exchange resin-triiodide ion bactericide, for use on a laboratory scale, was prepared in the following manner. Solid iodine was added to a 1 molar aqueous solution of potassium iodide (166 grams potassium iodide per liter of solution) so as to produce an iodide:iodine ratio of 3.5:1. Sufficient volume of this solution was used to provide a calculated excess over the capacity of the quantity of strong base resin used.

The resin beads (in the commercial chloride or sulfate form) were placed in a glass column and the potassium triiodide solution was passed through at a slow rate so as to saturate the resin's quaternary ammonium exchange sites with triiodide ion, after which the column was thoroughly washed with distilled water. Final washes of the column so prepared showed no detectable oxidant (iodine or triiodide ion), when tested with cadmium iodide-linear starch reagent, or iodide ion, when tested with silver nitrate solution. Resin triiodide combinations, prepared from Ionac A-540, Rexyn 201 or Amberlite IRA-400 strong base ion exchange resins in the chloride or sulfate forms, were effective in making complete kills of concentrations of 10^6 bacteria per milliliter and higher.

Bacteria tested included *Escherichia coli, Streptococcus faecalis,* and *Staphylococcus aureus.* Several other methods of preparation of the resin-polyiodide bactericide were also used in addition to that described in detail above, with comparable results. In one alternative, the reacted resin is washed with an aqueous solution of potassium iodide before washing with the distilled water, to insure conversion of any excess iodine or higher polyiodide ions to I_3^-, thereby insuring saturation of the resin ion exchange sites with triiodide ion.

An alternative procedure for the preparation of the disinfectant is to react the quaternary ammonium strong base resin in the chloride or sulfate form batchwise with the proper amount of the 3.5:1 iodide-iodine solution as described above, rather than in a column. After adequate time for reaction, the procedure is continued as above.

In still another alternative procedure, the quaternary ammonium strong base resin in the chloride or sulfate form is first reacted with a solution of potassium iodide or other iodide salt to convert it to the iodide form. After thorough washing to remove excess iodide salt solution, the disinfectant is prepared by either of the methods described above.

To achieve the stable resin-triiodide disinfectant with a reduction in the amount of washing required, slightly less than stoichiometric proportions of triiodide to basic sites can advantageously be used. For example, proportions giving 96 to 98% saturation (molar basis) of the basic sites with triiodide are desirable.

Example 2: Another mode of practicing the process is as follows. React a solution consisting of 16.6 parts by weight of potassium iodide and 7.26 parts by weight of iodine dissolved in 100 parts by weight of water through a weight of quaternary ammonium resin beads sufficient to yield a combination consisting of one quaternary ammonium site to one triiodide, I_3^-, ion.

Wash thoroughly with distilled water until the eluent tests negative for iodine, polyiodide or other oxidant with a cadmium iodide-linear starch reagent and negative for iodide ion with silver nitrate solution. In an alternative procedure, the reacted resin is washed with an aqueous solution of potassium iodide before washing with the distilled water to insure the conversion of any excess iodine or higher polyiodide ions to I_3^-, thereby insuring saturation of the resin exchange sites with triiodide ions.

Similar resin-polyiodide combinations, prepared from Ionac A-540, Rexyn 201 or Amberlite IRA-400 strong base ion exchange resins in the chloride or sulfate forms, were effective in making complete kills of concentrations of 10^6 bacteria per milliliter and higher.

Example 3: When prepared as described in Examples 1 and 2, the resin-triiodide column can be treated with distilled water or with solutions of up to 500 parts per million of sulfate, chloride or nitrate ions, without detectable concentrations of iodine, triiodide or other oxidant appearing in the eluent. The tests are made with a cadmium iodide-linear starch reagent which will detect concentrations of iodine or its equivalent of other oxidizing agents down to at least 0.2 part per million (200 ppb). Under some conditions, the sensitivity may extend to as low as 50 to 100 ppb oxidizing iodine.

The total concentration of iodine in all forms in the eluent (including the nonoxidizing I^-) when distilled water is passed through the column was less than 0.5 part per million (500 ppb) as determined by neutron activation analysis. Iodide ion release when distilled water is passed through the column produces no more than a faint turbidity of silver iodide when tested with silver nitrate solution. Release of iodide ion is no greater than that with distilled water when solutions of the following anions are passed through the column: 50 ppm of sulfate ion, or 100 ppm each of chloride or nitrate ions.

MEDICAL

Glassy Carbon Fibers for Blood Filter

J.C. Bokros; U.S. Patent 3,972,818; August 3, 1976; assigned to General Atomic Company describes a device for treating human blood prior to its return to a living human body. The blood filter employs a bed of fibers between about 1 and 100 microns in diameter, the outer surface of which is formed of impermeable carbon. Suitable fibrous substrates may be coated with vapor-deposited pyrolytic carbon, and the fibers may be supported between upper and lower screens which are likewise coated. It has been found that by utilizing a bed of fibers having an impermeable carbon surface a filter device can be constructed which does an excellent job of removing undesirable debris and microemboli from blood. However, because of the inherent compatibility of impermeable carbon with the components of human blood, a filter constructed in this manner causes no adverse effects in blood being so treated and reduces the rate of nucleation of new emboli downstream of the filter.

Shown in Figure 6.2a is an illustrative blood filter **11** which includes a housing made from an upper cap **13** and a lower body **15**. The illustrated housing is formed from glass, and the two parts are suitably joined together at their interface by adhesive or by suitable mechanical means so as to create an internal, closed chamber **17** of a generally cylindrical shape, the horizontal cross section of which chamber is circular. Instead of forming the housing parts from glass, a suitable metal such as stainless steel could be used, in which case the interior chamber-forming surface would be coated with impermeable vapor-deposited carbon.

An entrance conduit **19** is provided in the upper cap **13** of the housing, and an exit conduit **21** is provided in the lower body **15**. A vent conduit **23** is also provided in the cap of the housing to which a suitable check valve **25** can be attached which will permit the escape of air or other gases from the chamber **17** while precluding flow thereinto in the opposite direction.

The filtration effect of the blood filter **11** is performed by a bed of carbon fibers **27** which are supported within the chamber **17** between a lower screen **29** and an upper screen **31**. Instead of screens, other porous grid materials, e.g., lattices of fused glass frit, which are nonthrombogenic can be used. The screens are suitably held in place either adhesively, or by any other suitable means, for example, by grooving the internal wall of the chamber **17**. The lower screen **29** is spaced from the bottom of the chamber a sufficient distance to create a lower plenum **33** wherein the filtered blood can collect and flow smoothly to the exit conduit **21**. The upper screen **31** is preferably also spaced slightly below the top of the chamber, creating an upper plenum **35** so that the incoming blood can be fed to the entire upper horizontal surface of the fiber bed **27**. The screens are made of suitable materials which are compatible with blood, such as stainless steel wire that has been coated with vapor-deposited impermeable carbon.

The carbon fiber bed which effects the filtering is made up of fibers **27** having at least an outer impermeable carbon surface. Pyrolytic carbon is created by the thermal destruction of a carbon-containing compound, usually a hydrocarbon, in vapor form; however, carbon coatings can also be deposited by vapor-deposition without such destruction by using ion plating or the like, as where a carbon atmosphere is created under very low pressure conditions using, for example, electron beam heating.

For purposes of this process, such vapor deposited carbon is considered to be impermeable if it has density equal to at least about 70% of its theoretical maximum density (or at least about 1.55 grams per cm^3 for substantially pure carbon). Glassy or vitreous carbon is inherently impermeable and may be employed at slightly lower densities, and fibers made solely of such carbon are commercially available in the United States.

FIGURE 6.2: GLASSY CARBON FIBERS FOR BLOOD FILTER

a.

b.

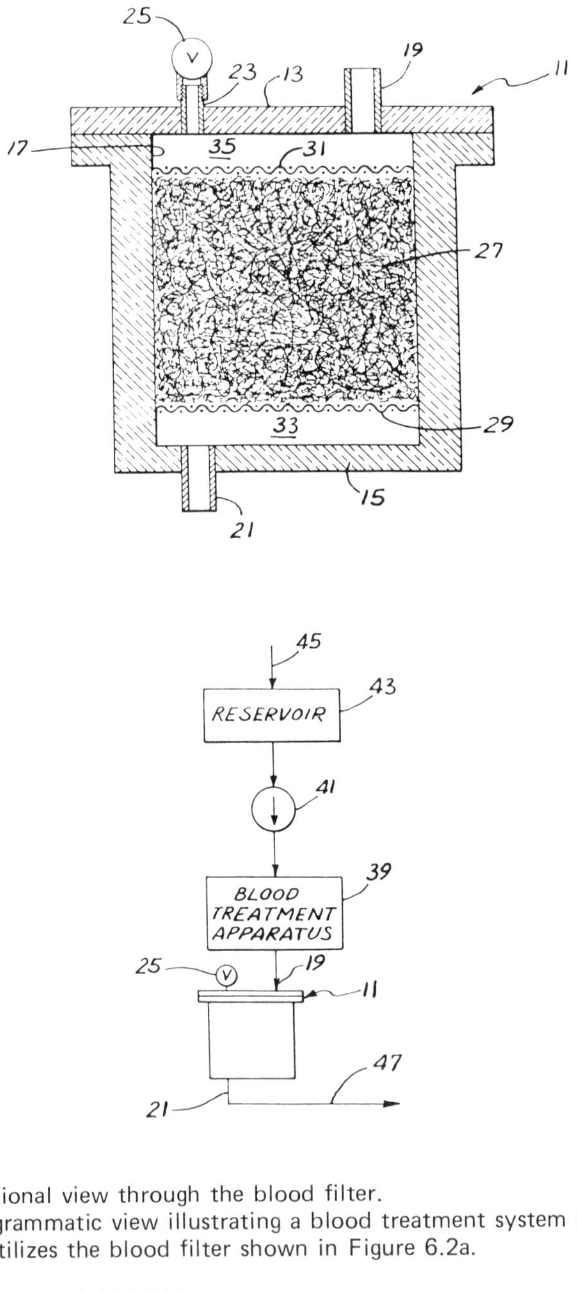

(a) Sectional view through the blood filter.
(b) Diagrammatic view illustrating a blood treatment system which utilizes the blood filter shown in Figure 6.2a.

Source: U.S. Patent 3,972,818

The size of the carbon fibers **27** may vary between about 1 micron and about 100 microns in diameter; however, preferably, fibers between 5 and 50 microns in diameter are employed. Fibers in this size range provide a tortuous path through which the blood must flow to traverse the filter from the inlet **19** to the outlet **21**, and undesirable microemboli are removed as the blood flows along this tortuous path. The length of the fibers will depend somewhat upon the processing method. The fibers should preferably be at least 0.1 inch (0.25 cm) in length.

Figure 6.2b depicts one illustrative system where the blood filter **11** might be employed. Diagrammatically illustrated is a blood treatment apparatus **39**, which might be a heart-lung machine. The heart-lung machine **39** would be fed with blood by suitable blood pump **41** which draws its suction from a reservoir **43**. The reservoir has an inlet line **45** which may be connected to the human body from which the blood is being drawn; however, the line might also be branched to provide for a subsidiary supply of blood from a blood bank, should it be needed. The heart-lung machine would discharge the blood to the entrance **19** to the blood filter **11**, where the blood would percolate downward through the carbon fiber bed **27**. The filtered blood leaves through the outlet **21** and is returned to the human body through the line **47**, driven by the pressure head supplied by the pump **41**.

The illustrated blood filter is made from a glass body **15** where a chamber **17** is formed, the internal diameter of which measures about 3 inches. The chamber includes a bed of carbon-coated fibers **27** having a depth of about 2 inches. The fibers are made of 10-micron diameter silicon carbide substrates which have been coated with an outer coating of vapor-deposited pyrolytic carbon, the thickness of the coating averaging about 25 microns. The fine fibrous substrates are coated using a mixture of methane and argon at atmospheric pressure (about 10 volume percent methane) and at a temperature of about 1100°C to produce laminar pyrolytic carbon having a BAF of 5 and a density of about 2 g/cm^3. The coatings are smooth in the as-deposited condition, so no further treatment is necessary.

The coated fibers range in size from 50 to 75 microns in diameter, and about 0.25 to 1.27 cm in length, and they are supported upon a lower screen of stainless steel wire coated with vapor-deposited carbon having screen openings measuring about 50 microns. The upper screen **31** is made of similar material and is formed to have openings of about 75 microns. Accordingly, the upper screen serves to remove relatively large emboli aggregates from the incoming bloodstream, for example, those over 200 microns in size, leaving the carbon fiber bed to remove the remaining microemboli. The blood filter **11** is, of course, designed to be a disposable item which would not be reused after the operation with a particular patient is completed.

Disposable Blood Filter Cascade

D.J. Rosenberg; U.S. Patent 3,765,536; October 16, 1973; assigned to Pall Corp. describes a disposable filter element which is capable of passing human blood at a high flow rate and useful in human blood transfusion systems, to remove undesired suspended particles, substantially without removal of normal and desirable blood components. The filter comprises in combination three filter elements in a cascade, the first comprising plastic netting and having a pore size within the range from about 800 to 4,000 microns, the second a woven square weave plastic mesh having a pore size within the range of 20 to 50 microns, and the third a fibrous unwoven mat having a pore size within the range of 10 to 30 microns.

The filter element of Figures 6.3a through 6.3c is composed of a housing **1** having first and second housing portions **2** and **3** defining a fluid chamber **4**. A fluid port **5** is at the base of housing channel portion **2**, and a fluid port **6** is at the base of housing portion **3**. It will be seen that these ports are coaxial. The port **5** serves as a fluid inlet, and the port **6** as a fluid outlet, and flow is only in the direction from port **5** to port **6**, because of the arrangement of the three filter sheets in the cascade.

The inlet port is housed in a spiked tube **7** which is designed to pierce a blood bag stopper for use in blood transfusion. Housing portion **2** has an internal projection or rib **8**, extending from one side **9** to the other side **10**, and housing portion **3** has a like internal projection **11** extending from one side **12** to the other side **13**. These serve as supports extending across the corrugations of the three-component filter sheet cascade and spacer composite **40**, serving as the filter.

Each housing portion **2** and **3** is generally channel-shaped, with opposed sides **9** and **10** extending outwardly from the base of portion **2**, the opposed sides **12** and **13** extending outwardly from the base of portion **3**. The bases of the housing portions are each provided with fluid flow channels **14**. Portion **2** has on each side a pair of locating fingers **15**, and portion **3** has on each side two pairs of locating fingers **16**, with a slot **17** therebetween into which fingers **15** fit. These locate the portions **2**, **3** on assembly. Sides **12**, **13** at their ends abut sides **9**, **10**, and are fused thereto so that the housing portions are bonded together as one piece. A pair of projecting members **20**, **21** on portion **3** extend parallel to and internally of sides **12**, **13** all the way to the internal wall of housing portion **2**.

The three-component filter sheet cascade and spacer composite **40** at each edge **22**, **23** projects into the sockets **24**, **25** defined by sides **12**, **13** and members **20**, **21**, and where the five sheet composite curves around the tips of the members it is held tightly at **26**, **27** against the internal wall of portion **2** and is bonded thereto. The bond is produced by fused integration of the members **20**, **21** to the housing portion **2** through the open pores of the three filter sheets **41**, **42**, **43** and the two spacers **44**, **45** forming a fluid-tight seal at **26**, **27** all along those sides of the composite, and bonding all five sheets together at those points. Such a bond can be obtained by ultrasonic welding, solvent softening, or heat fusion.

The filter sheet cascade and spacer composite **40**, best seen in Figure 6.3c, is composed of a first coarse filter sheet **41** of extruded polypropylene netting, pore size about 1,500 microns. The two spacers **44**, **45** are also made of extruded polypropylene netting of the same pore size. The second filter sheet **42** is made of open mesh square weave polyester monofilament fabric having a pore opening of 40 microns, a monofilament diameter of 40 microns, and 27% open area. The weft monofilaments and the warp monofilaments are backed together by heat setting at their points of crossing defining fixed 40-micron pores at their interstices. The third filter sheet **43** is a microporous

FIGURE 6.3: DISPOSABLE BLOOD FILTER CASCADE

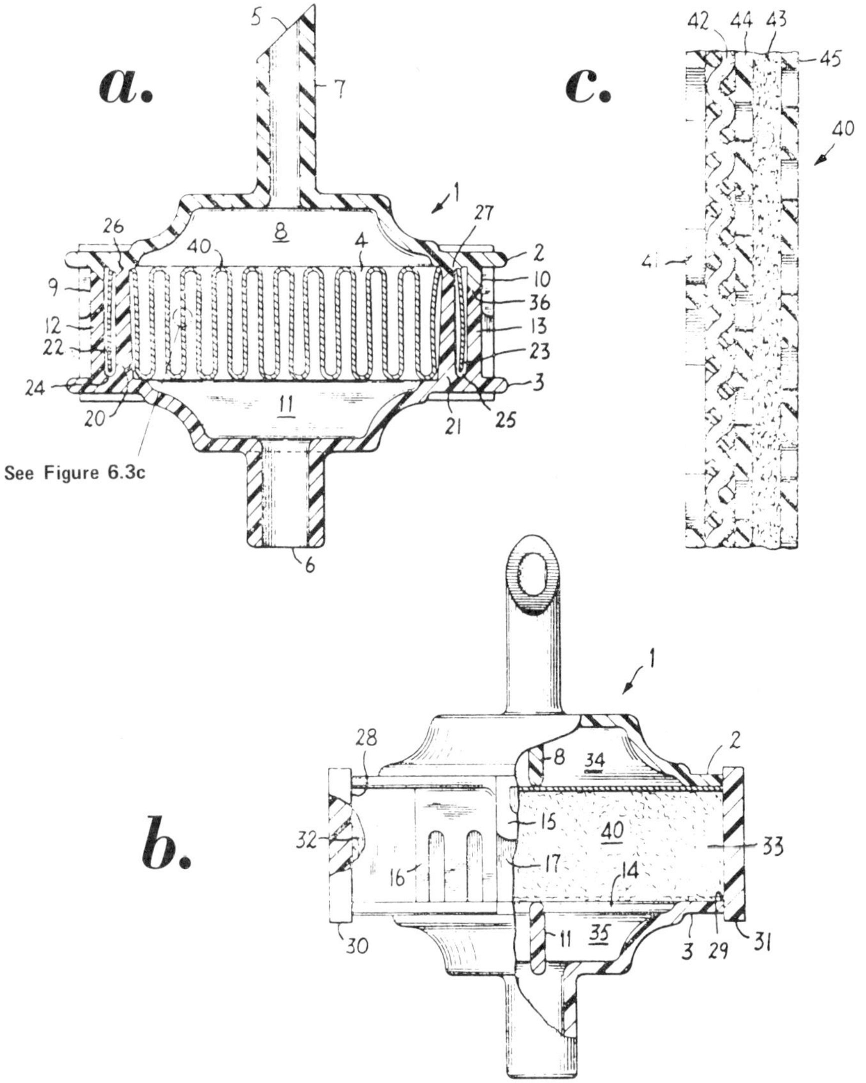

(a) Side view, in section, of a filter element in box form including a three-filter cascade.
(b) Side view, partly in section, of the filter element of Figure 6.3a, showing the side caps on the filter housings.
(c) Detail of a portion in section shown in Figure 6.3a of the filter sheet cascade.

Source: U.S. Patent 3,765,536

filter material composed of paper with a microporous layer of resin-bonded ceramic fibers thereon, prepared following the procedure of U.S. Patent 3,246,767, and having a pore size of 15 microns.

The three filter sheets of the filter cascade are in corrugated form, for an increased surface area in the limited space of fluid chamber **4**, and the tips of the corrugations abut and are held in place by the projecting sections **8** and **11** of the housing portions **2** and **3**. The edges **32, 33** of the filter sheet cascade run right to the ends of the sides **12** and **13**.

The housing channel portions are open at their sides and, as best seen in Figure 6.3b, define openings **28, 29** leading into the fluid chamber **4** of the housing **1**. The openings are closed off by side caps **30, 31** which are bonded to the housing portions **2** and **3** and also to the edges of the filter sheet cascade and spacer composite **40**, extending along the openings from end to end between the mating sections **9, 10, 12** and **13** of the housing portions. This closes off the other two side edges of the filter sheet cascade to fluid flow, and restricts flow between the two portions **34, 35** of fluid chamber **4** in the housing via the pores in each of the filter sheets in the filter sheet cascade **40**. Thus, all flow between fluid ports **5** and **6** of housing **1** must pass through each of the three filter sheets in the direction **41, 42, 43**, as is evident from Figure 6.3c.

The assembly of the box filter element of Figure 6.3a through Figure 6.3c is as follows. It will be seen on reference to Figure 6.3a that the side sections **9, 10, 12** and **13** of each housing portion have a special construction which ensures a fluid-tight seal between the housing portions when they are bonded together. The opposed sides **9, 10** of housing portion **2** meet and abut like sides **12, 13** of the other housing portion **3**. Portion **2** has one pair on each side of side-locating fingers **15**, and portion **3** has two pairs on each side of side-locating fingers **16** with receiving fingers **15** therebetween to ensure that the portions fit snugly together in the correct position to hold the filter sheet cascade 40 in place.

The sides **9, 10, 12** and **13** are each slightly longer than their combined length after they are bonded together. When portions **2** and **3** are fitted together, the sides **9, 10** are readily fused to sides **12, 13** respectively, to produce an integral one-piece structure at the seal **36** (Figure 6.3a and Figure 6.3b). Internally of the sides **12, 13** in housing portion **3** are the projecting members **20, 21** which extend all the way to the interior wall at **26, 27** of the housing portion **2**.

In assembly, the edges **22, 23** of the corrugated filter sheet cascade and spacer composite **40** are folded around the projecting members **20, 21** of housing portion **3** into the sockets **24, 25** between sides **12, 13** and the projecting members where they are held securely. Housing portion **2** is then fitted over the portion **3** and pressed down smartly against the filter sheet cascade and combined spacers, pinching the five sheets at **26, 27** against the tips of the projecting members and holding the five sheets firmly in place by the tight engagement between the inner wall of the housing portion **2** and the ends of the members **20, 21**.

The projecting members are then integrated through the pores of the spacers **44, 45** and the three filter elements **41, 42, 43** at **26, 27** to the wall of the housing portion **2**, forming a fluid-tight seal therebetween, and closing off both sides of the filter sheet cascade to fluid flow. The sides **9, 10** of housing portion **2** can also be bonded to the sides **12, 13** of the housing portion **3** by fusion, such as by ultrasonic welding, at the same time or thereafter, so that the two housing portions are sealed together, preventing fluid leakage to the outside of the filter assembly.

Next, the side caps **30, 31** are bonded across the openings **28, 29** into the housing portions and to the filter sheet cascade and spacer composite edges **32, 33**, bonding the filter sheet sides and spacers to the side caps, and completing the fixing of the filter sheet cascade and spacer **40** in place in the fluid chamber **4**, as well as the seals between the filter sheet cascade and the four side walls of the housing. This can be done using, for example, an adhesive, or a melt of adhesive or resin or other potting composition, or by fusing the end caps. The filter element is now complete, and ready for use.

The filter element is operated in line as follows. The spiked tip is plunged into the stopper of a blood bag, thereby permitting blood to flow from the bag through the inlet port **5** and enter the channel **14** flowing into chamber portion **34**. The blood then flows through the filter sheet cascade **40** via filter sheet **41**, filter sheet **42**, spacer **44**, filter sheet **43**, and spacer **45**, and enters the chamber portion **35**, where it emerges via channel **14** from the housing via port **6**. Line connections can be made at ports **5, 6** in any desired manner.

Stainless Steel Filter for Medical Injection Equipment

T.E. Weichselbaum and W.W. Gusky; U.S. Patent 3,933,652; January 20, 1976; assigned to Sherwood Medical Industries Inc. describe an improved filter for medical infusion and injection equipment which is manufactured by compacting stainless steel powder particles of an irregular or dendritic shape, in the substantial absence of a binder, to form a porous, self-supporting green article, and thereafter sintering the green article. The resulting filter is porous with the pores communicating with one another to form a network of tortuous or winding paths having lateral, oblique and vertical segments which function to trap contaminating particles of a solution passed therethrough. The porous filter is sealed in a tubular fitting by force-fitting the filter into the internal peripheral wall of an annular plastic bead in the fitting and heating the filter to cause the plastic bead to soften or melt and flow into the pores of the filter along a continuous peripheral ring.

Figure 6.4a illustrates a hypodermic needle **10** into which the filter device may be incorporated. The needle includes a needle cannula **12** (preferably stainless steel) connected to the forward or distal end of a thermoplastic tubular hub or fitting **16** having a porous, sintered, powdered stainless steel filter **18** therein. The filter element is manufactured according to the process as described below. It is manufactured from stainless steel particles having an irregular or dendritic shape.

The filter **18** is preferably cylindrical cup shape in form, having a closed rear end **36** and an open forward end **38** defined by a forward peripheral wall **39**. The cylindrical cup shape of the filter element maximizes the effective surface area of the filter which is exposed to the solution to be passed therethrough, consistent with the overall size limitations imposed by the fitting into which the filter is inserted, to maximize the permissible flow rate of the solution. For example,

FIGURE 6.4: STAINLESS STEEL FILTER FOR MEDICAL INJECTION EQUIPMENT

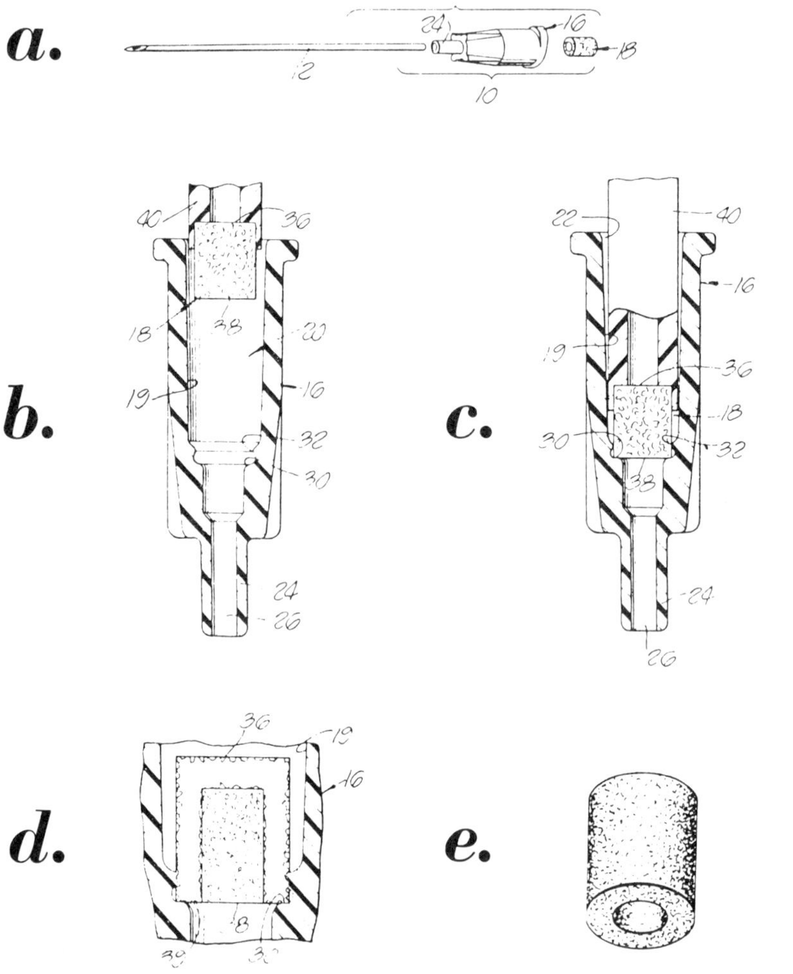

(a) Exploded perspective view of a hypodermic needle into which a filter may be incorporated.

(b) Sectional elevation view of the hub or tubular fitting of the apparatus shown in Figure 6.4a with the filter being inserted therein.

(c) Sectional elevation view similar to Figure 6.4b showing the filter in place in the fitting.

(d) Partial sectional elevation view showing in greater detail the seal between the filter and the fitting.

(e) Enlarged perspective view of the filter.

Source: U.S. Patent 3,933,652

in the process filters having a surface area of about 0.06 in^2 and a wall thickness of about 0.02 to 0.06 inch (preferably about 0.03 inch) are preferable. The filter is manufactured according to the following process.

(1) Metal powder having particles of an irregular or dendritic shape are screened to obtain particles of the desired size.
(2) A predetermined amount of the screened powder is blended with a predetermined relatively small quantity of a lubricant.
(3) The powder-lubricant blend is loaded into a suitable mold and compacted under a predetermined compaction pressure to obtain a self-supporting green filter.
(4) The green filter is removed from the mold and subjected to a two-step heating process to first burn off the lubricant and thereafter sinter the filter.

The specific details associated with the foregoing basic steps, including times, temperatures, compaction pressure, materials, etc, will now be described.

Two grades of filters are described. One grade is termed S3-5, and the second grade is termed S8-10. The filter grade identification indicates the efficiency of the filter. The S3-5 filter is designed to remove from the solution to be filtered 99% of all particles larger than 5 μm and 95% of all particles larger than 3 μm. The S8-10 filter is designed to remove 99% of all particles larger than 10 μm and 95% of all particles larger than 8 μm. These filter efficiencies have been achieved while permitting a relatively high rate of flow of solution through the filters.

The stainless steel powder from which the S3-5 filter has been manufactured has mean diameter particle sizes of between 44 and 74 μm, also identified as having sieve cuts of 200/325 mesh. The S8-10 filter has been manufactured from stainless steel powder whose mean diameter particle sizes are between 125 and 149 μm, having sieve cuts of 100/120 mesh. In both cases the shapes of the particles are sufficiently irregular so that the "branches" or "arms" of adjacent particles lock together when the powder is compacted to form a self-supporting article.

The screening step employed may be the two-step screening process where a larger mesh screen is first used to eliminate powder particles larger than the largest particle size desired (e.g., 74 μm for the S3-5 filter, 149 μm for the S8-10 filter) and a smaller mesh screen is then used to eliminate powder particles smaller than the smallest particle size desired (e.g., 44 μm for the S3-5 filter, 125 μm for the S8-10 filter).

Following the screening step, the stainless steel particles are mixed with a suitable lubricant and loaded into a suitable mold (e.g., a mold have a cup shape) and compacted into the desired shape. Any number of lubricants may be employed, such as a synthetic wax of general fatty acid origin.

The method by which a filter **18** manufactured according to the process may be sealed in a tubular thermoplastic connector **16** and the resulting filter assembly or device may best be understood by referring to Figure 6.4a through Figure 6.4c. The thermoplastic connector is of generally tubular shape and has internal peripheral wall **19** defining a longitudinal passage **20**. The outer or proximal end **22** of the passage is tapered outwardly (a female luer taper) to facilitate sealing reception of the tip of a syringe (or other tubular member) to which the proximal end of the fitting **16** is to be connected. The forward or distal end of the passage through the stem **24** may be cylindrical or tapered to receive the butt end of the needle cannula **12**, which may be epoxy bonded therein.

The intermediate portion of the peripheral wall **19** of a passage **20** is provided with a shoulder **30** which functions as a seat for the end of the filter **18**. An inwardly-projecting annular ring or bead **32** is provided on the interior wall **19** of the fitting **16** just above the shoulder or seat **30**. The internal diameter of the bead **32** is slightly less than the outer diameter of the filter **18** so that the end of the filter can be force-fitted into the bead in a manner and for a purpose decribed more fully below. The thermoplastic fitting or connector **16** is preferably constructed of a relatively rigid thermoplastic such as polypropylene, polycarbonate or polyethylene terephthalate.

The filter element is positioned and sealed or bonded in the tubular fitting **16** in the following manner. The filter element is inserted, open end first, into the passage **20** of the fitting **16** by means of a suitable tool such as the insertion tip **40** of a suitable vacuum tool, as shown in Figure 6.4b. The filter element is held on the tool by suction, and is pressed downwardly through the plastic annular ring or bead **32** until the forward peripheral wall **39** of the open forward end **38** rests or seats on the annular shoulder **30**. As noted above, the outer diameter of the filter element is slightly larger than the internal diameter of the annular bead so that the forward end of the filter is force-fitted in the annular bead to compress and deform the bead (see Figure 6.4c).

With the filter element in place (i.e., with the forward end wall **39** of the filter abutting the shoulder **30** in the fitting **16** and the forward end portion of the peripheral wall of the filter force-fitted in the annular bead, the filter element is heated to a temperature and for such a time to heat the thermoplastic bead, which melts or softens and flows into the pores in a continuous ring around the forward end portion of the outer peripheral wall of the filter element. The filter element may be heated by an electromagnetic field generated by induction heating coils. However, other means of heating the filter element may be employed as long as such heating means does not interfere with the compressive force exerted by the filter element on the annular bead.

Figure 6.4d is an enlarged sectional view illustrating the intrusion of the plastic bead into the pores around the outer peripheral wall of the forward end portion of the filter element after the bead has cooled and resolidified. It is to be noted that the seal or bond formed between the bead and the forward end portion of the outer surface of the filter element is a relatively thin continuous ring which does not significantly reduce the effective surface area of the filter which is exposed to the solution to be filtered. Thus, the seal will not inhibit the flow rate of the solutions which pass through the filter which is a significant advantage because it is essential to maintain the effective surface area of the filter element at a maximum to insure that the flow rate of the solution passed through the filter may be maintained at the desired level.

It will also be noted that the cylindrical cup shape of the filter element maximizes the effective surface area of the filter which is exposed to the solution to be passed through, consistent with the overall size limitations imposed by the fitting into which it is inserted, to maximize the permissible flow rate of the solution. The foregoing description of the overall configuration of the filter and method of bonding it within a tubular fitting or hub is more specifically described in U.S. Patent 3,817,389.

Gel Impregnated Media for Virus Removal

A process described by *C. Wallis and J.L. Melnick; U.S. Patent 3,816,304; June 11, 1974; assigned to The Carborundum Company* relates to a filter element which is suitable for concentrating or removing viruses from fluid containing viruses. The filter element comprises an annular filter media with a center passage and a contiguous layer of filter media impregnated with a gel. Viruses are removed from a fluid containing viruses by passing the fluid through a layer of gel which is formed by contacting the filter media with an aqueous solution containing colloidal particles of a polyvalent material.

By using the filter cartridge of the process, viruses can be concentrated from large quantities of water since the viruses are adsorbed into the gel layer which is impregnated in the filter cartridge while other microorganisms and particulate can pass through the cartridge.

Referring to Figure 6.5, the filter element **11** is annularly shaped with a center passage **13**. The filter media **15** can comprise a variety of materials constructed in a variety of ways to form an annular filter element. The filter media may be formed from woven or nonwoven material or a roving. In the former case the material is preferably spirally wound around a perforable support tube in successive layers. When a roving is employed, the fiber is preferably wound in helical convolutions spaced along the core in one direction and then in the opposite direction in a crisscross manner. This provides a honeycomb filter with diamond shaped openings between successive layers and results in passages extending from the outside of the unit to the inside.

The filter media is impregnated with gel which forms a layer **17** substantially concentric with the center passage **13**. Preferably the impregnated layer of gel is located upstream of an unimpregnated portion **19** of filter media **15**. With this type of construction the chances of losing gel to the fluid being filtered are reduced since the unimpregnated portion tends to adsorb sluffed-off gel. In another form of the process, as illustrated in Figure 6.5c, the layer of gel is located between radial portions of unimpregnated filter media. In this latter type of construction, the gel is firmly held in place even during backwashing.

The diameters of the pores in the filter media **15** are preferably from about 0.5 to 20.0 microns. The porosity of the filter element is sufficient so that particulate and other coagulants in the fluid pass through the filter element while sufficient surface area of gel impregnated fiber is present to adsorb viruses from the fluid.

The gel can be formed from any polyvalent metal salt capable of forming a colloidal suspension which attaches to the filter media. Examples of typically useful salts are those of tri- and tetravalent metals such as iron, aluminum, zirconium, tin, and thorium. In preparing the colloidal gel from polyvalent metal salts, methods of forming colloidal aggregates which are well-known may be used. These methods include mixing a metal halide with sodium carbonate, heating an aqueous solution of the hydrous metal oxide or adjusting the pH of a solution of metal ions to turbidity.

FIGURE 6.5: GEL IMPREGNATED FILTER MEDIA

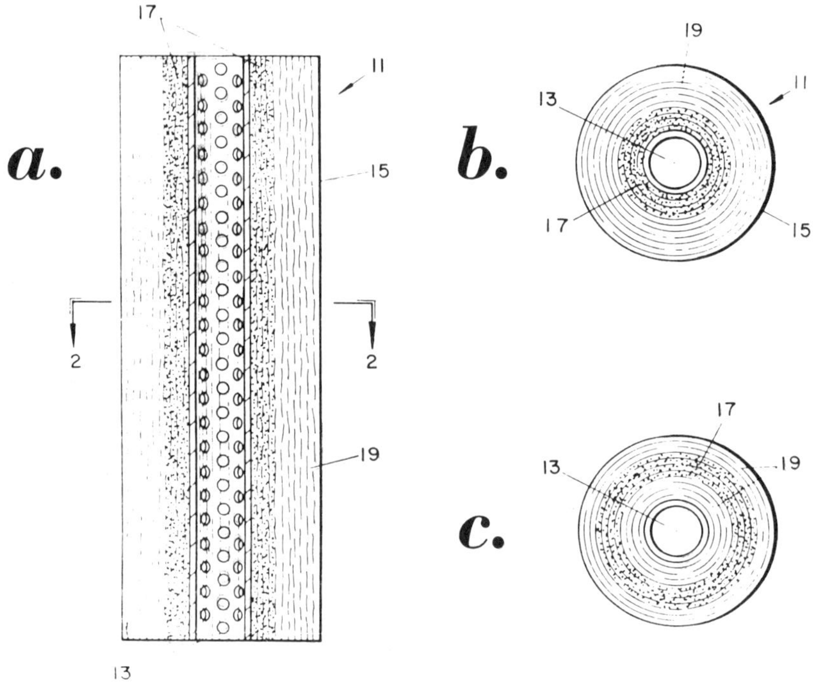

(a) Side elevational view in section of a filter cartridge.
(b) End elevational view along section **2–2** of Figure 6.5a.
(c) End elevational view in section with the gel in the middle portion of the filter element.

Source: U.S. Patent 3,816,304

The concentration of the colloidal suspension varies somewhat with the metal used but ranges upward from about 10 ppm concentration of metal ion per aqueous solution. The gel may be impregnated into the filter element **11** by a variety of methods. The gel can be impregnated by contacting a preformed filter element with a solution of the gel and applying the solution by brushing, injecting under a slight pressure, or soaking the cartridge so that the gel permeates through the filter media **15** forming a layer.

In another method, the gel is impregnated into a predetermined length of roving used in forming a honeycomb cartridge or a predetermined length of cloth used in forming spirally wound filter cartridge. Since only a portion of the entire roving or cloth is impregnated, a layer is formed as the roving or cloth is made into a filter cartridge. By varying the amount of roving or material impregnated and the location of the impregnated section in relation to the overall length of roving or material, the thickness and location of the gel within the filter cartridge can be varied. For instance, if an intermediate portion of the roving is immersed in a bath of gel prior to winding the final wound filter element has an intermediate layer of impregnated gel with outer portions of unimpregnated filter media.

When an aqueous solution containing viruses is passed through the filter element and the gel impregnated layer of filter media, the viruses are adsorbed into the gel. The viruses are not removed from the gel by surging water, but can be removed when desired by the addition of a suitable eluant such as polyethylene glycol at a pH of 12. The viruses in the eluant can then be counted. Thus, the filter element of the process is ideally suited for removing viruses from a fluid and concentrating the viruses removed so that a determination of how many viruses were present in the original fluid can be made. The following examples illustrate the process.

Example 1: An aqueous solution of ferric chloride, at a concentration of about 0.002 mol per liter, is mixed with an equal volume of an aqueous solution of sodium carbonate at a concentration of about 0.01 mol per liter. The colloidal suspension which forms is soaked into the center portion of a glass fiber filter element (K-27, The Carborundum Company) for about 10 to 30 minutes to form a gel impregnated layer about 0.25 inch thick.

Example 2: About 3.8 liters of water containing about 10,000 infectious particles of polio virus is passed through the gel impregnated filter of Example 1. The filter of Example 1 is supported in a Fulflo holder of The Carborundum Company so that water containing viruses first pass through the impregnated gel layer. No viruses were detected in the effluent from the filter cartridge. A polyelectrolyte method is used for concentrating and counting the viruses (see "Concentration of Virus from Sewage and Excretion on Insoluble Polyelectrolytes," *Applied Microbiology* by Wallis et al, December 1969, page 1007-1014).

GAS FILTRATION

AIR FILTERS—DUST COLLECTORS

Dust Filter Unit

A.J. Wiser; U.S. Patent 3,979,194; September 7, 1976 describes an improved unit for efficiently removing the large dust particles and chaff still present in the exhaust of existing apparatus. The dust filter unit may be readily attached to conventional air cleaners so as to enable them to meet any state or federal dust pollution standards.

Referring to Figure 7.1, numeral **10** designates the dust filter unit as a whole which comprises a cylindrical housing **12** open at its bottom, a pair of elongated filter cleaning brushes **14**, a cylindrical filter unit **16**, and a dust collecting trough **18** having a dust discharge pipe **20**.

While dust laden air or gases may be passed axially upward through the cylindrical housing of the unit **10** which may be used by itself to effect an efficient cleaning thereof, the only inlet of the unit is shown as mounted by bolts **13** on the exit or discharge cylinder **21** of an existing dust collector **22**. This is of the cyclone type and has an inlet **24** and the dust discharge opening **21** from which any externally protruding air outlet rain hood has been removed.

The filter housing **12** has diametrically opposed openings **26** for the complete exhausting of the filtered air and these openings are of substantial width, extend downwardly more than half the height of the housing, and are protected against rain, etc., by removable hoods **28** (Figure 7.1b). Projecting centrally above the top of the cylinder **12** and rotatably mounted therein by a sealed bearing **29** is a depending steel shaft **30** which is rotated by means of a gear or pulley **32** by a chain or belt **34** which is driven by a suitable electric motor.

The filter unit **16** is fixed to the inner side of the housing **12** by means of clamps **36** and the brushes **14** which rotate against the inner face of the filter unit **16** are mounted on pairs of vertically spaced, diametrically opposed radially extending arms **38** fixed to the shaft **30**.

FIGURE 7.1: DUST COLLECTOR

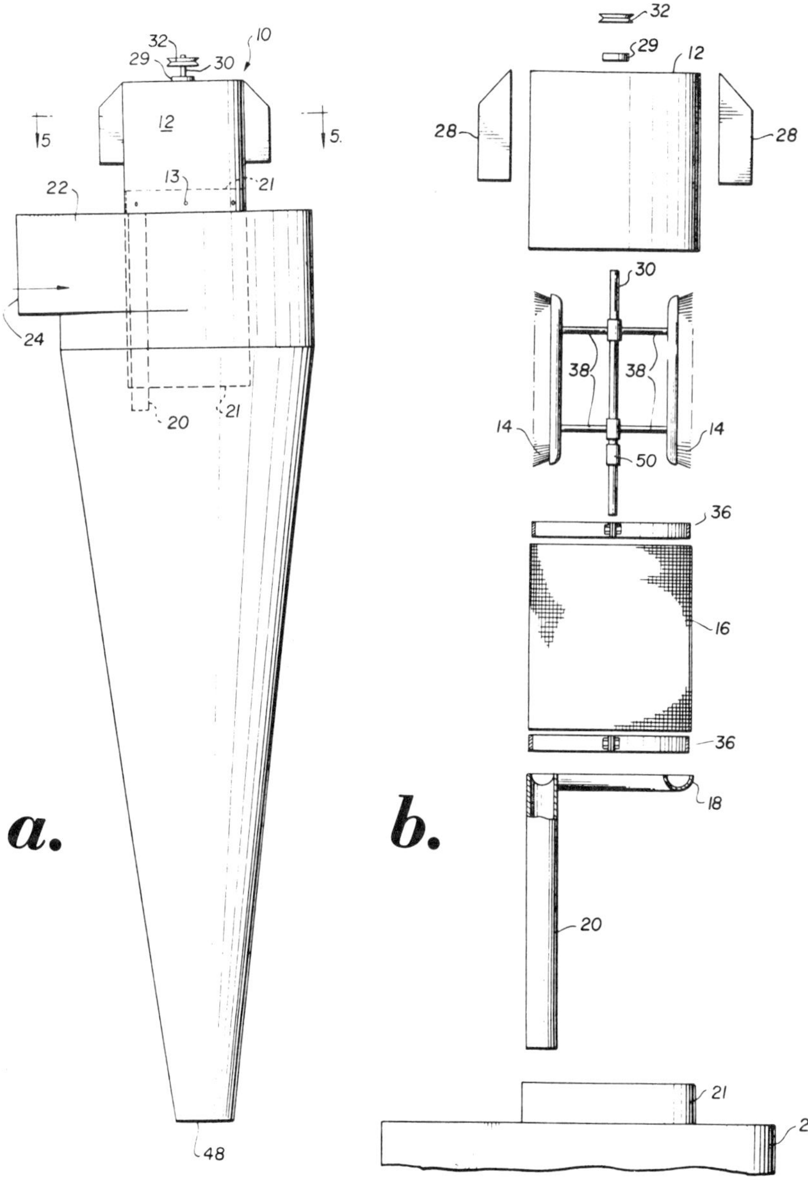

(a) Side Elevational View of Dust Filter Unit Mounted on the Exhaust Cylinder of a Cyclone Type Dust Collector
(b) Exploded View of Filter

(continued)

FIGURE 7.1: (continued)

c.

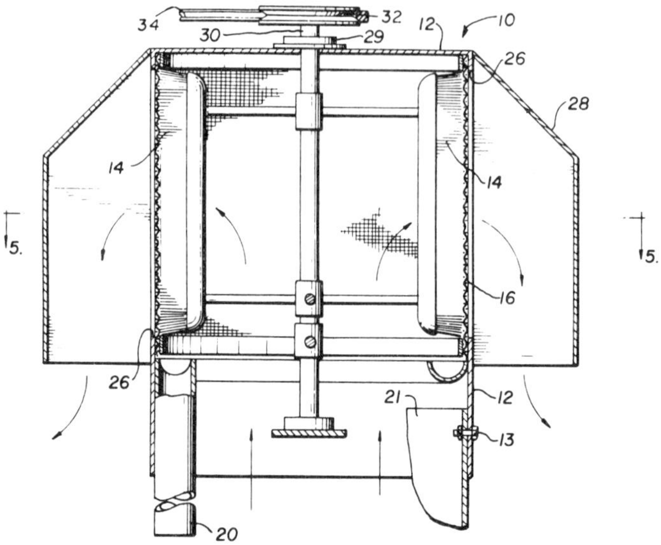

d.

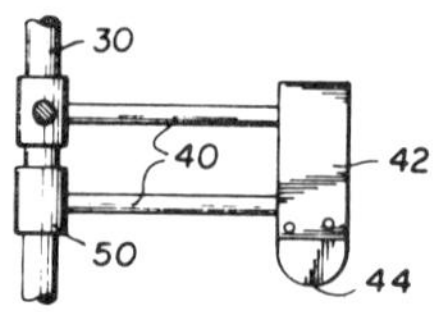

e.

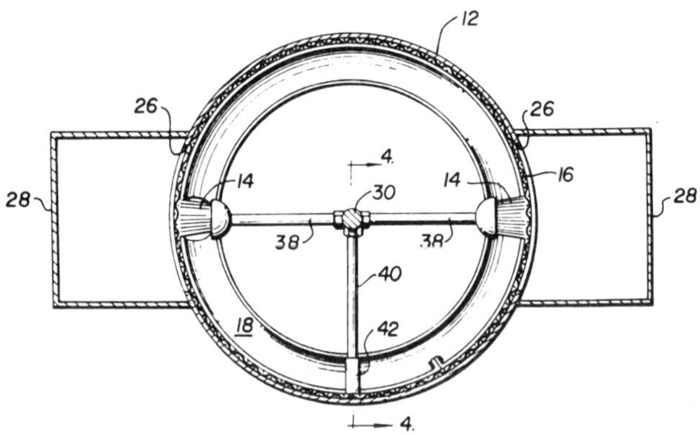

(c) Central Vertical Sectional View to an Enlarged Scale
(d) Fragmentary Elevational View of the Scraper and its Mounting
(e) Horizontal Sectional View Showing the Brushes, the Scraper and the Scraper Trough

Source: U.S. Patent 3,979,194

At right angles to the arms (Figures 7.1d and 7.1e), are a pair of vertically spaced radially extending arms **40** which support a scraper bar **42** having a rounded lower end **44** which conforms with and moves in the trough **18**. The scraper bar **42** collects dust particles which are brushed from the filter element **16** by the brushes **14** and fall into the semicircular trough **18** fixed to the lower end of the housing **12**. The dust particles then fall into the discharge pipe **20** to a point below the end of cyclone exhaust cylinder **21** where they have a better chance to centrifuge out.

It is apparent that when the dust filter unit **10** is mounted on an air exhaust cylinder or, as shown, on the exhaust cylinder **21** of any shape of a cyclone type dust collector, that the dust-laden air enters the cyclone by inlet **24** as shown by the arrows, and the dust follows the normal centrifugal pattern with the heavier particles moving to the outside and falling down through the dust exit **48**. The lighter particles move upwardly through the cyclone air exhaust cylinder **21** and pass through the filter unit **16** and out to the atmosphere through the openings **26** and hoods **28**, as indicated by the arrows (Figure 7.1c).

The dust particles which are trapped on the inner side of the filter element **16** are brushed off by the two brushes **14** which are rotated therewithin by the motor and fall into the semicircular trough **18** along which they are moved by the scraper bar **42** to the discharge pipe **20**.

The vertical position of the brushes or the scraper is readily adjusted on the supporting motor driven shaft **40** by means of a shaft collar **50** having positioning lock screws. It will be noted that the external configurations of the cyclone exhaust cylinder **21** and the bottom of the housing **12**, or of the trough **18** and the scraper, are immaterial as long as they are compatible.

Anisometric Compressed and Bonded Knitted Wire Mesh

D.B. Pall; U.S. Patents 3,780,872; December 25, 1973 and 3,690,606; September 12, 1972; both assigned to Pall Corporation describes filters and filter units which comprise as the filter sheet anisometric compressed and bonded knitted wire mesh composites composed of a plurality of sheets of knitted wire mesh, superimposed at random orientation with respect to each other, compressed or densified to a voids volume within the range from about 10 to 90%, and bonded together.

The sheets are taken in sufficient number, usually at least five and preferably ten or more, and as much as 1,000 or more, to form a self-supporting relatively nonresilient composite of high tensile strength and high breaking strength having an average pore diameter of less than 200 microns, and preferably less than 100 microns, that is relatively uniform in any unit area of the surface, and having an anisometric porosity, the through pores extending crosswise of the sheet greatly exceeding in number the through pores extending laterally of the sheet, which latter pores can be reduced virtually to zero in a highly compressed composite.

The composite is formed by superimposing a plurality of knitted wire mesh sheets, annealing the composite to avoid wire breakage during later processing, compressing the composite to the desired density and anisometricity by application of pressure in a direction approximately perpendicular to the plane of the layers of the composite, and bonding the sheet layers and wire filaments of the sheets together at their points of contact and/or crossing. The bonding holds the composite at the selected density, prevents relative movement of the wires in the composite, and in conjunction with the multilayer structure imparts the self-supporting nonresilient characteristic, together with high tensile strength and high breaking strength.

Figures 7.2a and 7.2b show an anisometric knitted wire mesh composite, made up of 10 layers **1** of a warp-type single-bar tricot knitted wire mesh, 12 needle ends per inch, of 4 mil stainless steel wire **2**, rolled and sintered to a voids volume of 85% and a thickness of 0.04 inch.

This composite can be cut into disks, and it can also be corrugated into a sheet having a plurality of corrugation folds. Such a sheet can be made into a cylindrical filter element. Figure 7.2c shows photographically, enlarged six times, a stack of sixteen layers of loose-weft-knitted mesh, of 0.004 inch diameter wire, ready for rolling and sintering. The looped wires of the mesh retain their knitted identity quite clearly.

Figure 7.2d shows photographically, enlarged six times, the same stack after rolling and sintering. The knitted pattern is still evident, but the consolidation of the mesh has resulted in a considerable reduction in the size of the mesh openings. The end view, Figure 7.2e, shows that the consolidation has in fact resulted in a sheet that is plate-like in character, with a smooth surface. The porosity is anisometric, with many through pores extending across the sheet, and few through pores extending laterally.

FIGURE 7.2: FILTERS COMPRISING ANISOMETRIC COMPRESSED AND BONDED MULTILAYER KNITTED WIRE MESH COMPOSITES

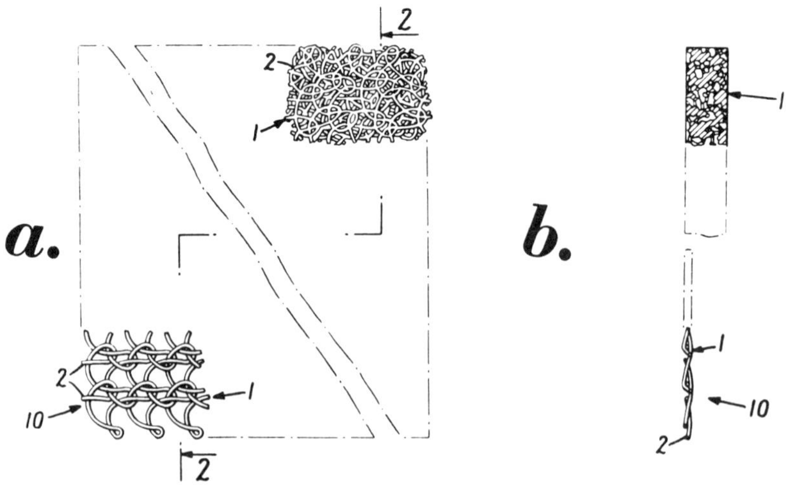

(a) Plan View of a Rigid Warp-Knitted Mesh Composite, Showing in Two Parts the Starting Mesh and the Composite, Respectively
(b) View in Transverse Section Taken Along the Line **2–2** of Figure 7.2a, and Looking in the Direction of the Arrow

(continued)

FIGURE 7.2: (continued)

c.

d.

(c) Photographic Top View Enlarged Six Times of a Loose-Weft-Knitted Mesh Stack, Ready for Compressing and Bonding to Form a Composite

(d) Photographic Top View Enlarged Six Times of a Weft-Knitted Mesh Composite

(continued)

FIGURE 7.2: (continued)

e.

(e) Photographic End View of the Composite of Figure 7.2d

Source: U.S. Patent 3,780,872

Example 1: Four anisometric knitted wire mesh compositions were prepared, made of 0.0011 inch diameter AISI 347 stainless steel wire, using a weft knit mesh that had from 12 to 18 needle ends per inch. Sixteen layers of this mesh were stacked at random orientation to make a composite, and sintered at 1200°C. The composite was cut into four pieces which were rolled to thicknesses of 0.007 inch, 0.0045 inch, 0.003 inch and 0.002 inch, respectively. The four layers were stacked in that order, and resintered to make the final anisometric composite.

The dirt capacity of the final composite was determined in accordance with the following test procedure. The composite described in the preceding paragraph was clamped in a flow jig fitted with gaskets 3½ inches o.d. and 3.06 inches i.d., and connected to a pressure build-up and collapse pressure apparatus.

Hydraulic fluid was run through the mesh at a flow of 40 gpm/ft^2. Direction of flow was such that the upstream face of the test piece was the 0.007 inch (highest voids volume) face. Standardized fine air cleaner (AC fine) test dust in a slurry was added through the dust valve in 0.2 gram increments at 4 minute intervals. The clean-up filter was not by-passed during this test.

Two minutes after each test dust addition, the pressure differential at rated flow through the apparatus was recorded. The initial pressure drop was 0.2 psig, and the weight of contaminant added in the same manner to develop at differential pressure across the mesh of 15, 40 and 90 psig was, respectively 85, 91 and 97 grams per square feet. After cleaning the composites, a suspension of glass beads in oil was passed through them. The maximum bead passed was 62 microns. This is the maximum particle rating.

These data represent a very high dirt capacity, considerably higher than a woven wire mesh of equivalent maximum particle rating. For a 325 x 325 x 0.0014 stainless steel square weave wire mesh, of nominal opening 43 microns, and maximum particle rating 51 microns, the weights of contaminant (AC test dust) were 20, 24, and 26 grams per square feet for differential pressures of 15, 40 and 90 psig, respectively for a 200 x 200 x 0.0021 stainless steel square weave wire mesh of nominal opening 74 microns and maximum particle rating 38 microns, the weights of contaminant were 56, 63, 68 grams per square feet to 15, 40, and 90 psig, respectively.

These two woven meshes are industry standards for removal ratings in the range of 43 to 83 microns. Thus, the knitted wire mesh composite of the process has a dirt capacity greater than that of woven wire mesh of the comparable or somewhat larger pore size.

Example 2: An anisometric knitted wire mesh composite was prepared, made of layers of 0.002 inch A151347 stainless steel wire weft knit mesh having from 12 to 18 needles per inch. Ten layers of this were stacked, annealed, rolled and sintered at 1200°C to a composite 0.006 inch thick.

Twenty layers were stacked, annealed, rolled and sintered at 1200°C to a composite 0.008 inch thick. Ten layers were stacked, annealed, rolled and sintered at 1200°C to a composite 0.0023 inch thick. The three composites were then stacked in that order, and resintered to make the final anisometric.

The dirt capacity and maximum particle rating of this wire mesh composite were determined in accordance with the test procedure of Example 1. The weights of contaminant to 15, 40 and 90 psig were 56, 60 and 78 grams per square foot, respectively. The maximum particle rating was 71 microns.

Example 3: An anisometric knitted wire mesh composite was prepared made from 64 layers of 0.0011 inch AISI 347 stainless steel wire weft knit mesh having 12 to 18 needle ends per inch, stacked, annealed, rolled and sintered at 1200°C to a composite 0.028 inch thick.

The dirt capacity and maximum particle rating of this composite were determined in accordance with Example 1, except that the flow was 50 gpm per square foot. The weights of contaminant to 15, 40 and 90 psig were 91, 107 and 120 grams per square foot, respectively. The maximum particle rating was 60 microns.

Example 4: The procedure of Example 3 was repeated except that the final composite was rolled and sintered to a composite of 0.018 inch thick. The dirt capacity and maximum particle rating of this composite were determined in accordance with Example 3. The weights of contaminant to 15, 40 and 90 psig were 64, 68 and 80 grams per square foot, respectively. The maximum particle rating was 31 microns.

Example 5: The composites of Examples 3 and 4 were stacked and sintered at 1200°C. The resultant composite was clamped in a flow jig as in Example 1 with the 0.028 inch thick portion upstream. The dirt capacity and maximum particle rating of this composite was determined as in Example 3. The weights of contaminant to 15, 40 and 90 psig were 81, 90 and 97 grams, respectively. The maximum particle rating was 30 microns.

Examples 6 through 19: A number of anisometric knitted wire mesh composites were prepared, from weft knitted wire mesh made of AISI 347 stainless steel wire, 0.002, 0.003 or 0.004 inch in diameter as noted in Table 1. These were stacked, using the number of layers noted in the table, sintered at 1150° to 1400°C, rolled to the thickness noted in the table and then resintered at 1100° to 1250°C. The bubble points and air flow at the pressure differential noted and the Rayl numbers were determined, and are listed. The Rayl number is a measure of flow resistance.

The tensile breaking strength, specific strength (ratio of breaking strength to weight per unit area) and Young's Modulus were determined for several of the examples given in Table 1. These data are given in Table 2.

TABLE 1

Ex. No.	Wire Diameter (in)	No. of Layers	Thick-ness (in)	Bubble Point (inches of water column) 1st	10th	Open	ΔP (inches of water column)	At Air Flow (ft/sec)	Weight (lb/ft^2)	Rayl No.
6	0.002	122	0.046	1.75	1.80	2.2	2.7	10	0.47	10
7	0.002	122	0.026	2.15	2.65	3.8	8	10	0.48	32
8	0.002	122	0.024	3.1	3.4	5.0	12	10	0.48	50
9	0.002	242	0.043	3.1	3.5	3.9	11.8	10	0.80	50
10	0.002	242	0.032	4.3	5.0	6.7	2.7*	10	0.80	150
11	0.002	242	0.027	5.2	5.9	10.4	12*	10	0.80	450
12	0.002	242	0.021	7.9	10.8	24.4	10*	2.5	0.80	2,250
13	0.003	56	0.017	1.0	1.4	2.6	2.7	10	0.45	10
14	0.003	70	0.024	1.0	1.3	2.5	3	10	0.52	11
15	0.003	152	0.050	1.8	2.0	3.4	7.1	10	0.75	32
16	0.003	152	0.044	2.0	2.4	4.2	12.1	10	0.75	50
17	0.004	32	0.020	1.0	1.2	2.2	4.0	10	0.52	11.5
18	0.004	70	0.038	1.5	1.9	3.3	10.6	10	0.90	35
19	0.004	82	0.046	1.8	2.4	3.7	14.3	10	1.0	50

*Pounds per square inch (psi).

TABLE 2

Example No.	Tensile Breaking Strength (lb/ft)	Specific Strength (ft)*	Young's Modulus (psi)
6	3,468	7,332	4.2×10^6
7	3,432	7,150	8.1×10^6
8	3,744	7,800	10.2×10^6
17	3,564	6,854	11.4×10^6
18	7,140	7,933	12.5×10^6
19	8,820	8,820	10.3×10^6

*lb/ft/lb/ft^2

For comparison with Examples 1 through 19, similar data are given in Table 3 for a number of sintered woven wire mesh sheets of comparable weights and pore size, made of the same stainless steel, AISI 347, as these examples.

It will be noted that all the above data are for material which has been resintered, thus at least partially annealing the mesh, and the variation in specific strength is not very large. This indicates that, for any alloy, the breaking strength is about proportional to the weight of material used, with the amount of compression performed on the material having at most a small effect. These anisometric materials were satisfactory as filters for air and for liquids.

TABLE 3

Woven Mesh Sample	Type Weave	Weight (lb/ft^2)	Breaking Strength (lb/ft)	Rayl No.	Specific Strength (ft)	Young's Modulus (psi)
A	Sintered Dutch twill	0.46	2,232; 2,760*	10	4,852; 6,000*	1.3×10^6
B	Sintered Dutch twill	0.70	3,744	34	5,349	4.6×10^6
C	Sintered square weave	0.60	3,120	35	5,200	4.9×20^6
D	Sintered square weave	0.60	3,360	50	5,600	5.3×10^6

*Directional properties

The anisometric wire mesh composites are especially suited for use as liquid filter elements because of their disproportionately high flow crosswise of the sheet, their unusually high dirt capacity, low flow resistance, and particle removal rating, and their high strength. A typical filter unit including an anisometric knitted wire mesh composite of the process as the filter element is described in the patent.

Pile Fiber Filter

R.C. Clark; U.S. Patent 3,977,847; August 31, 1976; assigned to Brunswick Corp. describes a method for removing particles from a gas stream which employs a depth filter which may be cleaned and is adapted to function at high temperatures and in corrosive environments.

According to the process, the gas is passed through a depth filter made essentially solely of metal. This filter is a fabric including a base and pile connected to the base. This fabric may be made by conventional techniques such as knitting, tufting or weaving. The pile is made from textile metal fibers having a diameter of less than about 50 microns and a rough, unmachined, unburnished, fracture-free surface.

These pile fibers lie down upon each other to form the depth filter media as the particle laden gas flows through the filter. Preferably the fibers have a diameter ranging between 1 and 25 microns. A suitable pile height may range between ⅛ and 1¼ inches. Stainless steel is the most preferred alloy, but other suitable metal alloys are Driver Harris 242 and Karma. However, any corrosion and temperature resistant alloy may be used. The density of the filter media formed by the fibers may range between 0.1 and 15% (density equals the volume occupied by the fibers, divided by the total volume of the depth filter, multiplied by 100).

In a preferred form, the pile fibers have long fibers interdispersed with short fibers. Preferably, the short fibers have a smaller diameter than the long fibers; for example, the long fibers' diameter may be from 2 to 4 times larger than the short fibers' diameter.

Because both short and long fibers are used, two layers of fibers are formed as particle laden gas flows through the filter. The first layer which the gas encounters is the least dense layer. Generally this layer has a density ranging between 0.1 and 3%. The second or denser layer has a density ranging between 2 and

10%. Preferably, there is a gradual increase in density along a transition layer of fibers formed between the low density and high density layers.

It has been very difficult if not impossible to clean a depth filter media. This problem is overcome and one can clean the depth filter media employed in this method by passing clean gas through the filter in a direction counter to the direction that the particle laden gas stream flows through the filter. Consequently the pile fibers open and extend outwardly from the base, permitting entrapped particles or particle agglomerates to be carried by the cleaning gas, out of the filter.

Preferably, the cleaning gas is heated to a temperature above the dew point of the particle laden gas. This keeps the fabric hot so that moisture or corrosive acids do not condense on the fabric. Heating the cleaning gas also avoids a potential problem caused by water and solid material forming a cement-like cake on the fabric which would be difficult or impossible to remove. To assist in cleaning, it is also required in some circumstances to agitate the pile fibers, for example, by vibration.

The method is particularly adapted to treat particle laden gas streams having a temperature in excess of 500°F and including particles having a diameter of less than 15 microns. Gas streams including particles having an average diameter ranging between 15 and 1 micron or less are most readily cleaned by the method.

Figures 7.3a and 7.3b schematically depict the method for removing particles from a stream of gas. The central feature is the use of an all metal fabric **10** having a base **12** such as a porous scrim and a plurality of metal pile fibers **14** connected to the base. The base may be woven and includes warp and fill yarns **19** and **21**, respectively. The pile fibers are wrapped around the fill yarns and held snug in position by the warp yarns **19**. The fabric is supported by a wire screen **17**.

FIGURE 7.3: PILE FABRIC FILTER

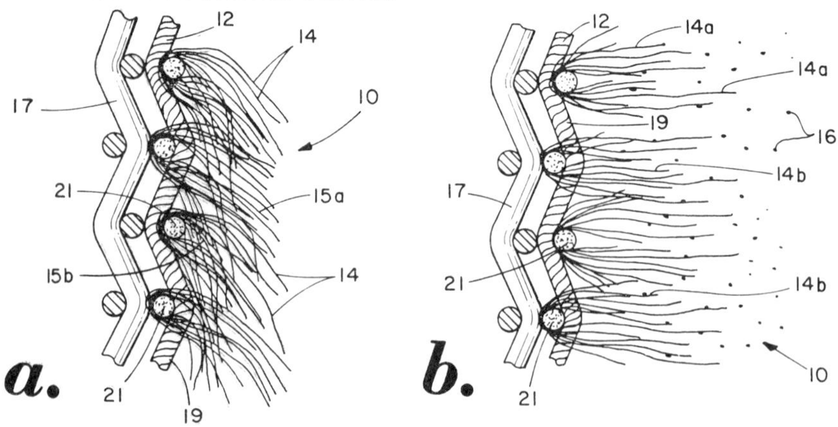

(a) Cross-Sectional View of the Depth Filter
(b) Cross-Sectional View of the Depth Filter of Figure 7.3a

(continued)

FIGURE 7.3: (continued)

c.

d.

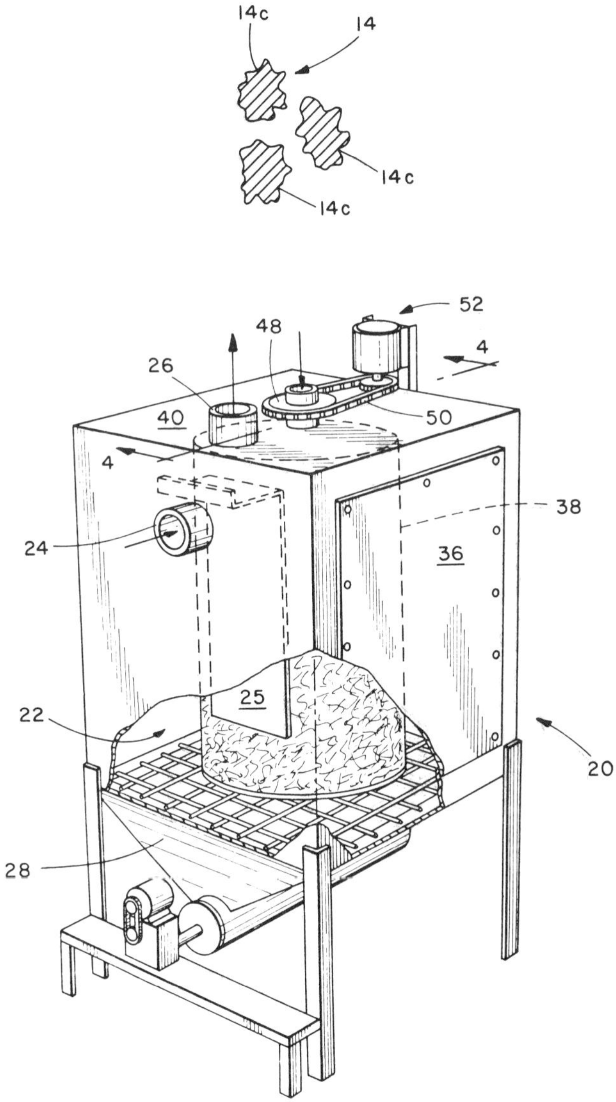

(c) Cross-Section of the Pile Fibers Illustrating Their Rough, Unmachined, Unburnished, Fracture Free Surface

(d) Perspective View, with a Section Broken Away, of the Apparatus of the Process

(continued)

FIGURE 7.3: (continued)

e.

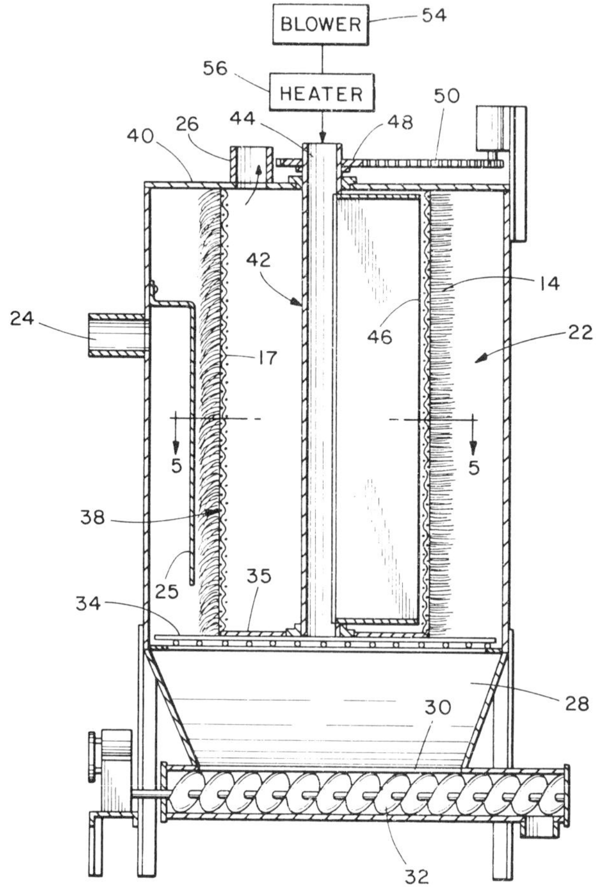

f.

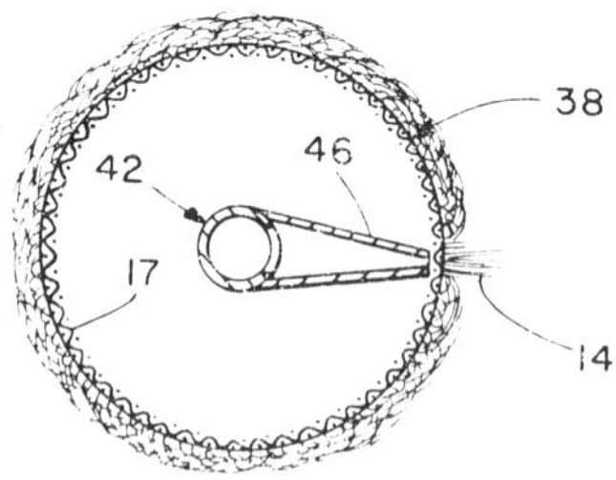

(e) Cross-Sectional View Taken Along Line **4–4**
(f) Cross-Sectional View Taken Along Line **5–5** of Figure 7.3e

Source: U.S. Patent 3,977,847

As best shown in Figure 7.3b, there are long fibers **14a** interdispersed in a random fashion among short fibers **14b**. Preferably, the long fibers have a larger diameter than the short fibers. As the particle laden gas stream flows through the fabric as illustrated in Figure 7.3a, the pile fibers lie down upon each other to form a depth filter media. Because both long and short pile fibers are employed, a low density layer **15a** and higher density layer **15b** are formed. Large particles tend to collect in the lower density layer, and small particles tend to collect in the higher density layer.

As shown in Figure 7.3c, the metal pile fibers have a rough, unmachined, unburnished, fracture free surface **14c**. This rough surface facilitates collection of fine particles, i.e., particles having a diameter of about 15 microns or less. These fine particles stick to the rough surface and act as sites at which large agglomerates grow by a gradual build-up of particles. In other words, small particles in the particle laden gas stream adhere to particles already attached to the rough surface to form large agglomerates which are removed from the filter media during cleaning and then collected.

Eventually the pressure drop across the depth filter media increases to the point where it is desirable to clean the depth filter media. Cleaning is achieved by simply passing a current of gas through the fabric **10** in a direction counter to the flow of the particle laden gas stream.

This is illustrated in Figure 7.3b. As the reverse gas current passes through the fabric, the pile fibers **14** open and extend outwardly from the base, permitting a major portion of the entrapped particles and particle agglomerates **16** collected by the fibers to release and become entrained with the reverse current of gas out of the fabric. Since the agglomerates are substantially larger and heavier than the particles carried in the incoming gas stream, these agglomerates fall downward due to gravity and are collected. In some instances, it is desirable to agitate the fibers, for example, by mechanical means which vibrate the fabric and thereby assist in removal of entrapped particles.

Figures 7.3d through 7.3f illustrate the apparatus **20** which is designed to carry out the method for removing particles from a gas stream. This apparatus **20** includes a chamber **22** having an inlet **24** and outlet **26**. At the inlet is a striker plate **25** which serves to remove large particles from the incoming gas stream.

Such large particles strike the plate and fall downward. The bottom of the chamber **22** tapers downwardly to form a hopper section **28** at which there is a trough **30** holding an auger **32**. A grate **34** provides an elevated floor above the hopper section **28**. In one side of the chamber **22** is a door **36** to permit entry, and disposed within the chamber is a tubular element **38** made of the all metal fabric described above.

This tubular member has its top attached to the roof **40** of the chamber **22** underneath the opening in the outlet **26**, and its bottom connected to the grate. The bottom of the tubular element **38** is closed off by plate **35** so that the particle laden gas stream entering the inlet will pass through the sidewall of the tubular element into its interior, with filtered gas exiting the apparatus **20** via the outlet **26**. Disposed inside the tubular member is a rotatable cleaning device **42**, including a hollow shaft **44** having a nozzle **46** attached to this hollow shaft.

At the top of the shaft is a sprocket **48** connected by a chain **50** to a motor drive **52**.

In operation, a particle laden gas stream flows into the inlet **24** of the chamber **22**, and through the wall of the tubular element **38**. As gas flows through the tubular element, the pile fibers **14** lie down upon each other as illustrated in Figure 7.3f. This forms the depth filter which collects particles. Simultaneously, the cleaning device **42** is revolving.

Cleaning gas is forced through the hollow shaft **44** and into the nozzle **46**. Thus, as the cleaning device advances, high velocity gas exits through a small increment of the wall of the tubular element **38**. This causes the pile fibers to open and extend outward from the base of the tubular element. This occurs only along the incremental portion of the tubular element adjacent the tip of the nozzle **46**.

Agglomerates formed in the depth filter are blown by the air stream out of the fabric, and these agglomerates fall into the hopper section **28** and are collected in the trough **30** and removed by the auger **32** from the chamber **22**. Purified gas leaves the chamber by the outlet **26**.

In the preferred arrangement, cleaning gas is forced by a blower **54** through a heater **56** into a hollow shaft. The heater heats the cleaning gas to a temperature above the dew point of the particle laden gas so that any moisture in the cleaning gas does not condense on the depth filter. This is desirable in order to prevent forming a calcified material which would be very difficult to remove from the depth filter and/or the formation of aqueous condensates of corrosive acids.

The volume of cleaning gas forced through the tubular member relative to the volume of particle laden gas entering into the chamber **22** is controlled by adjusting the nozzle size and/or velocity of cleaning gas relative to the process gas flow capacity of the apparatus. For example, the ratio of the cleaning gas to the volume of particle laden gas should range between 1:1,000 and 1:3.

In-Stack Filter Cell

A process described by *R.M. Martin; U.S. Patent 3,976,457; August 24, 1976; assigned to the U.S. Administrator for the Environmental Protection Agency* relates to a filter cell which meets the standards of the Environmental Protection Agency (EPA) and which may be used in an EPA reference method train for determining the amount of pollutants contained in an industrial stack or duct.

In 1967, the Abatement Branch of the National Center for Air Pollution Control formed a group to undertake stack sampling. At that time, there was no standard system of stack sampling equipment which was directly applicable to the needs of the branch. Thereafter, a system was designed to determine the amount of particulates or pollutants which are contained in an industrial stack.

This system is used as the standard system as referenced in the *Federal Register* Vol. 35, No. 247, December 23, 1971, and is known as Method 5. Essentially, all federally accepted data in support of New Source Performance Standards for particulate matter, and for determining compliance with particulate standards, are based upon this system. Commercially available equipment conforming to this EPA design is widely obtainable and is used for collecting the desired data. However, the sample box containing the fiber filter in this system contains much elaborate and delicate glass equipment thereby complicating the testing of the particulate matter. Furthermore, this sample box cannot be inserted through the standard 3-inch box or 4-inch box portholes which are contained in a great number of the stacks.

The process provides a filter cell having a substantially rectangular glass fiber filter having the approximate dimensions of 2" x 6" or 2" x 7". Therefore, the size of the filter area is equivalent to that of the standard 3-inch circular filter which is used in the EPA reference system, but which can be inserted into a typical 3-inch or 4-inch port.

The filter itself is composed of a substantially cylindrical housing having an inlet port communicating with the interior of the cell housing for drawing a flow of gas into the filter cell housing. A preweighed filter is disposed in this housing substantially parallel to the flow of the gas. After the gas flows through the filter and the particulate is trapped in the filter, it exits the cell housing via an outlet port placed under the filter.

After a predetermined period of time, the filter cell is removed from the stack, disassembled, and the filter itself is weighed to determine the amount of pollutants in the stack. This filter cell is to be used in conjunction with the EPA reference train and, therefore, the measuring instruments which are used in that train measure the isokinetic gas flow and other variables.

The device can be used in stacks which contain high moisture content, such as those on scrubbers, wet processes or combustion processes. Additionally, the filter cell is operated in the stack such that the temperature of the filter media is maintained at the temperature of the gas stream.

Figure 7.4a shows a filter cell **30** which is to be placed directly into a stack or duct to test the level of pollutants. This cell, preferably tubular in shape, consists of a top semicylindrical housing **32** and a bottom semicylindrical housing **34** which can essentially be a longitudinally cut stainless steel tube, although the exact constructional material is not crucial. An inlet port **10** which is connected to a nozzle **9** is in communication with the interior of the cell **30** so that the gaseous flow through the stack will be directed into the cell.

A pair of hose clamps **16** which are tightened by screws **20** encircle the cell to insure a tight, leakless seal. An outlet port **15** is connected to the bottom housing tube **34** and allows the gas to exit from the cell **30**. This outlet port is also attached to a probe in the EPA reference train which is in communication with a testing console to insure the isokinetic flow of the gas and to measure other variables.

Figure 7.4b shows the interior of the filter cell. As can be seen from the figure, a substantially rectangular filter **12** is provided which is aligned parallel to the flow of gas (dashed line in Figure 7.4b) as it enters a top chamber **11** in the top housing **32** from the nozzle **9** and the inlet port **10**.

FIGURE 7.4: IN-STACK FILTER CELL

(a) Side View of Filter Cell
(b) Exploded View of the Filter Cell Showing its Several Components

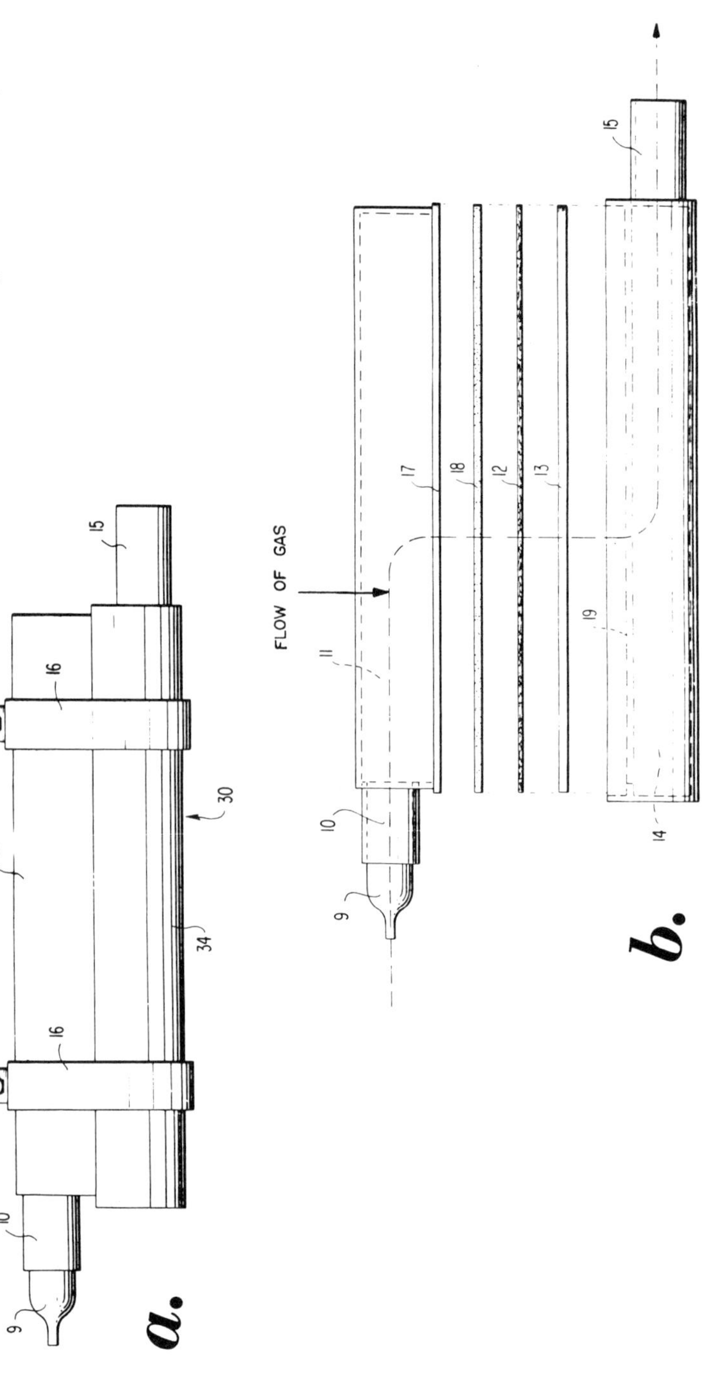

Source: U.S. Patent 3,976,457

The disposition of the inlet and outlet ports in relation to the filter to obtain gas flow parallel to the filter in an important feature of the device. This filter must be preweighed before it is inserted into the cell **30** to insure accurate measurement of the pollutants. The filter should be constituted of a glass fiber material and must have the same total area as the circular fiber filters utilized in the EPA standard reference train. This is accomplished by using a rectangular shaped filter having dimensions between 1 by 3 inches and 1 by 4 inches.

A perforated or sintered stainless steel rectangular plate **13** is included to support the filter **12** and allow the gas to flow into a bottom chamber **14** in the bottom housing **34** and out of port **15** to an acceptable metering device. To insure stability, the filter and the filter support **13** are supported by a filter support rest **19** which is disposed in the bottom chamber **14**.

A gasket **18** encircles the cell **30** to insure a leakless seal and can be constructed of Teflon, silicone rubber, or any like substance for low temperature applications or it can be constructed of asbestos such as that used for automobiles for use at elevated temperatures. If silicone rubber is used, the gasket should be provided with a central web that insures gasket seals on the edges.

To also insure that a minimum amount of leakage occurs, the inlet port **10** and the outlet port **15** are connected to the nozzle **9** and probe, respectively, with standard Swagelok fittings or the like. Additionally, during sampling, the filter cell may protrude out of the stack slightly and, if this occurs, the filter cell should be insulated by covering it with suitable insulating material, e.g., cloth such as asbestos cloth.

The filter cell is assembled and closed securely by use of the two hose clamps **16** which, upon tightening, force a flange **17** down onto the gasket **18** providing a leakless seal.

The above-described device operates in the following manner: Before the filter cell is assembled, the glass fiber filter **12** is weighed and then mounted on filter support **13** and the filter support rest **19** in the bottom chamber **14**. The entire cell is then placed directly within a stack and a sample of gas from this emission source is isokinetically drawn to the device in which particulate matter is collected on the filter. The filter cell is then removed from the emission source, disassembled, and the filter **12** is then weighed and calculations may be made which quantify the amount of emissions from the source.

Removal of Radioactive Radon Daughters from Air

A process described by *C.W. Stringer; U.S. Patent 3,853,501; December 10, 1974; assigned to Radon Development Corp.* relates to an improvement in the method for removing radioactive radon daughter products from the working environment in underground mines and ore processing plants by inserting a filter medium into a water washed filter system and circulating the water in which the radon daughter products are trapped through the filter medium to remove the daughter products, thus cleaning the water for reuse to wet the air filter.

The filter medium involves the use of dried water-insoluble granular substrate which has been wetted with a normally liquid hydrocarbon. The preparation of the filter medium involves the heating, preferably under vacuum, of the granular substrate for a time sufficient to remove substantially all of the moisture which may be on the surface, and in the pores, of the substrate. The substrate material is then treated with a normally liquid hydrocarbon to wet the surface of the substrate with the hydrocarbon.

The medium thus prepared, when placed in a suitable filter container, has been found to effectively remove the radon daughters from water such that the water can be recycled over an air filter to trap more radon daughter contaminants which are removed from the air by the water washed filter.

Considerable evidence indicates that excessive exposure to radon daughter products is associated with a high incidence rate of lung cancer. In view of the foregoing, it is not surprising that there is considerable concern for the radon gas and radon daughter levels in the working environment in mines and ore processing plants.

This concern has led to the establishment of allowable exposure levels and led to what has been called the working level (WL) which has been defined as that concentration of short-lived radon daughter products in a liter of air that will yield 1.3×10^5 million electron volts (mev) of alpha energy in decaying through RaC'.

This definition has become the standard by which engineers and regulatory agencies operate. Defining WL in picocuries per liter of air avoids the necessity of defining the equilibrium relationship between RaA, RaB, and RaC. A WL is the potential energy released by 100 picocuries each of RaA, RaB and RaC, but under this equilibrium condition, 300 picocuries of the daughters may not be equivalent to 1 WL.

Regulatory agencies have set forth limitations of working level to which miners particularly can be exposed, such regulations being promulgated to the extent that if the atmospheric concentration of radon daughters exceeds 1 WL but is less than 2 WL immediate corrective action has to be taken or the workers must be withdrawn from the environment contaminated with radon and radon daughters.

When concentrations higher than 2 WL exist, the workers are withdrawn immediately until the radon daughter atmospheric concentrations are reduced to 1 WL or less. Even more strict regulations may be imposed. Further discussion of the alpha radiation problem in underground uranium mines is found in "Controlling Employee Exposure to Alpha Radiation in Undergound Uranium Mines," R.L. Rock and D.K. Walker, Volume 1, Bureau of Mines, United States Department of the Interior, 1970.

OTHER PROCESSES

Sterilization of Gases in Fermentation Processes

W. Rothmayr and J. Dasek; U.S. Patent 3,881,896; May 6, 1975; assigned to Societe d'Assistance Technique pour Produits Nestle SA, Switzerland describe a filter for gaseous fluids which includes a packing comprising quartz fibers.

The filter may advantageously be used for sterilizing gases, for example, gases used in fermentation processes. As shown in Figure 7.5a, the filter comprises a cylindrical chamber **6a** containing a packing **8** consisting of quartz fibers retained between two screens **9**. These fibers, the diameter of which is generally between 5 and 15 microns, can be present as a nonwoven mass of such fibers or quartz wool; however, woven quartz fibers may also be used as packing.

Figure 7.5b shows a filtration system comprising an inlet pipe **1** feeding a gaseous fluid to a vessel or fermenter **2**. On the path of the pipe are arranged, from upstream to downstream, a first filter **3** having for example a packing of glass wool for eliminating impurities such as dust and other solid particles, a compressor **4**, a heater **5**, a filter **6** packed with quartz fibers and a cooler **7** for the gaseous fluid.

Retention of microorganisms present in the gaseous fluid by the permeable material comprising quartz fibers has been found to be especially efficient and the sterilizing effect of this material on a gaseous fluid is observed to be substantially superior to that of filtering materials normally used, such as glass wool or cotton fibers, with respect to the same microorganisms and under the same conditions of temperature and fluid flow rate.

FIGURE 7.5: APPARATUS FOR FILTERING GASEOUS FLUIDS

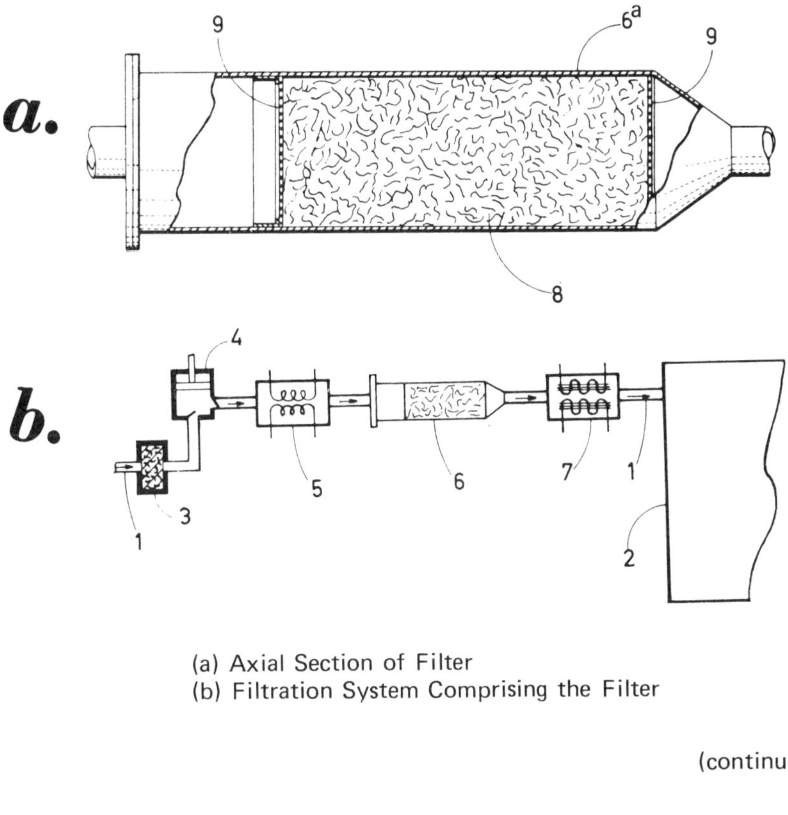

(a) Axial Section of Filter
(b) Filtration System Comprising the Filter

(continued)

FIGURE 7.5: (continued)

c.

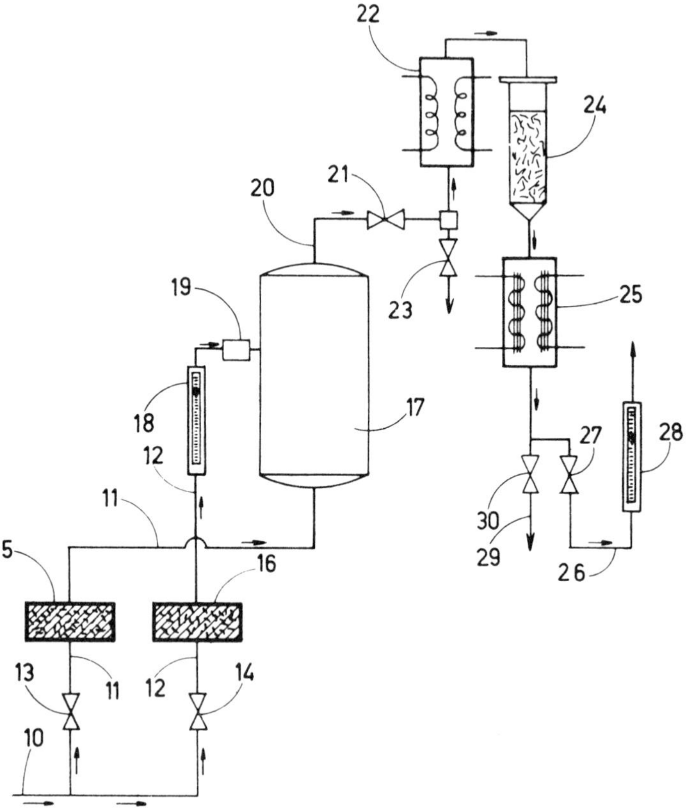

(c) Experimental Apparatus

Source: U.S. Patent 3,881,896

In particular, total sterilization of a gas at ambient temperatures, that is close to 25°C, which is very difficult, or impossible, to achieve with conventional filter materials, has been easily obtained by filtration of the gas through quartz wool. The sterilizing power of quartz fibers, which is substantially higher at ambient temperatures than that of traditional materials, is also better at high temperatures.

A particularly advantageous application of the device comprises sterilizing the gas or gases fed to a fermenter by filtration, through the quartz fiber filter. The compressed gas, for example, air for fermentations under aerobic conditions, ammonia as nitrogen source, or a gaseous paraffin such as methane as carbon source, is passed through at least one filter element packed with quartz fibers before entering the fermenter. According to the type of undesirable microorganisms present in the gas, it may optionally be heated to an appropriate temperature. This temperature is generally considerably below the temperature necessary to obtain the same sterilizing effect when using conventional filter materials, and may for example be obtained by simple adiabatic compression of the gas to be sterilized. The process is illustrated by the following examples.

Example 1: In order to observe the sterilizing effect of a filter, the experimental apparatus shown in Figure 7.5c is used. This apparatus comprises a feed pipe **10** for gaseous fluid branching into two pipes **11** and **12** connected in parallel and respectively having from upstream to downstream, flow rate and pressure control valves **13** and **14** for gaseous fluid and filters **15** and **16** packed with glass wool.

The pipe **11** is connected to the lower part of a reservoir **17** and the pipe **12**, provided with a flowmeter **18** of the rotameter type, is connected to the upper part of the reservoir by way of a spray chamber **19** containing a suspension of a microorganism. The gas streams carried by the pipes **11** and **12**, that is the main stream passing through pipe **11** and having the highest flow rate, and the stream passing through pipe **12**, contaminated with the microorganism present in the chamber **19**, are mixed in the reservoir and arrive, by way of the pipe **20** and valve **21**, at a heater **22** in the bottom of which is provided a device **23** for taking samples of contaminated gas. The heater **22** is itself connected to the filter **24** the sterilizing capacity of which is being observed. The exit opening of this filter **24** is connected to a cooler **25**. The exit pipe **26** of the cooler, which is provided with a control valve **27** and a flowmeter **28**, has a branch **29** provided with a valve **30** for taking samples of filtered gas.

The filter **24** is cylindrical, the diameter and length being respectively 108 mm and 161 mm. After packing of the filter **24** with 55 grams of quartz wool having an average diameter of 7 to 10 microns, the part of the apparatus including the filter, the heater **22**, the cooler **25** and the sampling ports is sterilized with a current of steam at 130° to 140°C during 30 minutes. Air is then circulated through the pipes **10, 11, 20** and **26** in order to dry the different parts of the apparatus and to equilibrate the temperature.

After an aqueous suspension containing *Bacillus stearothermophilus* spores has been introduced into the spray chamber **19**, the valve **14** is opened to introduce air into the pipe **12** and chamber **19**. The pressure within the reservoir **17** and the total air flow rate are adjusted with valves **27** and **13**. Samples are taken at different times and for varying periods at **23** (contaminated air) and at **29** (filtered air).

The spore suspension used to contaminate the gas stream to given concentrations, which are expressed as the number of spores per liter of air, is examined by counting the spores in a series of 0.1 ml samples placed on sterile nutrient agar plates which are incubated during 48 hours at 56°C and then counted. These counts give the concentration of the suspension in terms of the number of spores per liter.

The samples of air taken at **23** and at **29** pass through sterile Millipore membrane filters for specific periods of time. Counting of the spores collected on these filters is made under the same conditions as those described for the suspension samples. The trials have been effected at different temperatures and with various times; that is, the samples of filtered air were taken for different periods and after various periods of use of the filter. The results of these trials are summarized in Table 1. In this table, the parameters shown have the following meaning

Period 1: Period of use of the filter with a constant flow of contaminated air (95 l/hr) before the sample is taken.
Period 2: Length of time during which the sample is taken at **29** at a flow rate of 10 l/hr; the flow rate in the filter remaining at 95 l/hr.
Number of spores before filtration: Determined by counting spores in contaminated air samples taken at **23** during Period 1.
Number of spores after filtration: Determined by counting spores in the sample taken at **29**.
Temperature 1: Temperature of the air at the exit of the heater **22.**
Temperature 2: Temperature of the air at the exit of the cooler **25.**
Efficiency: $[(N_0 - N)/N_0] \times 100$, N_0 being the number of spores counted before filtration and N being the number of spores counted after filtration in the same volume of air.

TABLE 1

Period (minutes) 1	2	. . Number of Spores . . Before Filtration	After Filtration	Temperature (°C) 1	2	Efficiency (percent)
0	15	51,400	1	25	25	99.998
60	15	18,241	0	25	25	100
60	15	9,741	0	25	25	100
0	15	37,200	0	150	38	100
60	15	109,600	2	150	41	99.998
60	30	120,200	0	150	37	100
60	15	127,500	0	150	38	100
0	30	127,500	0	150	39	100
0	30	91,800	0	150	38	100

Trials effected under identical conditions with a filter having the same geometry and packed with 100 grams of glass wool have shown that its efficiency varies between 97.6 and 99.7% for filtration temperatures close to 25°C and only reaches 100% at temperatures close to 250°C.

These results demonstrate the extreme efficiency of the filter packed with quartz wool, which is 100% in numerous cases, and independent of temperature within the temperature ranges used.

Example 2: The efficiency of a cylindrical filter, the diameter and length of which are respectively of 108 and 161 mm is measured as described in Example 1. The filter is packed with 55 grams of quartz wool manufactured by Electrotermic, Lucerne, Switzerland. The results obtained are summarized in Table 2 on the next page, the parameters being those defined in Example 1.

These results also show a higher efficiency as compared to the filter packed with glass wool described in Example 1.

TABLE 2

Period (minutes) 1	2	.. Number of Spores .. Before Filtration	After Filtration	Temperature (°C) 1	2	Efficiency (percent)
0	15	212,500	36	25	25	99.983
0	10	13,500	1	25	25	99.992
1	5	29,200	5	25	25	99.982
0	10	47,700	9	25	25	99.981
0	15	48,500	1	150	36	99.998
0	30	85,850	2	150	36	99.997
0	30	170,000	8	150	36	99.995
15	30	10,625	0	170	144	100

Aerosol Filter

A process described by *R.C. Clark; U.S. Patent 3,979,253; September 7, 1976; assigned to U.S. Secretary of the Navy* relates to methods of forming aerosol filters and more particularly to aerosol filters which are formed with a uniform mat of a beta fiber reinforced glass paper filter media.

In the process, glass filter media is made by a formulation consisting of different sizes and types of glass fibers and yarns. The following method is used to obtain good dispersion of beta fibers in the construction of high efficiency particulate aerosol filters.

The method comprises removing the starch type sizing or surface coating from the beta glass fibers. The beta glass fibers 0.05 gram each, of ¼ inch and ½ inch having a diameter of from 3 to 4 microns are placed into a high temperature oven having a temperature of from about 325°C to about 375°C for high temperature exposure for a period of about 16 hours.

The applied temperature causes the degradation of the existing coating of a starch type sizing, thereby exposing a restored surface of glass. This restored surface will be highly susceptible to the solution of acid (HCl) and Silanox which is the next step in the process. These heat cleaned beta glass fibers are then added to 2½ liters of a 2.5 to 3.0 pH hydrochloric acid solution in a beaker to which has been added 0.02 gram of Silanox. Silanox is a hydrophobic fumed silicon dioxide treated with a silane.

The heated treated beta glass fibers, acid and Silanox form a slurry which is subjected to a sonic probe power of 3 amps DC for a time period of about 6 minutes. The sonic probe disperses any clumps or bundles of beta fiber into individual fibers and helps to open the fibers so that they will become coated with the Silanox within the slurry.

The Silanox conditions the beta fibers so that they are held apart thus preventing roping or tangling. The operation of Silanox to prevent roping or tangling is not fully understood. After having subjected the beta glass fiber slurry to the sonic probe power, other glass fibers such as 0.356 gram of A grade (2.1 microns), and 1.168 grams of AAA grade (0.6 micron) are added to the acid-Silanox solution and mixing by further use of the sonic probe is carried on for from about 5 to about 6 minutes depending on the solids concentration of the slurry.

Once the glass fiber solution has been mixed, it is poured into a hand sheet mold for forming the filter. The valve control for the hand sheet mold is opened to permit the solution to drain through the bottom. The filter is formed on the bottom of the hand sheet mold. The tube of the mold is opened and the filter is removed by use of a roller to which the filter clings as the roller is rolled along over the filter.

The filter is then placed on a sheet for final drying. The use and operation of the hand sheet mold for making small filters is well-known in the art. The different type of glass fibers mix uniformly and settle on the bottom of the hand sheet mold in such a manner as to form a highly improved filter. Filters formed by the above method have a thickness of 17 to 20 mils, weight of 64.37 square feet per pound, with a tensile strength of 1.4 pounds per inch of width. Such filters, used in a gas mask, filter 99.7% of 0.3 micron particulate matter at a flow rate of 16 liters per minute.

The cleaning, acid-Silanox mixture treatment of the beta glass fibers conditions the beta glass fibers such that the fibers are held apart thus preventing roping or tangling into a mat. The reaction between the beta glass fibers and the Silanox is not understood; however, the treatment is known to enhance dispersion of clumps of beta fiber into individual strands resulting in a highly improved filter media. In carrying out the above method, a typical glass filter media contains by weight the following table formulation

	Hand Sheet Weight, g	Machine Run Weight, g
Glass fiber, A grade (2.1 μ), 21.9%	0.356	8.21
Glass fiber, AAA grade (0.6 μ), 71.9%	1.168	26.96
Glass yarn, ¼" long (3 to 4 μ), 3.1%	0.05	1.16
Glass yarn, ½" long (3 to 4 μ), 3.1%	0.05	1.16

Using terms known in the art, the concentration of the above glass fibers in 1,000 ml of water is 4.5 grams of solids comprising the above percentages by weight. The hand sheet mold may be used for making small sized filters. Filters of much larger size such as for room air filters in hospitals and other clean places which require a high percentage of particulate filtering may be made by other machines such as a Fourdrinier paper machine.

A beta glass fiber mat is placed in a pulp beater with the beater plate spaced about ½ inch from the beater so that the beater will not chop or damage the glass fibers. The mat is beat for about 15 minutes so that the glass fibers are spaced and broken down to lengths of from ¼ to ½ inch of fiber. About

7 liters of 2.5 to 3.0 pH hydrochloric acid solution containing 0.46 gram of Silanox, is added to the glass fibers and mixed for about 6 minutes. The Silanox solution forms a spiniferous surface that attaches to the restored surface of the glass fibers. The restored surface of the glass fiber has a number of reactive Si—O—Si sites that can form bonds between the silanol groups of the glass.

Some Si—O—Si bonds must form between unadsorbed SiOH groups of adjacent adsorbed coupling molecules. The remainder of the glass fibers are added to the slurry acid solution and mixed for from about 10 to 15 minutes. The slurry is then diluted with water and added to the storage tank of the Fourdrinier paper machine. The diluted slurry is then pumped to the heat tank from which it is deposited onto the moving drain screen where the glass filter media takes its final sheet form as the liquid mix drains away.

Glass fiber filters made in accordance with each of the methods above without Silanox in the acid solution were not efficient because the beta glass fibers were not uniform due to roping or tangling into a mat. The small amount of Silanox added to the acid solution attaches itself to the beta glass fibers and holds the fibers apart to prevent roping or tangling. It has been determined that the percent by weight of Silanox necessary to prevent tangling and roping is the same for corresponding percent amounts of beta glass fibers used.

GEL FILTRATION AND OTHER CHROMATOGRAPHIC SEPARATION TECHNIQUES

GEL FILTRATION

Methacrylamide-N,N'-Methylenebisacrylamide Copolymers

A process described by *I.H. Johansson and M.K. Joustra; U.S. Patents 3,728,290; April 17, 1973; and 3,616,936; November 2, 1971; both assigned to Pharmacia Fine Chemicals AB, Sweden* relates to a method for preparing a separating medium for gel filtration in which grains of a rigid hydrophilic macroporous copolymer of acrylamide (or 1-vinyl-2-pyrrolidinone) with N,N'-methylenebisacrylamide has copolymerized in its pores a second copolymer of one or more acrylamides (or 1-vinyl-2-pyrrolidinone) within N,N'-methylenebisacrylamide in the presence of an initiator. The initiator is present in the pores of the rigid hydrophilic, macroporous copolymer when the copolymerizable substances of the second copolymer are introduced.

Example 1: 100 grams of fine grain chalk covered with calcium stearate and 5 grams of an emulsifier [dodecyl-phenoxy-poly-(ethyleneoxy)ethanol] were slurried in 1,000 ml of toluene in a reaction vessel placed in a water bath, the temperature of which was maintained at 50°C. 22.5 grams of methacrylamide and 27.5 grams of N,N'-methylenebisacrylamide were dissolved in 500 ml of distilled water at 50°C. 1 gram of ammonium persulfate was added to the monomer solution as an initiator and the solution together with the initiator added thereto was introduced immediately into the reaction vessel.

The two phase mixture was strongly stirred. After 15 minutes nitrogen gas was blown through the mixture to expel oxygen from the reaction vessel. After 1 hour solid beads had formed. Then another addition of monomer solution was performed, this time with no addition of initiator. 24.75 grams of acrylamide and 0.250 gram of N,N'-methylenebisacrylamide were dissolved in 50 ml of distilled water. Then the solution was added to the reaction mixture in the reaction vessel. After 4 hours the stirring was interrupted and the beads were washed with dilute acetic acid, toluene, acetone and water. The product was allowed to settle in a measuring glass, whereupon the yield was measured as 900 ml.

The rigidity of the obtained product was determined by subjecting it to flow tests. To these ends, the product was slurried in water and packed into a bed having a diameter of 5 cm and a height of 10 cm. A gradually increasing hydrostatic pressure drop (p) was applied over the bed, whereupon the linear flow (U) and the bed height (L) was measured for each value of p. The permeability of the bed was then calculated by applying the expression K = LU/p. If K is plotted on a logarithmic scale in a graph against the value of p/L a straight line is obtained. The point at which the line intersects the ordinate gives the value K_0. If the slope angle of the line is designated α the maximum flow (U_{max}) is then obtained for the gel bed from the equation

$$U_{max} = \frac{K_0}{\ln 10 \cdot tg\alpha \cdot e}$$

The exclusion limit, i.e., the upper limit of the molecular weight range within which a separation according to molecular size is obtained, was determined by a method in which some proteins of different molecular weights were chromatographed through a bed of the separating medium and that the K_{av} values of the proteins were calculated. K_{av} is a parameter which gives the penetration ability of the chromatographed substance in the separating medium and is defined as

$$K_{av} = \frac{V_e - V_0}{V_t - V_0}$$

V_e = elution volume
V_0 = void volume
V_t = bed volume

K_{av} = 0 means that the chromatographed substance is completely excluded from the pores of the separating medium. The measured K_{av} values for the different proteins were plotted in a graph against the logarithm of the molecular weight. The points in the graph were connected by a straight line and the approximate exclusion limit was obtained at the point where the line intersected the abscissa. In the present case U_{max} for beads having an average diameter of 150 micrometers in wet condition was calculated to be 315 ml/cm^2-hr and the exclusion limit was calculated to be approximately 600,000.

Example 2: The method was effected in a manner similar to that described in Example 1. In the first charge of monomers 22.5 grams of 1-vinyl-2-pyrrolidinone and 27.5 grams of N,N'-methylenebisacrylamide and in the second charge 18.7 grams of acrylamide and 0.2 gram of N,N'-methylenebisacrylamide were used. As the continuous phase n-heptane was used instead of toluene. The yield was 850 ml. Exclusion limit was approximately 1,000,000. The flow maximum (U_{max}) was 200 ml/cm^2-hr.

Example 3: The method was carried out in a manner similar to that described in Example 1. In the first charge 10 grams of acrylamide and 30 grams of N,N'-methylenebisacrylamide and in the second charge 42.8 grams of acrylamide and 0.8 gram of N,N'-methylenebisacrylamide were used. The yield was 900 ml. The exclusion limit was approximately 150,000. The flow maximum (U_{max}) was 230 ml/cm^2-hr.

Regenerated Cellulose as Reinforcing Agent

H. Determann and T. Wieland; U.S. Patents 3,598,245; August 10, 1971; and 3,597,350; August 3, 1971; both assigned to Pharmacia Fine Chemicals AB, Sweden describe a separation medium for gel filtration having an exclusion limit within the molecular weight range of from 10,000 to 10,000,000. The medium consists of grains of certain water-insoluble copolymers reinforced with a rigid hydrophilic macroporous material. The rigid hydrophilic macroporous material can be grains of regenerated cellulose having a bulk density of 2 to 25 g/dl, and a pore size in the range of from 2 to 2,000 μ.

The separation medium is characterized by the exclusion limit. This can be determined by packing the products of the process into beds and chromatographing a series of homologous substances of different average molecular weights on the bed. The degree of penetration, K_{av}, can then be calculated for the respective homologues. K_{av} is defined by the equation

$$K_{av} = \frac{V_e - V_0}{V_t - V_0}$$

wherein V_e is the elution volume for the homologue in question, V_t is the volume of the bed and V_0 the void volume of the bed. V_0 is determined as V_e for a substance which cannot penetrate the grains, e.g., bacteria.

Figure 8.1a illustrates K_{av} versus average molecular weight for dextrans chromatographed on products prepared according the the process, and Figure 8.1b illustrates a flow test of reinforced and nonreinforced products.

Figure 8.1a shows K_{av} versus average molecular weight for dextrans, used as homologous substances, chromatographed on the products prepared according to Example 4 (upper curve) and Example 2 (lower curve) of the following examples. The exclusion limit, i.e., the molecular size limit up to which the product in question can fractionate in accordance with molecular size, is obtained at the point where the curve intersects the abscissa, and is given in the following example to characterize the different products.

It has been found that if globular proteins are used as homologous substances an approximately twice as large exclusion limit is obtained. In order that a separation medium can be used for separating substance mixtures, at least one of the substances in the mixture should have a molecular size which lies beneath the exclusion limit. Every separation medium has a suitable fractionating area which lies under the exclusion limit. In order that the substances of a substance mixture can be fractionated at all it is necessary that at least one of the substances have a molecular size within the fractionating area.

Example 1: 12 grams of cellulose powder (Munktell 400) were dissolved in a solution of 7.2 grams copper-II-hydroxide and 0.75 gram copper-I-chloride in 300 ml of concentrated ammonium hydroxide solution. The solution of cellulose was emulsified in 400 ml of toluene, 8 grams of a reaction product of castor oil fatty acids and ethylene oxide (Cremophol EL) being dissolved as an emulsifier. Subsequent to stirring for 10 minutes with an impeller type agitator at a speed of 500 rpm the emulsion was poured out in the form of a jet, down into a well-stirred regenerating bath consisting of a solution comprising 750 grams of benzoic acid in 9 liters of toluene.

FIGURE 8.1: SEPARATION MEDIUM FOR GEL FILTRATION

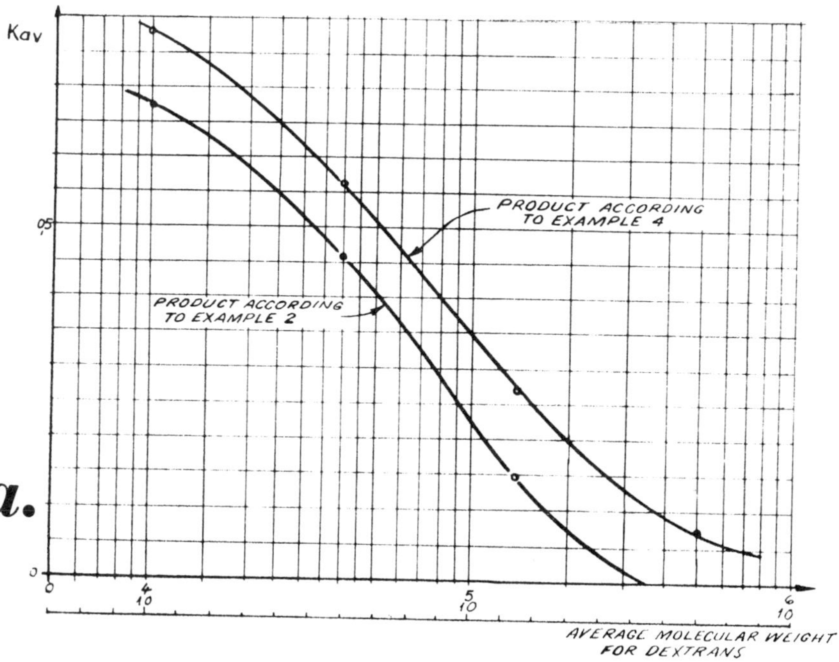

a.

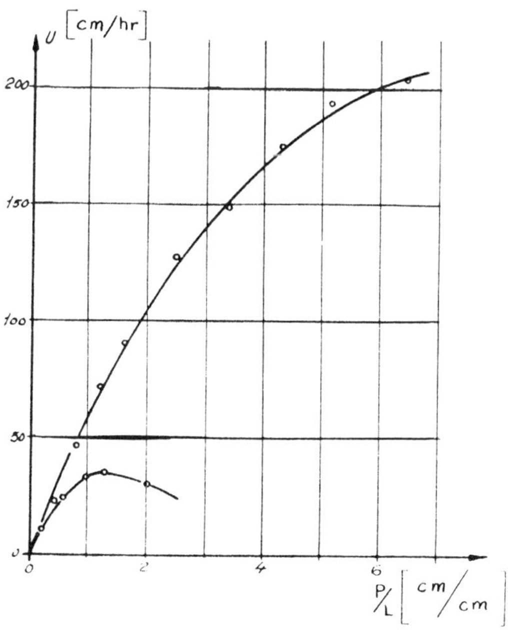

b.

Source: U.S. Patent 3,598,245

Ten minutes after the emulsion had been added, stirring was interrupted and the cellulose pearls and precipitated ammonium benzoate were permitted to sediment overnight. The remaining liquid was then removed by suction and the sediment was slurried in a solution of 120 grams of sodium hydroxide in 3 liters of water by agitation, until all the sodium benzoate had passed into solution. The cellulose pearls were then allowed to settle for approximately 3 hours, whereafter the then remaining liquid was removed by suction, washed with running water on a screen cloth having a mesh width of 0.040 mm. The pearls were then treated with 300 ml of 50% acetic acid for the purpose of removing copper salt rests, whereafter they were rewashed on the same screen cloth with water. The pearls were simultaneously passed through a screen cloth having a mesh width of 0.25 mm, for the purpose of separating coarse material.

The cellulose pearls obtained in accordance with the aforegoing were drained on a filter, whereafter 150 grams of pearls were mixed with 135 ml of water containing 10 grams of dextran with the limiting viscosity number 1.1 dl/g, and 7.5 grams of sodium hydroxide. The mixture was suspended in 600 ml of 1,2-dichloroethane, wherein 45 grams of cellulose acetate butyrate had been dissolved while stirring at 300 rpm with an agitator in a cylindrical reaction vessel. 15 ml of epichlorohydrin were then added. The reaction was allowed to proceed at 50°C for 16 hours.

700 ml of ethyl acetate were then added and, subsequent to sedimentation of the cellulose-containing pearls, the pearls were slurried five times in 300 ml of ethyl acetate. The pearls were then treated with 300 ml of 1 M sodium hydroxide solution, for hydrolyzing remaining acetate and butyrate groups. The product was then washed with running water on a screen cloth having a mesh width of 0.040 mm, treated with 300 ml of 50% acetic acid and rewashed on the screen cloth with water. Subsequent to sedimentation of the product in a measuring glass, a yield of 87 ml was read off. The product has an exclusion limit for dextran of approximately 100,000.

Example 2: Cellulose pearls obtained in accordance with Example 1 but with twice the quantity of cellulose were drained on a filter, whereafter 210 grams of the pearls were mixed with a solution of 12 grams of dextran having the limiting viscosity number 0.40 dl/g and 8 grams of sodium hydroxide in 200 ml of water. The mixture was suspended in 800 ml of 1,2-dichloroethane, in which 60 grams of cellulose acetate butyrate had been dissolved. 120 ml of glycerol-bis-epoxy-propyl ether (technical quality, Epicote 812) were then added.

The reaction was continued at 50°C for 16 hours. 1,000 ml of ethyl acetate were then added and, subsequent to sedimentation, the product was slurried five times in 5,000 ml of ethyl acetate. The product was then treated with 500 ml of 1 M sodium hydroxide solution, washed on a screen cloth with running water, treated with 500 ml of 50% acetic acid and then washed on the screen cloth with running water. 185 ml of a product having an exclusion limit for dextran of 350,000 were obtained.

Example 3: 210 grams of drained cellulose pearls prepared in accordance with Example 1 were mixed with a solution of 20 grams of dextran, having a limiting viscosity number of 1.1 dl/g, and 8 grams of sodium hydroxide in 200 ml of water. The mixture was suspended in 800 ml of 1,2-dichloroethane containing 60 grams of cellulose acetate butyrate. 24 ml of 1,4-butanediol-bis-epoxy-propyl ether were then added. The reaction was continued at 50°C for 16 hours. Subsequent to working up the product in accordance with Example 2, 96 ml of product were obtained with an exclusion limit for dextran of approximately 1 million.

Example 4: 140 ml of a product having an exclusion limit for dextran of approximately 1 million were obtained in a manner similar to that described in Example 3 from 20 grams of polyvinyl alcohol with a viscosity of 4 cp (4% solution at 20°C) and with a degree of hydrolysis of 89%.

Example 5: 100 ml of cellulose pearls obtained in accordance with Example 1 were mixed with a solution of 14 grams of dextran, having the limiting viscosity number 1.1 dl/g, and 5 grams of sodium hydroxide in 81 ml of water. The mixture was suspended in 600 ml of toluene, in which had been dissolved 24 grams of sorbitanlaurate (Span 20) and 24 grams of polyoxyethylene-sorbitanlaurate (Tween 20); 10 ml of epichlorhydrin were then added. The reaction was continued at 50°C for 16 hours. Subsequent to washing with running water on a screen cloth, 102 ml of a product having an exclusion limit for dextran of approximately 100,000 was obtained.

For the purpose of verifying the good flow properties of the products obtained in accordance with the process one such product was subjected to a flow test. This product was prepared in accordance with Example 2, although starting from cellulose pearls prepared in accordance with Example 1. The product was slurried in water and packed to form a bed having a diameter of 2.5 cm. A hydrostatic pressure drop (p) was then progressively applied over the bed, whereupon the linear flow (U) and the bed height (L) were measured at each value on p.

The result of the flow test is shown in Figure 8.1b, which shows U in cm/hr as a function of p/L for the product prepared according to Example 2 (upper curve) in comparison with a similar, nonreinforced product (lower curve) known in the art (Sephadex G-200) having an exclusion limit of approximately 200,000. Since the grains of this product were somewhat coarser than those of the products of the process, the latter having an average grain size of 0.140 mm, the flow data of the product was corrected in respect thereof, it being known that the flow is proportional to the square of the average grain size. It is evident from Figure 8.1b that a product produced in accordance with the process presents considerably better flow properties than nonreinforced gels of a similar type. The differences are still greater in beds of larger diameter, since the supporting effect of the column walls is then less.

Agarose Gel Product for Salting Out Adsorption

J.O. Porath and J.G. Rosengren; U.S. Patent 3,960,720; June 1, 1976; assigned to Exploaterings Aktiebolaget TBF, Sweden describe a process which concerns a gel product for chromatography based on agarose, and a method of using the product for hydrophobe salting out adsorption. The gel product is characterized in that the hydrophobe groups are bonded to the agarose with ether or ester bonding, and that the hydrophobe groups can be composed of aliphatic hydrocarbons, halogen or hydroxy-substituted alkyl groups, with 1 to 20 carbons, or contain 1 or more, up to 5, aromatic nuclei, unsubstituted or substituted with alkyl, alkyloxy, halogen or NO_2 or have the formula $-CH_2-CH(OH)-CH_2-OR$ where R can be alkyl, alkenyl, aryl, aralkyl, aralkoxy, alkaryl, optionally substituted with hydroxy, halogen or nitro, and where the alkyl parts of R contain 1 to 20 carbons. The method of separating the amphipatic (containing hydrophobic as well as hydrophilic groups) substances is based on the fact that they are adsorbed by an amphipatic gel in which the salt concentration in the solution is adjusted beyond the level at which the ion concentration least corresponds to the ion concentration in an aqueous solution of 1 M NaCl, and desorption takes place through lowering the ion concentration and/or lowering the pH and/or lowering the polarity of the solvent. The amphipatic substances are primarily proteins or peptides.

The chromatographic uses of the process are illustrated in the Figure 8.2. Figure 8.2a shows the adsorption at 280 nm in eluate of serum adsorbed on a gel according to the process as a function of the elution volume. Figure 8.2b shows an adsorption curve at 254 nm of an extract of brown beans fractionated through a bed of gel product. A number of different studies are given in Example 5 of adsorption conditions for a series of gels used.

Example 1: To 6% agarose crosslinked with dibromopropanol in the form of beads swelled in water with a diameter of 30 to 200 microns in an amount of 200 ml drained gel was added 200 ml 5 M NaOH, 50 ml benzyl chloride and 1 gram sodium borohydride. The suspension was heated to 80°C during agitation and refluxing. The reaction was stopped after 5 hours and the gel product was transferred to a suction funnel, and was washed successively with water, ethyl alcohol and water.

Example 2: To 250 ml drained 6% agarose gel crosslinked with epichlorohydrin there was added 250 ml 5 M NaOH solution. 20 grams solid NaOH was dissolved and 60 ml ethyl bromide and 1 gram $NaBH_4$ was added thereto.

The reaction occurred in nitrogen gas in an autoclave for 6 hours at 80°C. After cooling the gel was transferred to a filter and was washed with water, alcohol and water. After drying the gel swells in ethanol in contrast to the original gel.

Example 3: 25 ml of 6% agarose crosslinked with epichlorohydrin and desulfated was mixed with 25 ml 5 M NaOH, 10 ml allyl bromide and 50 mg sodium borohydride. The suspension was kept at 80°C for 3 hours with refluxing and was then transferred to a filter, was washed with water, ethyl alcohol and finally water. 6 grams of the gel discolored 1.2 ml of a bromine solution with 10 ml Br_2/250 ml. According to elementary analysis the dry gel contained 39.8% Br.

Example 4: The Effect of Different Salts on the Adsorption Capacity of Benzyl-Sepharose 6B According to Example 1 — The experimental conditions were: Column dimensions—column length = 1.5 cm, total volume V_t = 1.21 ml; Buffer—0.05 M tris-CHI buffer, pH 7.5 with an addition of extra salts according to the table below.

Trials were carried out as front analyses with cytochrome C in the concentration 1 mg/ml. The breakthrough volume of the front = V_e. The table shows the hydrophobic salting out ability of various salts. They further support the potential possibilities of the hydrophobic salting out chromatography since the choice of the salt significantly affects the interaction of the various proteins with the gel. This can be exploited to increase the separation capacity of the gels.

FIGURE 8.2: HYDROPHOBIC SALTING OUT ADSORPTION PROCESS

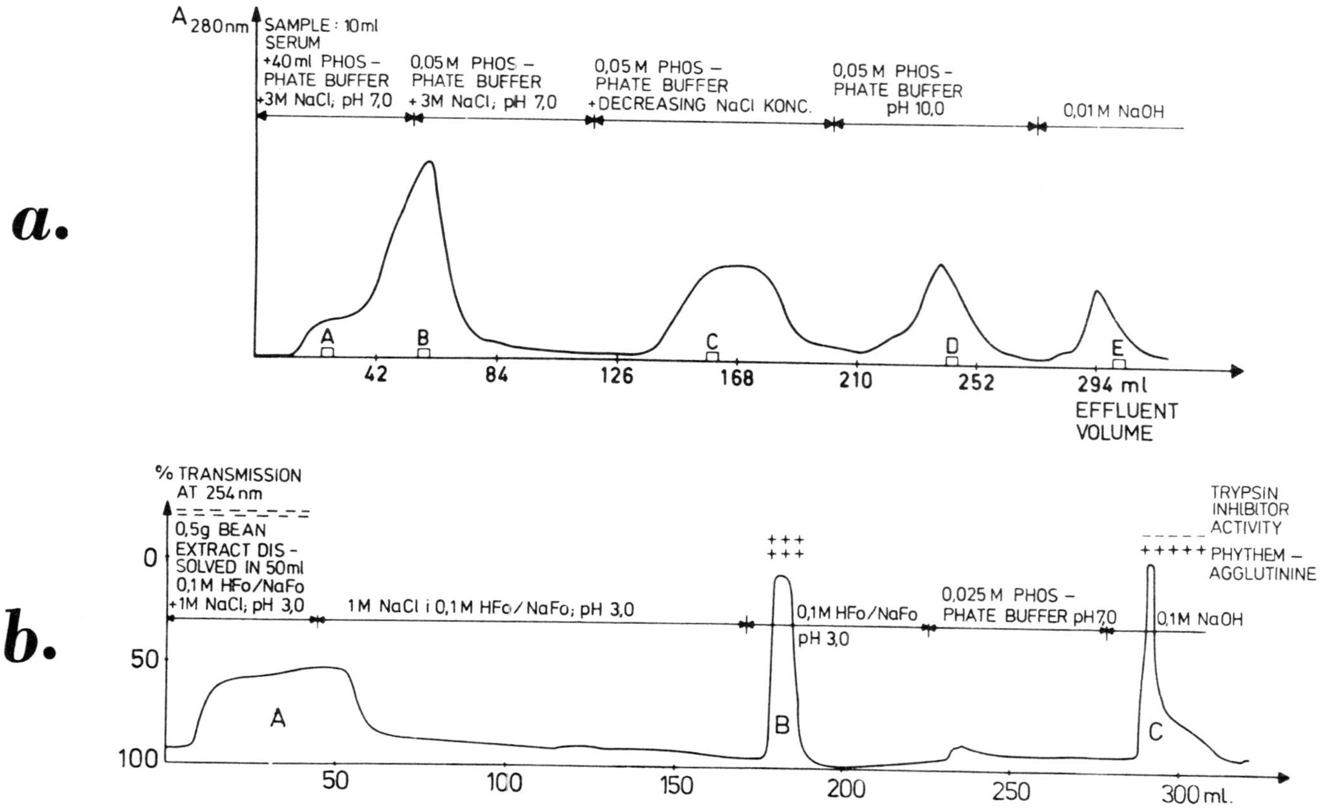

Source: U.S. Patent 3,960,720

Salt	Conc, M	Ion Strength of the Added Salt, M	V_e/V_t
NaCl	3	4.5	9.9
KCl	3	4.5	9.5
KBr	3	4.5	8.25
KF	3	4.5	8.0
LiCl	3	4.5	6.9
NH_4Cl	3	4.5	4.4
NH_4-formate	3	4.5	5.15
$MgCl_2$	1.5	4.5	5.8
$CaCl_2$	1.5	4.5	12.0
$BaCl_2$	1.5	4.5	14.5
Na_2SO_4	1	3	4.0
Na-trichloroacetate	0.5	0.5	8.7
NaCl/glycine	3/2	6.5	18.3
NaCl/N,N-dimethylformamide	3/2	4.5	2.12
Pure tris buffer	0.05	0	2.7
0.05 M phosphate buffer	0.05	0	1.47

Example 5: The following table illustrates some trials carried out with front analysis of cytochrome C by Example 4. The values for pH and ion strength are given in the table. The capacity values under different conditions are given as V_e/V_t. At pH 7, 0.05 M phosphate buffer was used and at pH 3, 0.1 M formate buffer.

Gel of Example	Hydrophobic Substituent	Capacity, V_e/V_t — Molarity NaCl, pH 7.0: 0	1	3	Molarity NaCl, pH 3.0: 0	1	3
	Control (agarose)	–	–	–	–	0.73	0.72
2	Ethyl	–	–	–	–	0.88	2.2
3	Allyl	–	–	0.95	–	5.2	25
1	Benzyl	0.90	1.6	10	3	20	55

Example 6: Crosslinked benzylated agarose according to Example 1 was packed in a column with the dimensions 1.4 x 8 cm. The gel was equilibrated with 0.1 M formate buffer pH 3.0 containing 3 M NaCl. 1 liter urine concentrate was pH adjusted to 3.0 and was pumped through the benzylated agarose. At elution with various solutions the proteins in the urine concentrate were separated in different fractions.

Metal-Containing Polymers

A process described by *W.J. Leonard, Jr.; U.S. Patent 3,779,952; December 18, 1973; assigned to Shell Oil Company* relates to porous, metal-ion-chain-bridged polymeric products. These materials are useful as catalysts, as substrates for gel permeation chromatography because of their pore size distribution and as molecular sieves.

These salts are rigid and highly porous and comprise polymerized ethylenically-unsaturated carboxylic acids in combination with from about 0.2 to 1.0 mol per equivalent of polyacid of one or more Group III-b to V-a metals. Such salts are prepared by the process of:

(A) contacting at moderate temperature a dilute solution of polymerized ethylenically-unsaturated carboxylic acids in polar non-aqueous solvent, and at least about 0.2 mol, per equivalent of polyacid, of ions of at least one Group III-b to V-a metal,

(B) recovering the resulting gelatinous precipitate of swollen metal-polyacid product, and

(C) removing residual solvent from the gelatinous product of step (B) to form a free-flowing powder of the rigid porous metal-ion-chain-bridged polyacid salt.

The polymeric acids used in preparing the metal-containing polymers consist essentially of polymers or copolymers derived solely from hydrocarbon carboxylic acid monomers of up to 10 carbon atoms possessing at least one carboxy group and at least one ethylenic linkage, i.e., a nonaromatic carbon-carbon double bond. The acids may also contain additional carbon-carbon unsaturation in the molecules, such as aromatic unsaturation. Suitable acids include, for example, vinylhydrogenphthalate, vinyl benzoic acid, acrylic acid and maleic acid. Preferred polymeric acids are polymers of lower alkenoic acids which contain from 1 to 2 carboxy groups, from 1 to 2 ethylenic linkages as the sole carbon-carbon unsaturation and contain from 3 to 6 carbon atoms including the carboxy carbon atom(s).

Illustrative of monocarboxylic alkenoic acids of this type, from which suitable polymerized acids can be prepared, are acrylic acid, methacrylic acid, crotonic acid, 3-pentenoic acid, sorbic acid, 2-methyl-2-butenoic acid and ethacrylic acid, while illustrative dicarboxylic alkenoic acid from which suitable polymers can be made include maleic acid, fumaric acid, citraconic acid, mesaconic acid, itaconic acid, 2-methyleneglutaric acid and 2-pentendioic acid.

In general, polymers of monocarboxylic acids are preferred over analogous polymers of dicarboxylic acids, and further preferred are those of monocarboxylic alkenoic acids where the ethylenic linkage is conjugated with the carboxy group and is a terminal ethylenic linkage. Most preferred are polymers of acrylic acid and the alpha-alkyl derivatives where the alkyl has from 1 to 3 carbon atoms, particularly 1 carbon atom, i.e., methacrylic acid.

It is essential that the acid be polymerized prior to combination with the metal ions if the desired porous product is to be obtained. Suitably the acid is polymerized to an average molecular weight of at least 500. The upper limit on the average molecular weight of the polymeric acid is not critical. Polymerized acids having an average molecular weight of up to about 2,000,000 can be used. Polymerized acids of average molecular weight of from about 2,000 to about 500,000 are preferred, giving excellent porous products and having generally low viscosities which provide processing and handling advantages.

The polymerized acid is employed as a dilute solution in a suitable inert solvent as defined below. The concentration of the polyacid in solution is suitably less than 2 equivalents of acid per liter. Preferably, the polyacid concentration is from 0.1 to 1.5 equivalents per liter, most preferably it is from 0.2 to 1 equivalent per liter. The following examples illustrate the process.

Example 1: (A) A solution of acrylic acid in dioxane containing 0.1 weight percent azobisisobutyronitrile and 0.06 weight percent azobiscyclohexanenitrile was held 16 hours at 90°C to give a polyacrylic acid having an average molecular weight, as determined by viscosity measurement, of about 1,000,000. A 2,000-milliliter portion of this polyacrylic acid solution containing 0.55 equivalent of acid was heated to 50°C and maintained at that temperature. Gradually, with stirring, a solution of 0.65 mol of nickel(II) hexanoate in 2,000 ml of dioxane, was added to the polyacrylic acid over a period of 1.5 hours. A gelatinous mass of pale green precipitate formed and was centrifuged from the bulk of the solvent. The precipitate was washed several times with dioxane to remove unreacted metal salts. Occluded solvent was removed to dryness at 100°C to leave 29 grams of a pale green powder which had a pore volume of 0.16 cc/g and a surface area of 125 m^2/g as determined by nitrogen adsorption. The median pore diameter was 70 A. Analysis of this product showed it contained:

	Percent by Weight
Ni	17.4
C	42.2
H	5.3
O (by difference)	35.1

Thus, the final product contained 0.51 equivalent of nickel(II) per equivalent of acid.

(B) A 0.36-gram sample of the product of part (A) was used to catalyze the hydrogenation of olefins. The catalyst, 6.0 grams of n-heptane solvent, 0.18 gram hexene-1 and 0.18 gram of cyclohexene was charged to a bomb, which was heated to 300°C and then pressured to 605 psig with hydrogen. After 50 minutes, 15% of the olefins had been hydrogenated.

Example 2: Using the general procedure of Example 1, a copper/aluminum-polyacrylate product was prepared. Over a period of 1.5 hours, 1,000 ml of a dioxane solution containing 0.049 mol of cupric bistetraisopropoxyaluminate, $Cu[Al(O\text{-}i\text{-}Pr)_4]_2$, was added to 3,000 ml of a 0.13 molar dioxane solution of polyacrylic acid (average molecular weight 10,000). The temperature was 50°C. A gelatinous mass formed which was separated by centrifugation and vacuum at 100°C for 5 hours to yield 41 grams of a fluffy pale blue porous product. This product had a pore volume of 0.32 cc/g and a surface area of 179 m^2/g and contained 7.1% by weight copper and 7.0% by weight aluminum or 0.19 equivalent of copper and 0.65 equivalent of polyacid.

Example 3: (A) Using the general procedures of Example 1 a copper polyacrylate having pores with a volume median pore diameter of 500 A was formed. 300 ml of dioxane containing 0.0134 mol of copper(II) acetate $[Cu(OAc)_2{\cdot}H_2O]$ was added over a period of 5 minutes to 50 ml of dried dioxane containing 1.0 gram of polyacrylic acid. The temperature was maintained at about 80°C during the addition of the cupric salt to the polymer. The resulting mixture contained a light gelatinous blue precipitate.

After cooling slowly to 25°C, the precipitate was filtered, washed with fresh dioxane and heated to dryness at 60°C for 10 hours to remove residual solvent. Approximately 1.4 grams of a copper polyacrylate product was obtained. This material was insoluble in water and all common organic solvents such as acetone, methanol, carbon tetrachloride and dimethylsulfoxide. It had a pore volume of 0.6 cc/g and a BET specific surface area of 145 m^2/g. Analysis of the product showed it contained 30 weight percent Cu.

(B) Adsorption isotherms for hexane and cyclohexane showed essentially no difference in capacity for either hydrocarbon. The large pores obtained in this material indicate its utility in macromolecular separations, e.g., gel permeation chromatography.

Example 4: (A) 3 grams of polyacrylic acid as prepared in Example 1 were dissolved in 300 ml of dimethylformamide. The polymer solution was heated and maintained at 120°C during addition of a solution of 0.027 mol of $Cu(OAc)_2 \cdot H_2O$ in 35 ml of dimethylformamide. The combined solution was distilled under a nitrogen blanket at 310 mm mercury with occasional addition of dimethylformamide to maintain volume. The distillate contained acetic acid. The mixture was cooled and the 4.0 grams of a green gel precipitate which formed was separated by centrifugation, rinsed and dried. The dried polymer was a powder and had a pore volume of 0.22 cc/g and a BET specific surface area of 257 m^2/g. The volume median pore diameter was less than 30 A. Analysis of the product showed it contained:

	Percent by Weight
Cu	28.0
C	35.4
H	4.0
O	31.8

Thus, the final product contained 0.96 equivalent of Cu(II) per equivalent acid.

(B) Adsorption isotherms for hexane and cyclohexane were measured. The volume capacity of hexane was approximately equal to the N_2 pore volume and twice that found for cyclohexane. These isotherms show that size separations can be performed using polymers prepared in this way.

THIN LAYER CHROMATOGRAPHY

Colloidal Silica

B. Englebrecht, H. Walker, J. Blome and F. Jordan; U.S. Patents 3,726,405; April 10, 1973 and 3,883,435; May 13, 1975 describe an adsorbent which comprises a mixture of adsorption active powder and from 1 to 50% by weight of silicic acid in the form of a silica sol having a content of from 1 to 15% of colloidal silicic acid, based on the weight of adsorption active powder. Preferably, the silica sol contains from 5 to 20% by weight of silicic acid. Examples of suitable adsorption active powders include silica gel, aluminum oxide, kieselguhr and magnesium silicate. The preferred adsorption active powder is silica gel. The preferred silica gel is a standard thin layer chromatography silica gel having a particle size of less than 50 μ.

The silica sol used as binder is a sol in which low molecular weight polymeric silicic acids are incoherently colloidally dissolved. Such silica sols have heretofore been used in, for example, the preparation of nonadhesive cellulose hydrate films or for fining wine and fruit juices. They are commercially available and represent odorless liquids which, in thin layer, are almost clear or slightly opalescent and bluish when viewed from above. They may be prepared, for example, according to the processes described by R. Griessbach in the *Chemiker-Zeitung,* Volume 57 (1933), pages 253 et seq. A typical preparation is by peptization of purified SiO_2 gelatins.

The products of this process may be used in layer chromatography, either previously activated or not. Activation is obtained, for example, by heating to temperatures of from 100° to 200°C.

Particularly good results with respect to abrasion resistance, minimum splintering tendency, resistance to being written on, resistance to water, etc. are obtained when using adsorbents consisting of a mixture of 1 part by weight of silica gel DC and 2 parts by weight of a 4.5% silica sol solution or of a mixture of 2 parts by weight of aluminum oxide DC with 3 parts by weight of a 5% silica sol solution. The following examples illustrate the process.

Example 1: 100 grams of silica gel DC were mixed by stirring with an amount of silica sol corresponding to 8 grams of silicon dioxide and with a total of 200 grams of H_2O. The spreadable mass was spread on previously cleaned glass sheets according to E. Stahl, *Dunnschichtchromatographie,* Berlin-Heidelberg-Gattingen 1962, and the coated sheets were air-dried. The resulting layer was abrasion-resistant and could be used in layer chromatography, either activated previously or not.

Example 2: 6 grams of a 15% silica sol and 14 grams of water were added to 10 grams of silica gel DC. Upon thorough mixing the mass was spread on aluminum films in conventional manner. By exposing the films to air, the water was evaporated. The resulting layer was abrasion-resistant and could be used in layer chromatography.

Example 3: 100 grams of silica gel DC were mixed by stirring with an amount of silica sol corresponding to 10 grams of silicon dioxide and with 200 grams of H_2O. The mass was spread on previously cleaned glass sheets in a known manner and the coated sheets were air dried. The resulting layer was abrasion-resistant and could be used in layer chromatography, either previously activated or not.

Example 4: To 100 grams of aluminum oxide DC was added a silica sol solution in an amount sufficient to provide 1,5% by weight of SiO_2 per 100 grams of aluminum oxide. 120 ml of water was added thereto. The spreadable mass was spread on previously cleaned glass sheets in a known manner, and the coated sheets were air dried. The resulting layer was abrasion-resistant and could be used in layer chromatography, either previously activated or not.

Silicic Acid, Cationic Starch and Glass Fibers

A process described by *E.W. Malcolm; U.S. Patent 3,647,684; March 7, 1972; assigned to The Dexter Corporation* is concerned with an improved differential restraining phase or chromatographic medium particularly applicable to thin layer chromatographic techniques. The medium advantageously contains the adsorbent material as a major component thereof while taking the form of a thin, soft, flexible structure of substantially uniform and homogeneous composition capable of being accurately reproduced so as to provide constant and controlled adsorbent characteristics which can be relied upon to give reproducible and accurate results. The chromatographic medium is provided with structural and dimensional integrity by the incorporation of water-dispersible fibers, preferably of an inorganic, chromatographically dormant nature which need not be entirely of submicron diameter and therefore lend additional and enhanced processability to the manufacturing operation and added strength to the medium. The improved structure is generally produced in sheet form in a manner similar to that described in U.S. Patent 3,253,978.

In other words, the medium is preferably formed on a papermaking machine as continuous, fibrous webs. The sheets are generally formed as medium to heavy basis weight material, i.e., about 50 to 300 pounds, with continuous lengths of such webs preferably having a basis weight of about 100 to 200 pounds. The term basis weight as used herein refers to the weight in pounds of 480 sheets of the medium having planar dimensions of 24 x 36 inches.

Although most adsorbents suitable for use in column or thin layer chromatography can be utilized in the process, the material considered most generally useful is silicic acid (hydrated silica). Alumina is also employed to a substantial degree, although less than silicic acid. In general, the adsorbent may include organic or inorganic materials depending upon their adsorption activity and the type of separation required of the medium. A typical chromatographic medium will have the following compositions:

Materials	Percent by Weight
Silicic acid	75.0
Glass fibers:	
0.5 micron diameter	14.75
2.0 micron diameter	8.0
9.0 micron diameter	2.0
Organic cationic agent	0.25

The cationic starches generally exhibit the most beneficial results and consequently are the preferred organic additives. Such starches, typically used at concentrations of 0.25% by weight, are generally amine modified starches which are considered to be starch ethers using an amino or imino group to provide a positive charge. The starches may be derived from conventional sources such as corn, wheat, potato, tapioca, waxy maize, sago or rice. Exemplary of such materials are the amine modified starches.

The process utilized to manufacture the chromatographic medium broadly involves the admixture in water of the adsorbent material and fibers to form a dilute, substantially homogeneous, aqueous dispersion or slurry. The thorough mixing of the components can be effected by adding a dispersion of the silicic acid to the fibers suspended in water within a suitable mixing device such as a paper mill beater. A solution or dispersion of the organic cationic material is then added to the adsorbent-fiber dispersion and the pH and temperature of the resultant admixture is appropriately adjusted prior to formation of the sheet structure on a papermaking wire, such as a conventional Fourdrinier wire.

The sheet material is made in accordance with conventional papermaking techniques and is a nonwoven web structure wherein the integrity of the sheet is maintained primarily by the physical interengagement of the individual fibers as deposited on the web-forming wire of the papermaking machine. As is known, the solids or dispersions of solids are admixed and thoroughly dispersed in an aqueous medium by means of a paper mill beater. The resultant mixture with its pH and temperature adjusted is then conveyed to the head box of a papermaking machine where it is further diluted and fed onto the continuous fiber accumulating, paper-forming wire, such as a Fourdrinier wire. The cationic material may be added to the fiber dispersion at the mixing stage or after the fiber slurry has been fed to the head box and as part of the diluting operation.

Although substantially all commercial papermaking machines, including the regular Fourdrinier or rotary cylinder machines, may be used, it is desirable where very dilute fiber furnishes are employed to use an inclined fiber collecting wire, such as that described in U.S. Patent 2,045,095. The fibers flowing from the head box are retained on the wire in a random three-dimensional reticulate network or configuration with slight orientation in the machine direction, while the dispersing aqueous phase quickly passes through the wire and is rapidly and effectively removed.

It has been found in accordance with the process that optimum chromatographic performance is obtained when the fiber furnish containing the cationic starch is maintained at an acid pH and a slightly elevated temperature, e.g., a pH of about 3 and temperature of about 120°F. This is particularly surprising since sheet materials produced without the cationic material show no substantial effect as a result of temperature change and heretofore the pH of the furnish was believed to have no noticeable effect on chromatographic performance.

The temperature utilized may be broadly categorized as warm since the fiber furnish or slurry need only be at a temperature of from about 20° to 70°F above room temperature, i.e., within the range of about 90° to 140°F. For consistently acceptable results operating temperatures within the range of 95° to 130°F are conventionally employed with a typical operating temperature in excess of 140°F being particularly undesirable at low pH levels as the hot acid solutions more readily attack the solids within the furnish under such conditions. The following examples illustrate the process. Unless otherwise specified all parts are given by weight.

Example 1: A fiber slurry was made by initially dispersing about 90 grams of glass fiber having an average diameter of 0.2 to 0.5 micron (Code 104 microglass), 40 grams of glass fiber having an average diameter of 0.5 to 0.75 micron (Code 106 microglass) and 15 grams of glass fiber having an average diameter of 1.6 to 2.6 microns (Code 110 microglass) in about 18 liters of water adjusted to a pH of 3.0 with HCl. To this dispersion was added 10 grams of ½-inch long glass rovings having an average diameter of about 9 microns and 400 grams of fluorescent silicic acid (SilicAR TLC-4F).

The resultant slurry was mixed for about 15 minutes in a standard laboratory Valley beater after which a portion was added to a sheet mold and adjusted to a temperature of 120°F and a pH of 3. To this portion was then added a 1% aqueous solution of cationic starch (Cato 15) in an amount sufficient to produce a starch concentration of 0.25% by weight based on the dry weight of the solids in the slurry. A hand sheet was formed, dried between blotters at a light pressure of approximately 10 psi and then dried on a laboratory drum dryer at approximately 230°F. The hand sheet exhibited a tensile strength of approximately 700 g/inch.

The chromatographic performance of the sheet was evaluated by visual observation of the developed chromatogram after using the following procedure. The sheets were conditioned in a circulating air oven for 15 minutes at 70°C. Samples of the standard three component Stahl dye mixture in both microliter amounts and in amounts approximately five times the maximum recommended amount for thin layer chromatograms were applied to a starting line on the sheet and the sheet developed by standard chromatographic ascending procedures using benzene as the developing liquid. The solvent front was allowed to advance 10 centimeters from the starting line at which point the material was removed from the developing tank and allowed to air dry.

The sharpness and length of the yellow component of the Stahl dye mixture was then examined together with the degree of separation or resolution of the chromatogram. Hand sheets produced as set forth above showed excellent separation of the three dyes present in both test samples of the Stahl dye mixture.

Example 2: The procedure outlined in Example 1 was followed except the silicic acid was replaced by an equal amount by weight of aluminum oxide, and the concentration of cationic starch in the slurry was 0.3% instead of 0.25%. Since the chromatographic aluminum oxide employed does not adequately separate the yellow component of the Stahl dye mixture, the red component was used as a measure of chromatographic performance. After running the chromatogram it was found that adequate separation occurred between the blue and red components with uniform spot size in both test samples.

Example 3: Using a standard laboratory beater with 18 liters of water at 100°F and a pH of 3, a slurry of glass fibers was prepared from 51 grams of Code 104 microglass, 12 grams of Code 106 microglass and 6 grams of Code 110 microglass. The fibers added to the beater were defibered and lightly brushed to clear the glass fibers after which 6 grams of ½-inch rovings and 300 grams of fluorescent silicic acid (SilicAR TLC-4F) were added and defibered to give a uniform suspension. Sufficient cationic starch was added to give a starch concentration of 0.05% by weight and the material in the beater was then used to make hand sheets having a basis weight of 175 pounds per 2,880 square feet. Sheet mold conditions were maintained at a temperature of 100°F and a pH of 3.

After formation and removal from the wire, the wet sheet was dried without wet pressing. Chromatographic evaluation was made in accordance with the procedure set forth in Example 1 and separation of the components in the Stahl dye mixture was judged acceptable but minimum.

Sheets made in the same manner except that the sheet mold conditions included a temperature of 70°F and a pH of 7 gave spot sizes ranging from 150 to 200% longer indicating poor and unacceptable chromatographic performance.

Example 4: Sheet material was produced on an experimental, sloped wire, Fourdrinier papermaking machine. About 250 gallons of water were supplied to the beater with the pH adjusted to 3 with HCl and a temperature adjusted to the 80° to 90°F range. To the beater was added 23 pounds of Code 104 microglass, 10 pounds of Code 106 microglass and 4 pounds of Code 110 microglass. The microglass was then given a light brush to clear the glass fibers. Next, 3 pounds of ½-inch glass rovings were added together with 100 pounds of fluorescent silicic acid (SilicAR TLC-4F). The beater furnish was then defibered for 20 minutes to thoroughly admix the components and 5 gallons of a 0.1 lb/gal cationic starch solution (Cato 15) was then added to the beater. The suspension after mixing for 5 minutes was then pumped to the machine chest with the addition of sufficient water to yield a 3% consistency in the tank.

The water in the headbox was adjusted to a pH of 3 with HCl and held at a temperature of approximately 120°F. Normal procedures were followed to form the wet fibrous web. After formation and water removal, the material was removed from the Fourdrinier wire and passed through a nip (approximately 10 pli nip pressure) and dried on the standard drying drums. Machine speed was approximately 25 feet per minute and the basis weight of the finished product was approximately 90 pounds per 2,880 square feet. Chromatographic evaluation of the material showed the material to have satisfactory chromatographic performance.

The material had an MD tensile of approximately 300 grams and a CD tensile of approximately 200 grams. The density of the product was approximately 0.36 g/cc, and the porosity was approximately 1.7 cfm as measured on the Frazier porosity instrument. The material possessed adequate strength to be handled and processed in the normal paper converting operations.

Thin Layer Medium for Molecular Diffusion

H.J. Bixler; U.S. Patent 3,878,100; April 15, 1975; assigned to Marine Colloids, Inc. describes a thin layer medium for use in molecular diffusion or affinity separation processes having a layer of hydrated gel material, the surface of which is scored or cross-scored at intervals of 0.05 to 1 millimeter.

The medium of the process is useful in a variety of molecular diffusion or affinity separation procedures including chromatographic and electrophoretic processes, and including radioimmunoassay procedures. While the medium is preferably in the dry form for convenience in shipping and storing, being hydrated simply by immersion in water shortly before use, it can also be maintained in hydrated form for shipping and storing by packaging in moisture-tight containers.

Example: There was prepared an aqueous solution containing 1% by weight of agarose, 0.5% sodium alginate, 0.25% sorbitol and 0.15% polyethylene glycol (molecular weight 400). A 5-milliliter portion of this solution was spread on a polyethylene terephthalate support film approximately 2⅝ x 3¼ inches in dimensions to form a layer about 1 mm thick. It was allowed to stand at room temperature until it gelled, then dried at 40°C in a forced hot air oven.

Using a steel ruler and a scalpel parallel cuts or score lines were made in the surface of the dry film at closely spaced (ca 0.5 mm) intervals and then an additional set of score lines crossing the first set at approximately right angles was made, the second set having approximately the same spacing as the first set. In each case the score lines penetrated approximately two-thirds of the thickness of the dried gel layer to form in fact a layer of dried agarose particles separated from each other but bonded to the underlying dried layer or film of agarose.

The medium was useful for chromatographic separation. It was hydrated by immersion in water at room temperature for about an hour, after which, when supported in an inclined (ca 45°) position, there was applied to the raised end a solution in an aqueous phosphate buffer of a mixture of Dextran Blue 2000 and of DNP aspartic acid dicyclohexylamine salt. Rapid separation of the components of the mixture occurred along the length of the medium under the influence of gravity. Similar results can be achieved by using the medium in horizontal position in a conventional electrophoretic device.

OTHER PROCESSES

Activated Polysaccharide Matrices

A process described by *P. Cuatrecasas and I. Parikh; U.S. Patent 3,947,352; March 30, 1976* relates to the activation of certain polysaccharide matrices for use in affinity chromatography procedures, and, more especially, to the sodium metaperiodate ($NaIO_4$) activation of polysaccharide matrices. The process also involves the coupling of various biologically active molecules as well as polyhydrazide derivatives.

Probably the most widely used technique for the covalent coupling of protein to insoluble matrices, finding considerable application in immunology and enzymology, is the cyanogen bromide method described by Axen et al, *Nature*, 214, 1302-1304 (1967); see also Cuatrecasas et al, *Proc. Natl. Acad. Sci. U.S.*, 61, 636-643 (1968). Briefly, cyanogen halides react rapidly with the hydroxyl groups of carbohydrates to form cyanate esters. The reaction occurs most rapidly at a pH between 10 and 12. The cyanate may react further with hydroxyl groups to form an imidocarbonate intermediate.

It has been postulated from studies of low molecular weight model compounds that the cyclic imidocarbonate is formed as a major product following the reaction of cyanogen halides with polymeric carbohydrates. The highly reactive intermediate can be isolated, and agarose beads which contain this reactive group are commercially available. Imidocarbonates react with amines to form N-imidocarbonates, isoureas or N-carbamates. However, the actual products formed in the reaction between cyanogen halide-activated agarose and the amino group of alanine, for example, are difficult to study owing to the insolubility of the carbohydrate and the relatively low concentrations of the products. Recent studies on the isoelectric point of products of the reaction between CNBr-activated polysaccharides and a primary amine indicate that an isourea is the major product, Svensson, *FEBS Lett.*, 29, 167 (1973).

Nevertheless, CNBr is a toxic chemical, and awareness of the hazards associated with CNBr activation reactions is important. CNBr, which sublimes rather rapidly at room temperature, is a powerful lachrimator, and is highly toxic. Allowing the chemical to stand for months at room temperature can result in the formation of explosive compounds. Filtrates collected after activation contain large amounts of cyanide ion and should never be allowed to come to neutral or acidic pH. Disposal of the filtrates should be performed with care and only after the gaseous products are dissipated from the filter flask in a fume hood for several hours. The importance of adequate ventilation during the activation procedures has been emphasized. Another disadvantage of the CNBr method for coupling amino group-containing ligands to agarose lies in the relative instability of the bond(s) formed between the ligand and the matrix. As a result, in most cases, during washing, storage and usage, a small amount of ligand is leaked out. This problem of ligand leakage is of great importance, especially during isolation of small (microgram) quantities of certain hormone receptors. The leakage of a ligand from the adsorbent becomes still more serious because the ligand which is released from the adsorbent in most cases has greater affinity toward the receptor protein(s) than the matrix-bound ligand. Besides the problem of ligand leakage, substitution does not work well with cellulose.

While nearly all of the earlier methods for coupling ligands or proteins to agarose depend on the initial modification of the gel with CNBr, an alternative method exists which is rapid, simple and safe and which promises to result in chemically stable ligand-agarose bonds. This method depends on the oxidation of cis-vicinal hydroxyl groups of agarose (or cellulose) by sodium metaperiodate ($NaIO_4$) to generate aldehyde functions. Compare Sanderson et al, *Immunology,* 20, 1061-1065 (1971). These aldehydic functions react at pH 4 to 6 with primary amines to form Schiff bases (aldimines); these are reduced with sodium borohydride ($NaBH_4$) to form stable secondary amines.

It has now been determined according to this process that the reductive stabilization of the intermediate Schiff base is best achieved with sodium cyanoborohydride ($NaBH_3CN$) since this reagent, at a pH between 6 and 6.5, preferentially reduces Schiff bases without reducing the aldehyde functions of the agarose. Since sodium cyanoborohydride selectively reduces only the Schiff bases, it shifts the equilibrium of the reaction to the right and thereby drives the overall reaction to completion.

Moreover, very useful derivatives can be prepared according to the process by allowing the periodate-oxidized agarose to react with a bifunctional, symmetrical dihydrazide most advantageously succinic dihydrazide. This reaction is quite advantageous since carbonyl groups react with hydrazides more completely than with primary amines. The matrix is thus converted to a hydrazido/hydrazone form and can be reduced with $NaBH_4$ or, preferably, $NaBH_3CN$, to produce unsymmetric hydrazides while the unreacted hydrazido groups are unaffected.

The stable hydrazido-agarose derivatives can be stored and used at will for coupling proteins or ligands that contain carboxyl or amino groups with carbodiimide reagents according to known procedures or by conversion of the acyl hydrazide to the acyl azide form, also according to known procedures. A somewhat similar polyhydrazide-agarose has been prepared by coupling adipic dihydrazide to CNBr-activated agarose; Lamed et al, *Biochimica et Biophysica Acta,* 304, 231-235 (1973).

The polyhydrazide-matrix is also coupled to aldehyde or keto group containing ligands followed again by reductive stabilization. Thus, the glycolipid, ganglioside, and certain glycoproteins are coupled to polyhydrazide-agarose (and cellulose) after periodate oxidation of suitable sugar component on the ligands.

Accordingly, polyhydrazide-agarose and -cellulose have been prepared utilizing the process by reacting succinic dihydrazide with $NaIO_4$ oxidized matrix followed by reduction with $NaBH_4$ or $NaBH_3CN$. These polyhydrazides were easily converted to the corresponding acyl azides and coupled to proteins or peptides in mild fashion. By controlling the pH of the coupling reaction, the protein or peptide can be directed to couple solely through its α-NH_2 group. And, again, the polyhydrazide matrix was also used to couple glycopeptides, glycoproteins, glycolipids and certain oligosaccharides after prior oxidation of the ligand with $NaIO_4$. Use of $NaBH_3CN$ economized the quantity of ligand. Further, both brain gangliosides and thyroglobin were successfully coupled.

Although the periodate oxidation activation method for agarose generally results in a lower degree of ligand substitution than can be achieved with the CNBr methods, the facility and safety of the method and greater stability of the ligand-agarose bonds formed, offer considerable advantages. In many cases, enzyme purification by affinity chromatography is achieved satisfactorily with ligand substitution of 0.05 to 1.0 μmol per milliliter of agarose, which can be readily achieved utilizing the procedures of the process.

Other solid supports, such as crosslinked dextrans and cellulose, unlike agarose contain very large numbers of cis-vicinal hydroxyl groups and are thus very well suited for activation and coupling by the periodate oxidation procedures. Cellulose, because of its stability to the conditions of oxidation, can be substituted to a very high degree with the periodate method. [The table below illustrates the high degree of incorporation of radioactive ligands into gels activated by the periodate/hydrazide procedures. In the procedure, the gels were oxidized with 0.5 M sodium metaperiodate for 2 hours and allowed to react with excess succinic dihydrazide (SDH) as described in the text. SDH incorporation was calculated from elemental analysis.] The more porous Sephadex gels (G-75 and G-200P shatter after periodate treatment (5 mM, 2 hours) and therefore are not recommended.

	Incorporation		
Matrix	**SDH (μmol/ml)**	**Alanine (μmol/ml)**	**Albumin (mg/ml)**
Agarose	1.2	0.25	0.20
Particulate cellulose	210	1.1	0.21
Beaded, porous cellulose	500	1.1	0.16

The following specific procedures will illustrate the process.

Sodium Periodate Oxidation of Agarose: To a suspension of 100 ml of agarose in 80 ml of water was added 20 ml of 1 M $NaIO_4$. The suspension, placed in a 500-ml tightly closed polyethylene bottle, was gently shaken on a mechanical shaker for 2 hours at room temperature. The oxidized agarose was filtered and washed on a coarse sintered-glass funnel with 2 liters of water.

Reductive Amination of Oxidized Agarose by Use of Sodium Borohydride: The washed, oxidized agarose was allowed to react with a solution of ligand containing an amino group. For example, an ω-aminoalkyl agarose derivative was prepared by adding the oxidized gel, 100 ml, to 100 ml of 2 M aqueous diaminodipropylamine (Eastman Chemicals) at pH 5.0. After 6 to 10 hours of gentle shaking at room temperature, the pH was raised to 9 with solid Na_2CO_3. 10 ml of freshly prepared 5 M $NaBH_4$ were added in small aliquots to the magnetically stirred suspension and kept at 4°C over a 12-hour period with precautions to avoid excessive foaming. The reduced agarose derivative was washed in a 250-ml sintered-glass funnel with 2 liters of 1 M NaCl without suction over a 4-hour period. After stopping the outflow (with a rubber stopper), the agarose derivative was incubated in an equal volume of 1 M NaCl for 15 hours while on the funnel, and washed with an additional 2 liters of 1 M NaCl over a 2-hour period. The filtrates of the wash were occasionally checked for the presence of free diamine by the TNBS test or with ninhydrin. The extent of substitution of diamine, as determined by reaction with ^{14}C-labeled acetic anhydride, was 2 to 3 μmol per milliliter of agarose.

In the TNBS (2,4,6-trinitrobenzenesulfonic acid) test, 2 to 3 drops of 1.5% ethanolic solution of TNBS are added to a mixture of 0.5 ml of filtrate and 0.5 ml of saturated sodium borate. The presence of diamine is indicated by formation of intense yellow color (420 nm) and of hydrazide by brick red color (500 nm).

Reductive Amination of Oxidized Agarose by Use of Sodium Cyanoborohydride: The product of reaction of the $NaIO_4$-oxidized agarose with the amino ligand was also subjected to reductive amination with sodium cyanoborohydride ($NaBH_3CN$). Although sodium borohydride is efficient, the use of $NaBH_3CN$ can be most advantageous, e.g., when the quantity of the amino ligand is limited. Since the latter reducing agent drives the reaction toward completion, relatively more efficient use of the amino ligand results. This method is also preferable when the ligand to be coupled is sensitive to the higher pH values (pH 9 to 10) necessary with $NaBH_4$.

A 1 to 50 mM solution of the amino ligand, a designation which includes proteins as well as smaller molecules, in 0.5 M phosphate buffer at pH 6, containing 0.5 mM sodium cyanoborohydride (Alfa Chemicals) was prepared at room temperature and centrifuged at 3,000 *g* for 10 minutes to remove insoluble material. The pH of the solution was adjusted to 6. The solution was then added to an equal volume of periodate-oxidized agarose which had previously been washed with 1 to 2 volumes of 0.5 M phosphate buffer at pH 6. The suspension was gently shaken for 3 days in a closed, capped polyethylene bottle at room temperature with a mechanical shaker. The gel was extensively washed as described above.

The unreacted aldehyde groups on the agarose matrix were then reduced with a solution of 1 M $NaBH_4$ (1 ml per each milliliter of agarose gel) for 15 hours at 4°C. The substituted agarose was washed extensively again. The substitution of ligand was about 2 μmol per milliliter of agarose when 50 mM diaminodipropylamine was used.

Hydrazidosuccinyl-Agarose: Preparation of Succinic Dihydrazide (SDH) — 64 g (2 mols) of hydrazine (99%) were stirred magnetically in a 1-liter Erlenmeyer flask with 100 ml of absolute ethanol at room temperature. Diethyl succinate (35 g; 0.2 mol) was added dropwise over 3 to 4 hours from a dropping funnel. The temperature of the reaction mixture was raised to 45° to 50°C. The clear reaction mixture was allowed to stand overnight at room temperature, and the resultant crystalline succinic dihydrazide (SDH) was filtered, washed with a liter of ice-cold absolute ethanol, and dried. The dihydrazide (MP 168° to 170°C with decomposition), obtained in 90% yield (26.5 g), was used without further purificaton.

Use of Periodate-Oxidized Agarose: A suspension containing 100 ml of oxidized agarose and 100 ml of 0.1 M SDH (pH adjusted to 5.0) was gently shaken for 1 to 2 hours at room temperature and the pH was raised to 9 with solid Na_2CO_3. 10 ml of freshly prepared aqueous $NaBH_4$ (5 M) was added in small aliquots, with gentle stirring at 4°C, over a 12-hour period (avoiding excessive foaming). The reduced agarose derivative was washed on a 250-ml sintered-glass funnel with 2 liters of 1 M NaCl without suction over a 3 to 4-hour period. The agarose derivative was incubated with an equal volume of 1 M NaCl for 15 hours at 24°C in the same funnel, by stopping the outflow with a rubber stopper. It was washed with an additional 2 liters of 1 M NaCl over 2 to 3 hours. The substitution of SDH, calculated from the total nitrogen content of the hydrazido-agarose, varied between 1 and 2 μmol per milliliter of packed agarose.

Activation and Use of Hydrazido-Agarose: The hydrazido-agarose derivative was stable for months when stored at 4°C in the presence of 0.02 M NaN_3. Carboxylic group-containing ligands are coupled to the hydrazido-agarose with the use of carbodiimide reagents by established procedures, Cuatrecasas, *J. Biol. Chem.*, 245, 3059 (1970) and Hoare et al, *J. Biol. Chem.*, 242, 2447 (1967). Primary or secondary amines are also coupled readily to the hydrazido-agarose after formation of agarose-azide derivatives with $NaNO_2$ and dilute hydrochloric acid.

Porous Inorganic Fibers

A process described by *H. Dutz, H.-O. Mulfinger and J.F. Pfrommer; U.S. Patent 3,570,673; March 16, 1971; assigned to Jenaer Glaswerk Schott & Gen, Germany* involves a separation column for liquid-chromatography in which the stationary phase for the desired separation effect is provided with the necessary uniform channels of 0.5 to 100 μ over a length up to 15 meters and at a cross section up to several hundred square centimeters. Furthermore, the flow resistance is uniform over great lengths. In the process, as stationary phase or as carrier of the liquid stationary phase, one uses a bundle of inorganic fibers or organic fibers, which is surrounded by a chemically resistant sleeve of glass, metal or synthetic material.

During the production of separation columns, the fibers are drawn in known manner from a glass, in which upon heat treatment a phase-separation occurs. The soluble phase is then eliminated so that a porous glass is produced. The fiber size lies suitably in the range of 5 to 200 μm. The size is adjusted according to the particular purpose of use, particularly also to the bending radius, the desired weight rate of flow and the required separation capacity.

The cover or envelope may also be produced by drawing the bundle of fibers wrapped in a foil of synthetic material into a tube and filling the remaining circular gaps with thermoplastic or Duroplastic. The following examples illustrate the process.

Example 1: There is produced a bundle of fibers from the tempered, but not leached starting glass with 7 mol percent Na_2O, 23 mol percent B_2O_3, 70 mol percent SiO_2 with a diameter of approximately 2 to 15 mm and a length of 0.1 to 15 meters. This bundle is drawn into a shrinkable, chemically resistant tube of synthetic material and subsequently thereto the tube is shrunk solid on the bundle by means of a suitable heat treatment. If the separation column is later to be operated at high pressures, then it is necessary to surround it by a protective sleeve of metal. The separation column produced in this way and consisting still of nonporous glass is now subjected to the leaching process. For this purpose, the column is attached to an acid-resistant pressure pump, with the aid of which the acid, the rinsing water, the lye and the wash water are pressed through. The advantage of this recommended method consists therein, that the fibers surrounded by a shrunken tube indeed break, but their position cannot be changed thereby, not even during later use.

Example 2: A bundle of fibers is grasped at the starting end and at the other end tightly with a metal sleeve, whereby one sleeve is provided with a ring. This bundle of fibers is drawn through a closely fitting chemically resistant tube of metal, glass or synthetic material, which may subsequently be narrowed. Thereafter, the starting end and the other end of this body are cut off, so that the metal sleeves are removed. Then starts the above described leaching process.

Example 3: The method described here is adapted both to the production of separation columns from leached out glass fibers as well as to the production of separation columns from solely tempered but not yet leached out glass fibers. In addition, the bundle of glass fibers is wound about spirally with a foil of synthetic material and the foil is subsequently fastened. This bundle is now introduced into a loosely fitting tube of metal, glass or synthetic material. It was found to be practical to construct the pipe in a U-shape.

The space between the bundle of fibers and the tube is filled with a fluid synthetic resin. After the filling, one may give to metal and synthetic material tubes a desired shape, as for example, spirals and coils. Thereafter the synthetic material is hardened. This last mentioned method when using metal tubes has the additional advantage over the first named case that it may be employed without special pressure envelope and also for high operational pressures.

Extraction Chromatography

R. Kroebel and A. Meyer; U.S. Patent 3,960,762; June 1, 1976; assigned to Bayer AG, Germany have found that absorbents based on synthetic resins can be obtained by adding an extractant, an extractant mixture or a solution of an extractant, in a solution promoter to a mixture to be polymerized comprising monomeric monovinyl and polyvinyl compounds, the extractant, the extractant mixture or the solution of the extractant in a solution promoter representing a solvent for the monomeric monovinyl and polyvinyl compounds, but not for the polymer; and by carrying out the polymerization reaction in the presence of the extractant, the extractant mixture or a solution of the extractant in a solution promoter. When absorbents are prepared in this way, the extractant or the extractant mixture is included in the polymer. All monovinyl- and polyvinyl compounds can be used in this process.

Suitable extractants include any of the compounds commonly used as extractants in extraction chromatography. A summary of these compounds can be found in Extraction Chromatography by H. Eschrich and W. Drent, Eurochemic Technical Report ETR-271 (June 1971). These extractants are substances from a variety of different classes of compounds such as esters, particularly acid and neutral esters of the acids of phosphorus, also sulfides, oxides, aliphatic and aromatic sulfoxides, amino oxides, phosphino oxides and arsino-oxides; also primary, secondary and tertiary amines and ammonium and sulfonium salts, as well as ketones, ethers, carboxylic acids, sulfonic acids, mercaptans, oximes, lactams, hydroxylamines, quinolines and pyrazoles.

In the context of the process, solution promoters are compounds of the kind which represent a solvent both for the extractants and also for the monomeric monovinyl compounds and polyvinyl compounds used. Compounds such as these are optionally halogen-substituted, linear or branched hydrocarbons, cycloaliphatic and aromatic hydrocarbons, and also such compounds as ketones, ethers and alcohols.

The absorbents are generally prepared by dissolving the monomeric monovinyl and polyvinyl compounds in the extractant, in the mixture of extractants or in the solution of an extractant in a solution promoter; adding a polymerization catalyst and polymerizing the resulting solution by a conventional technique.

To prepare bead-form absorbents, a mixture consisting of the monomeric monovinyl and polyvinyl compounds, the catalyst and the extractant is generally introduced into a mixture of water and a suspension stabilizer, the resulting mixture is then dispersed by stirring into droplets. The suspension stabilizers are known compounds, the following being examples: methylcellulose, hydroxyethyl cellulose, carboxymethylcellulose, gelatin, gum arabic, styrene maleic acid anhydride, polyvinyl alcohol, bentonite, sodium lauryl sulfate and Mersolats.

The absorbents obtained by the process give off hardly any extractant to the mobile phase. The elution curves show virtually no tail formation. The accompanying Figures 8.3a and 8.3b are comparisons of the elution curves of absorbents produced in accordance with the process with those obtained in the elution of conventional absorbents.

The absorbents used in Figures 8.3a and 8.3b for comparison purposes were obtained by conventional methods in accordance with Beranova H., Novak M., *Coll. Czech. Chem. Comm.* 30, 1073-1081 (1965), by impregnating 100 grams of a macroporous copolymer of 8% of divinyl benzene and 92% of styrene with 67 ml of tri-n-butylphosphate (Figure 8.3a) and by impregnating this macroporous copolymer of 25% of divinyl benzene and 75% of styrene with 150 ml of tri-n-butylphosphate (Figure 8.3b), while the absorbents prepared by this process were produced in accordance with the specification of Example 1, by copolymerizing 36.8% of styrene, 3.2% of divinyl benzene and 60% of tri-n-butylphosphate (Figure 8.3a) and 30% of styrene, 10% of divinyl benzene in the presence of 60% of tri-n-butylphosphate (Figure 8.3b).

Batches of 100 ml of the absorbents were introduced into a 2.4-cm diameter, 24-cm long column and 100 ml of an uranyl nitrate solution having a concentration of 100 g/l in 3 N HNO_3 was poured over them as the mobile phase.

The charged absorbents were then washed with 150 ml 3 N HNO_3 and eluted with water. The curves denoted by the reference I in Figures 8.3a and 8.3b are the elution curves of the absorbents prepared in accordance with the process. It can clearly be seen from these curves that the tail formation characterizing the elution curves of conventional absorbents is substantially eliminated in the case of the absorbents produced in accordance with the process. The curves relating to the absorbents produced in accordance with the process also show a distinct increase in capacity.

FIGURE 8.3: ABSORBENTS FOR EXTRACTION CHROMATOGRAPHY

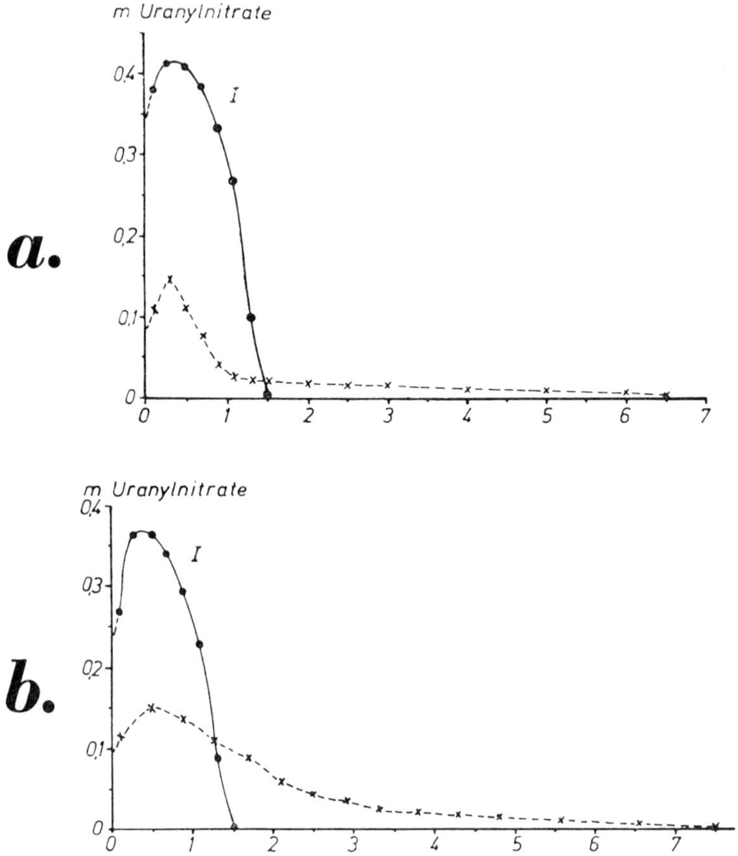

Source: U.S. Patent 3,960,762

Example 1: A solution of 13.4 parts of 60.8% divinyl benzene, 136.6 parts of styrene, 1.5 parts of benzoyl peroxide and 350 parts of tri-n-butylphosphate, was introduced into a starting mixture of 1,500 parts of water and 130 parts of a 2.5% methylcellulose solution. The mixture was heated with stirring for 3 hours to a temperature of 80°C, left at this temperature for a period of 10 hours and then heated for 1 hour to 90°C. After cooling, the aqueous phase was separated off and the resulting bead polymer was briefly washed with water.

Example 2: The procedure was as described in Example 1, except that a solution of 32.9 parts of 60.8% divinyl benzene, 167.1 parts of vinyl toluene, 2 parts of benzoyl peroxide and 300 parts of tri-n-butyl phosphate were added to the starting mixture.

Example 3: The procedure was as described in Example 1, except that 49.3 parts of 60.8% divinyl benzene, 150.7 parts of styrene, 3 parts of azobisisobutyronitrile and 300 parts of dodecylamine were added to the starting mixture.

Example 4: The procedure was as described in Example 1, except that 65.7 parts of 60.8% divinyl benzene, 134.3 parts of styrene, 2 parts of azobisisobutyronitrile and 150 parts of chlorododecane and 150 parts of phosphoric acid-bis(2-ethylhexylester) were added to the starting mixture.

Example 5: The procedure was as described in Example 1, except that 65.7 parts of 60.8% divinyl benzene, 134.3 parts of styrene, 3 parts of azobisisobutyronitrile, 240 parts of 1-chlorododecane and 60 parts of bis-(2-ethylhexyl)-amine were added to the starting mixture.

Example 6: The procedure was as described in Example 1, except that 32.8 parts of 60.8% divinyl benzene, 67.2 parts of styrene, 1 part of azobisisobutyronitrile and 150 parts of tricaprylamine were added to the starting mixture.

Example 7: A solution of 130.1 parts of 61.5% divinyl benzene, 269.9 parts of styrene, 4 parts of benzoyl peroxide, 20 parts of tri-n-butylphosphate and 80 parts of n-octane was added to a starting mixture of 500 parts of water and 130 parts of 2.5% methylcellulose solution. The mixture was heated with stirring for 4 hours to a temperature of 85°C, left at this temperature for a period of 10 hours and then heated for 1 hour to 90°C. The resulting bead polymer was separated off from the aqueous phase and washed briefly with water.

Example 8: A solution consisting of 216.9 parts of 96.8% divinyl benzene, 83.1 parts of styrene, 3 parts of benzoyl peroxide, 160 parts of tri-n-butyl phosphate and 40 parts of 1-oxo-1-dodecylphospholine, was added to 1,500 parts of saturated sodium sulfate solution. The mixture was heated with stirring for 3 hours to 85°C, left at this temperature for 10 hours and then heated for 1 hour to 95°C. After cooling, the resulting bead polymer was separated off from the aqueous phase and washed briefly with water.

POLYMER FILTRATION AND OTHER APPLICATIONS

POLYMER FILTRATION

Continuous In-Line Filter

J.E. Fogarty, Jr.; U.S. Patent 3,971,721; July 27, 1976 describes a filter system which includes a path of flow for a flowable material to be filtered and a continuous filter element moveable transversely to such flow. The continuous filter element passes through a channel of predetermined and uniform cross section and carries a plurality of seal members at intervals along its length, each seal member being of the same cross-sectional size and shape as the channel so as to form moving seals within the channel. The filter member is moved continuously or on demand through the channel, presenting a continuously renewed filter area to the flowable material.

The filter is adapted to operate satisfactorily under temperatures and pressures in excess of those ordinarily encountered in plastics operations. Specifically the seals between flanges and seal members or between channel and seals satisfactorily withstand pressures in excess of 5,000 psi at temperatures in excess of 200° or 300°F. The carrier members operate to carry the filter element through the filter zone under the same operating temperatures and pressures and to convey the filter element through the zone both when such element is fresh and free from collected contaminants and when such filter element has been used and is clogged with solid contaminants.

Figure 9.1a illustrates a filter assembly including an assembly block **11** having a channel **12** extending therethrough and a filter element **13** passing through the channel. Extending transversely through the filter assembly **11** is a passage or opening **15** adapted for the passage of a fluid to be filtered through the assembly, the direction of flow being from front to back in the view illustrated.

Figure 9.1d shows the entry passage **15** and the opposite mating exit passage **16** not visible in Figure 9.1a. The assembly is adapted to be connected into a flowing system, as shown in Figure 9.1d, by means of suitable pipes or the like to cause a flowing fluid to pass through the assembly, entering through opening **15** and exiting through opening **16** and being filtered within the assembly.

FIGURE 9.1: CONTINUOUS FILTER FOR PLASTIC MATERIALS

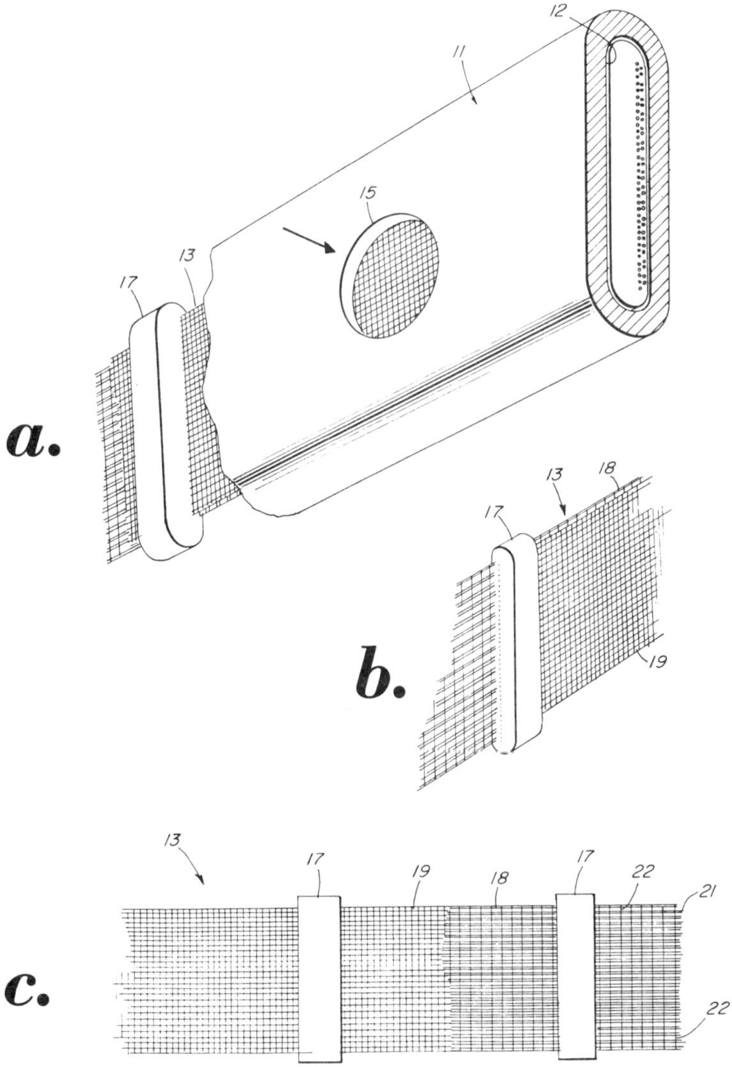

(a) Perspective View of a Filter Assembly
(b) Perspective View of a Filter Member
(c) Front View of the Filter Member According to Figure 9.1b

(continued)

FIGURE 9.1: (continued)

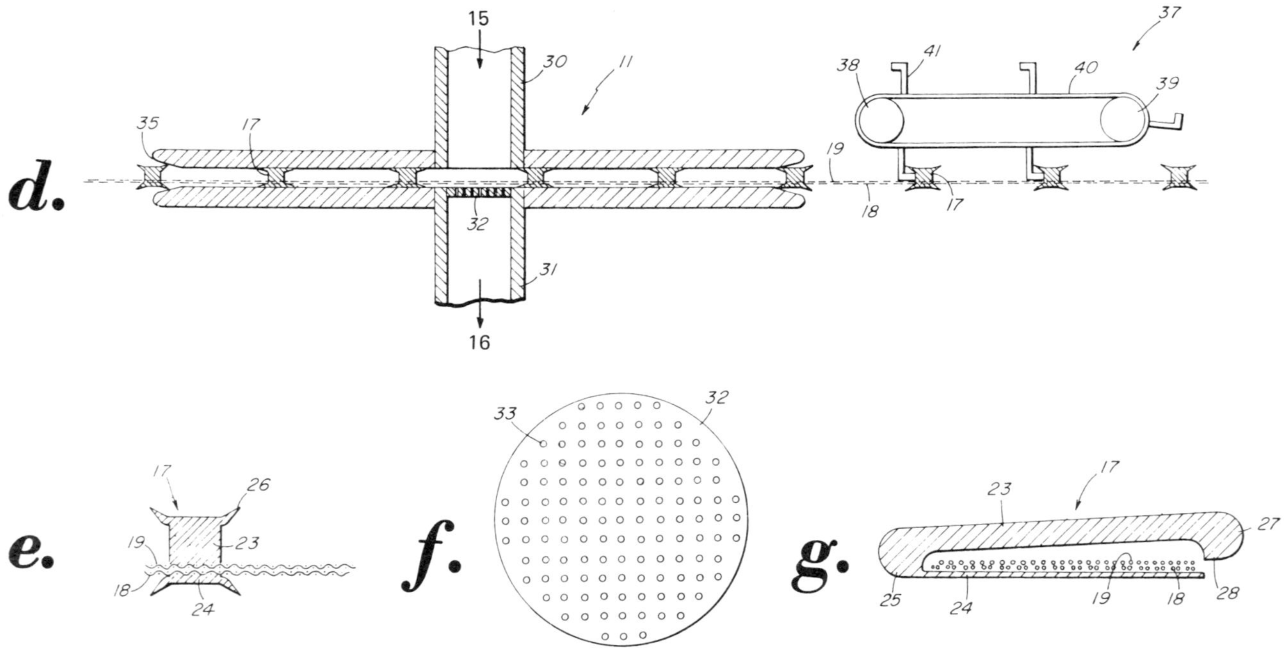

(d) Side View, Partially Schematic, of a Filter System, Partially in Section
(e) Side View Partially in Section of the Filter Member of Figure 9.1b
(f) Front View of a Backer Plate According to the Filter System of Figure 9.1d
(g) End View of a Seal Member for the Filter Member of Figure 9.1b or 9.1c

Source: U.S. Patent 3,971,721

The filter element **13** has a support member **18** desirably in the form of a wire mesh on which may be secured a filter screen or web **19**. Permanently affixed to and carried by the carrier **18** and filter web are a plurality of spaced seal elements **17**. As illustrated, carrier **18** and web **19** are held together by seal elements **17** as shown in further detail in connection with Figure 9.1g.

The carrier member is characterized by having at least high longitudinal strength to provide permanent mounting means for the seal member and to permit the filter element, including the seal member, to be drawn through the filter assembly **11**. Because seal elements **17** fit tightly within channel **12**, a high tensile strength carrier **18** is employed.

According to one form of the process, the carrier member comprises a plurality of longitudinal strands **21** of a high strength material. A suitable high strength material may, for example, be a metallic wire such as a stainless steel wire or the like. Brass coated steel wire is satisfactory. Joining together the longitudinal strands **21** of carrier **18** are a plurality of transverse strands **22** which may, but need not be, high strength strands.

According to the process, a plurality of high strength steel (brass coated) strands **21** are secured together and held in their proper spacing by means of a multiplicity of steel wires or other fibers **22** extending across the width of the carrier member **18** and secured to each of the longitudinal strands. The preferred mesh for all-around use has brass coated steel transverse fibers, but for low temperature uses polyester transverse fibers can be employed, as can other materials. The longitudinal strands provide strong longitudinal strength while the transverse strands provide permanent, controlled spacing between the longitudinal strands throughout the entire length.

Secured to one side of the carrier **18** is a filter member or screen **19**. This filter screen preferably extends the full length of the carrier and is secured to the carrier along its entire length. According to a preferred embodiment, the carrier is overlaid with a filter screen of desired mesh size and the two are secured together by means of the seals **17**, as shown further in Figure 9.1g.

For certain filtering operations the filter screen **19** is not necessary and the necessary filtering can be obtained by means of the carrier member **18** alone. In general, however, the filter screen **19** is employed and consists of a suitable filter material of the desired screen size for proper filtering. In the plastics industry it is usual to employ a filter member having a screen size corresponding to about a 40 mesh screen or finer and a screen of 100 mesh or 200 mesh is quite usual, but one of the advantages of the process is the wide freedom of choice to use virtually any desired filter porosity merely by choosing an appropriate filter.

The nature of the seal member **17** is more clearly illustrated in Figures 9.1b, 9.1e and 9.1g. Referring to Figure 9.1b and 9.1e, it is observed that filter member **13** includes a carrier mesh **18** having a filter screen **19** overlaid across the upstream surface thereof. The carrier and filter screen are held together along their full length by means of a plurality of seal members. Ordinarily the seal members are spaced approximately regularly, but if desired the several seal members may be irregularly spaced at programmed intervals. Between seal members are a succession of filter areas where the fluid being filtered presses the filter screen **19** against the carrier **18**. The carrier and screen are offset very close to the carrier or to the downstream side of the filter member.

Referring to Figure 9.1g, it is observed that the seal member **17** is a generally U-shaped member having a body or upper jaw **23** and a base **24** extending from the foot or closed end **25** of the body **23**. At the opposite or open end **27** of body **23** is a downwardly extending tooth **28** defining a recess between the tooth and the foot **25**. The carrier **18** and filter screen **19** are placed within this recess, resting on base **24**. The recess is then filled with a brazing compound, the base is pressed firmly toward body **23** and the seal assembly is then furnace brazed.

For less rigorous conditions other binding means may be used; an epoxy glue is fine for low pressure, low temperature uses. The seal **17** (see Figure 9.1b) firmly and permanently holds the carrier **18** and screen **19** together. Base **24**, as shown, is relatively thin with the result that the wire members ride through channel **12** positioned close to the downstream side of the channel.

The volume between successive seal elements on the upstream side of the filter screen serves as a trap or storage volume to hold and remove from the system the solids or other matter removed from the flowing stream.

At the top or upstream side of seal **17** a pair of pressure lips **26** extend from body **23** of the seal. These lips normally extend parallel with carrier **18** or are aimed slightly upward or away from the carrier. When the seals ride in channel **12**, fluid pressure urges these lips firmly against the upper or upstream wall of channel **12**, thereby helping to tighten the sealing effect. Another pair of lips **26** may, if desired, extend from base **24** of seal **17**, similarly sealing against the downstream wall of channel **12**.

In Figure 9.1d is shown a filter assembly generally designated **11** connected into a conduit or pipe for the flow of a fluid. For example, an intake pipe **30** leads from a source of fluid such as flowable plastic to the assembly and a discharge pipe leads to further apparatus such as an extrusion die, an injection mold, a storage container, a filament former or the like. Filter element **13** extends through channel **12** transversely to the direction of flow. Seals **17** fit tightly against the walls of the channel, substantially sealing the assembly against fluid leakage. A backer plate **32** is positioned essentially flush with the downstream wall of the channel and has a multiplicity of holes **33** therethrough to permit essentially unimpeded flow. Such a backer plate is conventional in the plastics industry.

Carrier wire **18** is positioned almost against the downstream wall of channel **12**, and when a fluid is being filtered, carrier **18** is pressed against backer plate **32**, largely preventing the fluid from by-passing the filter screen **19**. At the ends of channel **12**, lips **35** of the assembly block **11** spread slightly to receive pressure lips **26** of seal elements **17** on the entry end of block **11**, and at the exit end to release seals **17** gradually. Drive means **37**, of suitable design, pulls filter element through channel **12**. One suitable drive means has drive rolls **38** and **39** with belt **40** passing therearound. A plurality of drive hooks **41** on belt **40** are shaped to seize seal elements **17** and move them along either continuously or on demand. Either drive roll may be powered, and since the driving force is relatively high both rolls may be powered.

In the process a fluid to be filtered is introduced into the filter assembly through pipe **30**, usually under several thousand pounds pressure, where it fills the cavity defined by the walls of the assembly **11** and a pair of seals **17**, one on either side of the filter zone. The fluid accordingly is forced through the filter element and emerges in filtered condition into pipe **31**.

A plurality of seals **17** is at all times positioned within channel **12** of the filter assembly **11** with the result that the seal assembly which seals the filter against leakage includes always at least two seal elements and generally at least two such elements on each side of the filter zone. As the filter element is moved through the channel a new seal element enters the operating or filtering area and is backed up by a second seal element further removed from the filter zone. On the exit side of the assembly **11**, as successive filter elements **17** pass out of the assembly, a successive element is already in channel **12** to restrict the pressure acting against the emerging element. Accordingly, there is no surge or spurt of filter fluid upon the emergence into the open air of one of the seal elements.

Generally lubrication of the surfaces between the seal elements and the walls of channel **12** is not required, particularly if the fluid being filtered itself acts as a lubricating agent. In the case of filtering of plastic material, if the entire filter element is maintained at a sufficiently high temperature to keep the plastic in a flowable condition until after it has emerged from the filter assembly, the plastic will serve as a lubricant. In most cases a separate lubricant is avoided in order to avoid contamination of the fluid being filtered.

In addition to the plastics industry and the rubber industry, the filter of the process can be employed in essentially any situation where it is desired to filter a fluid including gases and/or liquids.

Metallurgically Integral Filter

J.L. Bidler; U.S. Patent 3,581,902; June 1, 1971; assigned to Minnesota Mining and Manufacturing Company describes a filter leaf which is entirely made of sintered powdered metal to form an integral rigid structure which does not have shortcomings of prior art filters made of powdered metal. Briefly, the filter leaf is metallurgically integral and comprises sintered laminae of shaped powdered metal, the laminae broadly comprising an impervious structural member, a filter element and a pervious support member interposed or sandwiched between the structural member and the filter element, with a drainage space or area between the structural member and the filter element, the drainage space preferably communicating with a discharge pipe.

The term metallurgically integral as used here means that the contiguous surfaces of the filter leaf components are bonded together by interatomic diffusion, as a result of the sintering of the assembly, the sintered filter leaf thus in effect being one piece of metal.

In a preferred case, the filter leaf before sintering comprises a laminae of five discs of green, dried, compacted powdered metal with the peripheral edges of the discs sealed with a ring of green undried powdered metal. After being so-assembled, the whole assembly is sintered as a unit to form a metallurgically integral rigid filter leaf. The middle or center disc serves as a structural core and it is substantially impervious and has on its two faces an array of spaced-apart protuberances, such as studs integral with the core. Each of the filtering elements is supported by a relatively more permeable support disc of powdered metal which is interposed or sandwiched between the filtering element and the protuberances or studs.

The spaces between the core and abutting support members form a drainage space or area which communicates with a central opening passing through the assembly. A series of such sintered filter leaves can be stacked in spaced relation, for example, by interposing separator rings between filter leaves, on a discharge pipe or channeled rod passing through the central openings of the series, and the stack mounted within a housing to form a filter assembly useful in removing solid particles from liquids or gas.

Figure 9.2 illustrates a powdered metal filter as follows. Figure 9.2a is an exploded isometric view of a filter leaf before it is assembled and sintered. Figure 9.2b is a sectional view in elevation of a sintered filter leaf. Figure 9.2c is an enlarged sectional view of the peripheral portion of the filter leaf of Figure 9.2b. Figures 9.2d and 9.2e are side elevational views of a discharge rod upon which a plurality of the sintered filter leaves of Figure 9.2b can be mounted. Figure 9.2f is a sectional view of Figure 9.2e taken along the plane **6-6** indicated. Figure 9.2g is an elevational view in partial section of a filter assembly in which has been mounted a series or stack of the filter leaves of Figure 9.2b.

Referring to Figure 9.2a, reference number **11** designates a circular or disc-like green structural core of shaped powdered metal, the two faces of which are provided with a plurality of integral protuberances or studs **12** (only a portion of which are shown for brevity), the core having a central opening **13**. Members **14** and **16** are circular or disc-like green filter elements or media, having a smooth planar face and central openings **17** and **18**, respectively. Disposed between core **11** and filter elements **14** and **16** are two circular green disc-like supports **19** and **21**, which also have planar faces and central openings **22** and **23**, respectively.

Supports **19** and **21** have outside diameters smaller than core **11** and filter elements **14** and **16** to accommodate seal rings **15** and **20** within which the supports are to be disposed, these rings being made of green, undried powdered metal. By applying a light but sufficient pressure to the top and bottom peripheral portions of the assembly, the seal rings will be squeezed or extruded around the periphery of the discs to seal the edge of the assembly as shown in detail in Figure 9.2c.

Figures 9.2b and 9.2c illustrate the filter leaf of Figure 9.2a as assembled and sintered, the contiguous parts of the various discs being metallurgically bonded as a result of the sintering to form an integral rigid unit, generally designated **25**, the several central openings of each of these components forming a common central opening **24** with the internal filtrate drainage areas being designated by reference number **26**. The peripheral edges of the discs are also metallurgically bonded by a ring **27** resulting from the pressing of rings **15** and **20**, as described above. When a plurality of these filter leaves are assembled, they are held in spaced relation by spacer rings **28** disposed on each side of each of the filter leaves. Such spacer rings have a central opening **29** of the same diameter as openings **13, 17, 18, 22** and **23** after the assembly has been sintered.

A series of filter leaves can be mounted on a discharge rod or shaft illustrated in Figures 9.2d, 9.2e and 9.2f. The outer diameter of rod **31** is such that the stack of filter leaves can be snugly mounted thereon. The outer surface of the discharge rod has a plurality of longitudinally aligned channels **32** which progressively increase in depth toward the bottom of the rod, such channels being in communication with the drainage areas **26** of the series of filter leaves **25** mounted on the rod. The lower ends of channels **32** communicate via passages **30** with a common drain area **35** in the bottom of the rod **31**, the lower end of which is threaded at **48** and the upper end being provided with a tapped hole **50**.

FIGURE 9.2: POWDERED METAL FILTER

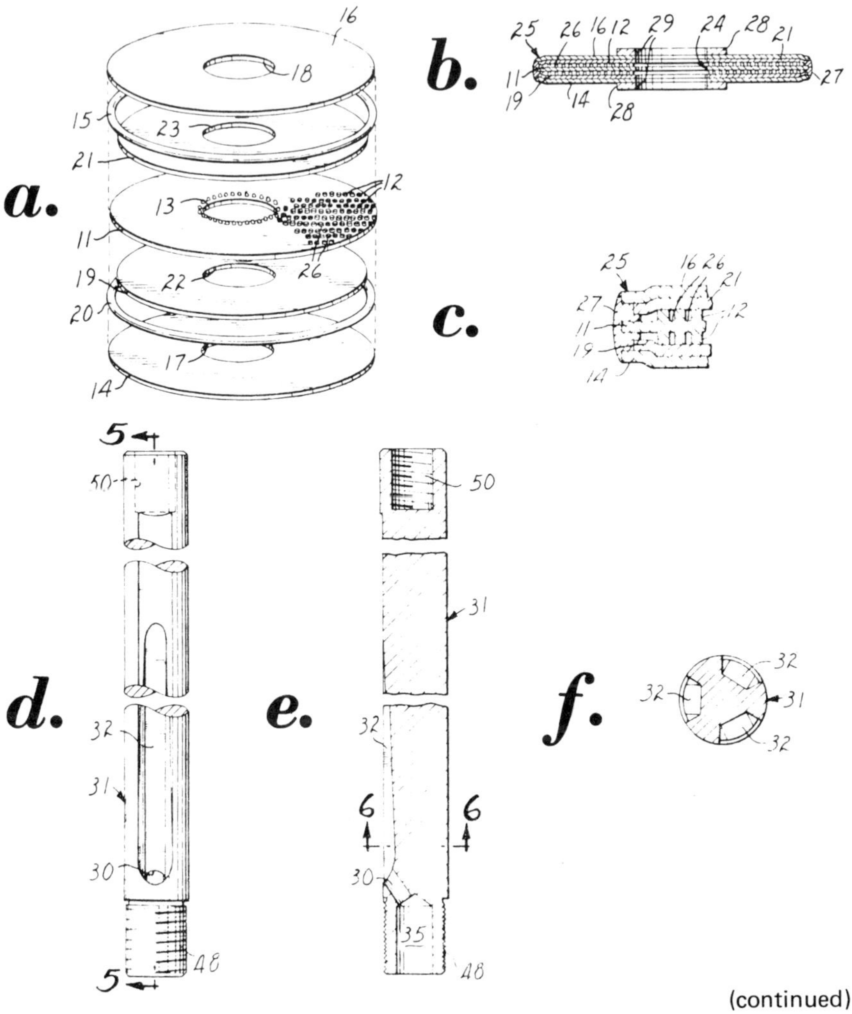

(continued)

FIGURE 9.2: (continued)

g.

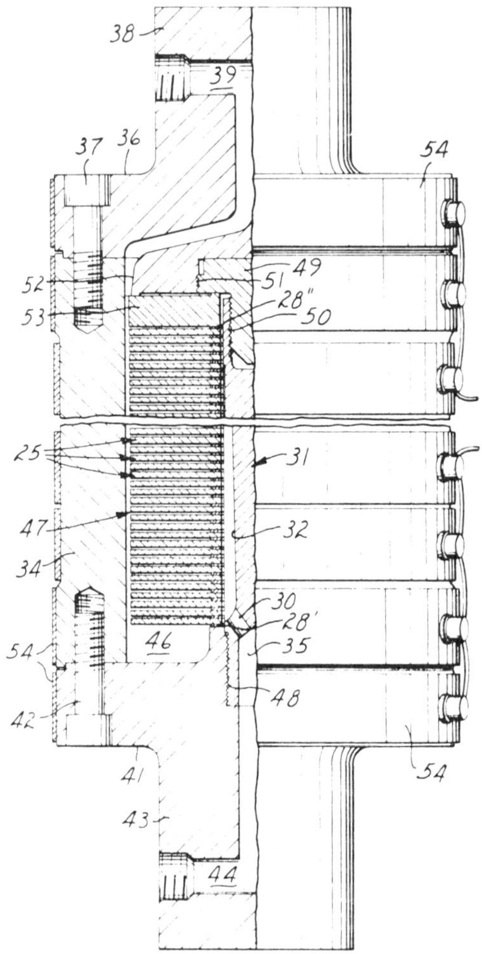

Source: U.S. Patent 3,581,902

Referring to Figure 9.2g, a filter assembly is shown comprising a cylindrical housing or casing **34** surmounted by a top cover or cap **36** which can be fastened thereto by means of bolts **37** or the like. Surmounting the cap is a flange **38** provided with inlet port or passage **39**. The lower end of casing **34** is mounted on a supporting bottom closure or baseplate **41** and fastened thereto by means of bolts **42** or the like. The baseplate has a lower flange portion **43** with a discharge port or passage **44** therein. As assembled, casing **34** with its top cap and baseplate define an internal chamber **46** which communicates at its upper end with inlet passage **39** to receive liquid or gas laden with solid particles or the like which are to be separated or filtered. Centrally disposed within chamber **46** is a series of stacked filter leaves **25** of Figure 9.2b, this assembly being generally designated as **47**. The lower threaded end **48** of discharge rod **31** is threaded into a tapped hole provided in the upper portion of baseplate **41**, so as to support assembly **47** within chamber **46**, the lower-most spacer ring **28′** of the assembly resting on an internal central shoulder of baseplate **41**. The upper threaded hole **50** of discharge rod **31** is threadedly connected to a nipple **49**, which in turn is threadedly connected at **51** to a distributor cap **52** which surmounts assembly **47**, the distributor cap serving to distribute around such assembly the solid-laden liquid or gas introduced via passage **39**. Surmounting the series of stacked filter leaves, and the uppermost spacer ring **28″**, is a retaining ring or collar **53**.

The entire filter leaf is made from sintered powdered metals using, in part, conventional powder metallurgy techniques. The preferred powdered metals used in this process are those of austenitic chromium-nickel stainless steel, these alloys generally containing 16.0 to 26.0 weight percent chromium, 6.0 to 22.0 weight percent nickel, 0.03 to 0.25 weight percent carbon, and occasionally some other elements added to develop certain specific properties, such as 1.75 to 4.00 weight percent molybdenum or small amounts of titanium, tantalum, and niobium to minimize formation of chromium carbides, especially in welding. Standard types of these steels have been assigned numbers and specifications by the American Iron and Steel Institute. These are generally known in the art as stainless steels of the AISI 300 series, types 301, 302, 304 and 305 generally referred to as "18-8" stainless steel, and the "workhorse" type 316 generally referred to as "18-8 Mo." All of these AISI stainless steels of the 300 series are applicable in this process.

A center structural core was made by mixing 3,200 grams of -325 mesh stainless steel powder (Anchor 316L), 150 grams of 4,000 cp methylcellulose (Methocel), and 410 ml of a 15 volume percent solution of glycerin in water, in a Brabender sigma blade mixer for 1 hour, after which the material had the consistency of modeling clay. The claylike material was then formed into a sheet 0.125 inch thick by passing it through a rubber mill.

After drying for 24 hours at room temperature, the green sheet was placed between two 1/16-inch thick perforated metal plates, having 9/64-inch round, staggered holes at 3/16-inch centers, and pressed until the material filled the holes in the perforated sheet to form an array of studs 1/16 inch in height. The perforated plates were removed and a 7.50-inch disc with a 2.00-inch center hole was cut from the studded green sheet using a suitable die.

The intermediate, large pore size, support discs were made from a mixture of 2,850 grams of -50+100 mesh 316L stainless steel powder (Anchor 316L), 150 grams of molybdenum powder (Fisher No. of about 4 microns), 150 grams of 4,000 cp methylcellulose, and 650 ml of a 15 volume percent solution of glycerin in water, in a sigma blade mixer for 1 hour, at which time the material had the consistency of modeling clay. A sheet 0.050-inch thick was made by passing the material through a rubber mill, after which the sheet was dried at room temperature, and 7.25-inch diameter discs with a 2.00-inch center hole were cut from the sheet using a suitable die.

The filter elements were made by mixing 2,400 grams of -100+200 mesh and 600 grams of -325 mesh 316L stainless steel powder (Anchor 316L), 150 grams of 4,000 cp methylcellulose, and 525 ml of a 15 volume percent solution of glycerin in water, in a sigma blade mixer. The resulting claylike material was sheeted, dried and cut into 7.50-inch diameter, 0.035-inch thick discs with a 2.00-inch center hole in the same manner as the support disc.

Both faces of the center structural disc and the intermediate support discs, and the inner face of the filter element discs were wetted with a 1 weight percent solution of 4,000 cp methylcellulose in water and stacked in the proper order (see Figure 9.2b) over a 2.00-inch post. An extruded rod, made by forcing the undried, claylike -325 mesh stainless steel through a 0.300-inch die, was shaped in the form of a ring around a ⅛-inch peripheral portion of both sides of the center structural core before the intermediate support and outer filter element discs were positioned.

The resulting assembly was compressed around the outer peripheral portion, using a pressing and retaining ring, to force the -325 mesh extruded rings over the edge of the inner structural disc. All discs were thereby cohered together using the cohesiveness of the methylcellulose. After room temperature drying, the peripheral portion of the assembly consisted of a -325 mesh bead sealing the edge of the assembly.

The green assembly was placed on an inert sagger and heated slowly to 1350°C in a hydrogen atmosphere and held for 2 hours. The resulting sintered filter leaf was an integral composite with all discs metallurgically bonded together. The filter elements were 0.035-inch thick, 50% porous, and had an absolute micronic rating of 20 microns. The intermediate support members were 0.050 inch thick, 50% porous, and had an absolute filtration rating of 65 microns. The center structural core was 0.180 inch thick across the embossed projections and had 0.062 inch of continuous material between the embossed projections. The filter leaf was 7.00 inches in diameter with a 1.90-inch hole through the center.

The pressure drop across two of the filter leaves prepared as described above was determined by filtering molten viscous polyester and for purposes of comparison the pressure drop of two commercial prior art filter leaves made of powdered metal was also determined. Each of the prior art filter leaves comprised an inner stainless steel screen sandwiched between two filter elements (supported only by the screen) each measuring 0.062 inch in thickness and 7 inches in diameter and made from sintered porous sheets of powdered stainless steel. The filter leaf assembly had a weld about its periphery and a machined center hub of wrought steel. These prior art filter leaves had substantially the same filter area and micronic rating as the filter leaves of this process made as described above. The following table sets forth the results of this comparison.

Filter Leaves of this Process				Prior Art Filter Leaves			
Number 1		Number 2		Number 3		Number 4	
Pressure Drop (psi)	Flow Rate (lb/hr)	Pressure Drop (psi)	Flow Rate (lb/hr)	Pressure Drop (psi)	Flow Rate (lb/hr)	Pressure Drop (psi)	Flow Rate (lb/hr)
318	13.7	330	20.6	560	17.3	900	16.1
407	22.1	475	31.7	817	25.8	1,200	21.6
583	29.4	690	39.5	1,208	32.4	1,625	29.1
768	35.6	865	47.9	1,503	39.2	2,050	33.2

(continued)

..... Filter Leaves of this Process				.... Prior Art Filter Leaves........			
...Number 1 ...		...Number 2 ...		...Number 3 ...		....Number 4 ...	
Pressure Drop (psi)	Flow Rate (lb/hr)	Pressure Drop (psi)	Flow Rate (lb/hr)	Pressure Drop (psi)	Flow Rate (lb/hr)	Pressure Drop (psi)	Flow Rate (lb/hr)
743	38.1	905	48.8	2,063	46.5	2,415	38.9
1,010	48.0	950	55.0	2,510	50.3	2,740	42.3
		1,165	70.3	2,495	53.5	3,145	49.5
		1,340	65.3	2,882	61.0	3,725	57.3
						4,050	57.6

As the data of the above table show, the pressure drop across the filter leaf of this process is as much as about 2.5 to 3 times less than that across the prior art filter leaf at equal flow rate.

The air permeability of the filter leaf of this process was 2.8 times greater than that of the prior art filter leaf and the flexural stiffness of the former was 2 to 3 times greater than that of the latter.

Filter leaves made by the method of the process were used to filter a viscous, molten polyester polymer at 550°F until a maximum pressure differential of 4,500 psi was reached, with excellent results.

Filtration of Polymer Melts Using Spinning Pack Assembly

A process described by *W.L. Phillips, Jr.; U.S. Patent 3,896,028; July 22, 1975; assigned to E.I. Du Pont de Nemours & Company* provides a particulate-metal filter medium which, in the filtration of polymer melts, has a pack-life about twice that of commonly used silica sand. It further provides a filter medium which is exceptionally porous, highly resistant to compaction at melt-filtration pressures, substantially inert to polymer melts (e.g., nylons and polyesters) and relatively inexpensive.

The particulate filtering medium of this process is formed from an alloy containing 30 to 40% (preferably 35 ± 1%) by weight chromium, 3 ± 1% by weight silicon, minor amounts, at ordinary levels, of carbon and nitrogen (i.e., 100 to 400 parts per million by weight carbon and 600 to 1,000 parts per million by weight nitrogen), the remainder being essentially iron. The particles are irregularly shaped and rough-surfaced to provide a shape factor less than about 0.58.

A bed of the particles is characterized by a porosity of at least 60% under no applied load and by compaction of no more than about 20% under an applied load of about 5,000 psi (352 kg/cm^2).

Figure 9.3a is a greatly magnified view of particles **10** of the process showing their irregular, rough and somewhat fibrous appearance. These particles are preferably prepared by impinging jets of water or steam onto an extruded stream of molten metal alloy, thereby explosively disintegrating the alloy into particles generally as illustrated in Figure 9.3a. Alternatively, these irregular particles may be prepared by the "Coldstream Process" in which a high pressure air stream shoots the molten alloy through a venturi nozzle at a target in an evacuated chamber. Temperature of the alloy drops instantaneously on exit from the nozzle whereby the chilled alloy shatters on impact with the target.

FIGURE 9.3: PARTICULATE METAL FILTER MEDIUM FOR POLYMER MELTS

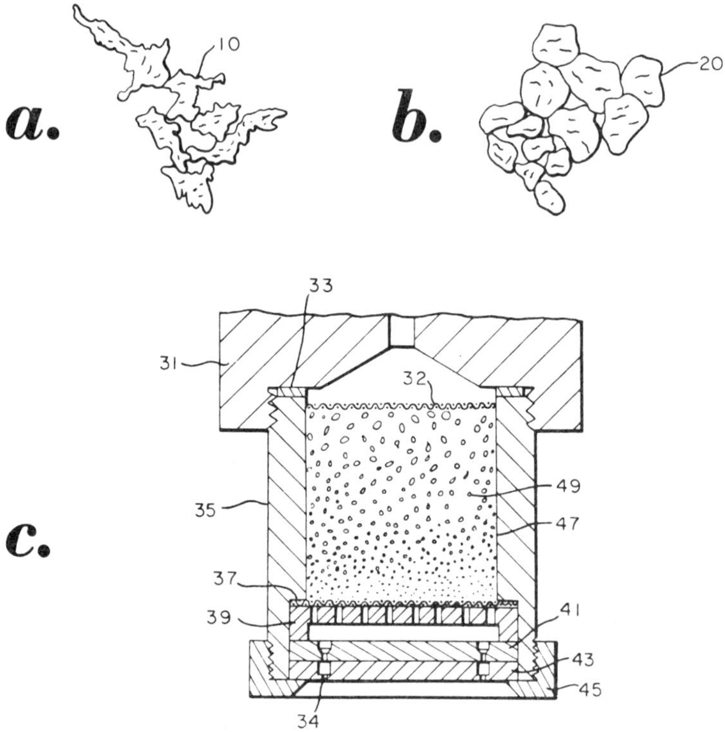

(a) Magnified View of Particles of the Process, Showing Their Irregular Shape and Rough Surfaces
(b) Similarly Magnified Particles of Silica Sand as Commonly Employed for Melt Filtration
(c) Cross-Sectional View of a Typical Spinning Pack Utilizing a Particulate Filtering Medium

Source: U.S. Patent 3,896,028

Figure 9.3b generally depicts, at similar magnification, particles **20** of commonly used filtering media, more specifically silica sand. Such particles approximate spheres in shape, although uneven bumpy surfaces often are present. These known shapes provide, for a given volume and average particle size, much less free space (porosity) than do the particles of this process. Even particles formed according to Figure 9.3a of known metal alloys accelerate the degradation and gelation of the polymer melt to too great a degree and/or are insufficiently resistant to compaction at pressures required for filtration of polymer melts.

Figure 9.3c shows a typical spinning pack assembly using particulate shear-filtration media. The spinning pack assembly includes a filter body **35** threaded into spinning block **31** against gasket **33**. A screen **37**, a distribution plate **39**, a metering plate **41**, and a spinneret **43** are retained in one end of body **35** by support ring **45** threaded onto the body. The cavity **47** of the body is at least partly filled with a particulate shear-filtration medium **49** and then, preferably, capped with a coarse screen **32**. Flow is from spinning block **31**, through the shear-filtration medium **49**, and ultimately through outlet capillaries **34** of spinneret **43**. Ordinarily cavity **47** is filled with two or more layers of particles **49** of graded mesh sizes, the layer with the finest particles resting against screen **37** and the layer with the coarsest particles being adjacent to screen **32**.

The chemical composition of the alloy of which the particles of this process are composed must be very carefully controlled in order to realize the improved properties. It should contain 30 to 40% by weight chromium and 3 ± 1% silicon, minor amounts of carbon and nitrogen, with the remainder being substantially of iron. The silicon is required to impart sufficient fluidity to the molten alloy that it will disintegrate on contact with jets of water or steam into a high yield of particles of mesh sizes commonly used for melt filtration (e.g., between about 25 and 200 mesh). Above 40% chromium, brittleness increases rapidly. Below 30% chromium, the particles become less inert to polymer melts, and resistance to compaction diminishes. Optimum results occur when the chromium content is 35 ± 1% by weight.

Example: Using a pack assembly substantially as shown in Figure 9.3c, yarns of polyethylene terephthalate are extruded in two tests identical except for the filtering media employed. Each yarn has 44 filaments and a total denier, after cold-drawing, of 70. The feed polymer has a relative viscosity of 20.5. (Relative viscosity is determined as the solution/solvent ratio of absolute viscosities at 25°C. The solvent is hexafluoroisopropanol. The solution results from dissolving 0.8 gram of polymer in 10 ml of solvent.) Temperature of molten polymer on entrance to the filter medium is 285°C, the cross-sectional area of the filter cavity is 3.8 square inches (24.5 cm^2), and flow rate of molten polymer through the filter medium is constant at 3.02 lb/hr (1.37 kg/hr).

The first test employs Fe/35% Cr/3% Si particles as the filtering medium. The filter cavity is filled, in order, with a coarse supporting screen, a 0.125 inch (0.32 cm) thick sintered stainless steel disc of porosity equivalent to 20 mesh, 10 ml of 100/140 mesh Fe/35% Cr/3% Si particles, 20 ml of 50/70 mesh Fe/35% Cr/3% Si particles, 30 ml of 25/50 mesh Fe/35% Cr/3% Si particles, and enough (about 20 ml) of coarse No. 10 aluminum oxide particles to fill the remaining space. The Fe/35% Cr/3% Si particles are characterized as follows, where E_0 is the uncompacted porosity and C is the percent compaction.

Mesh	Time to Gel, hr	Shape Factor	E_0	C (5,000 psi)
100/140	236	0.29	70	17
50/70	238	0.49	72	20
25/50	240	0.53	74	22

The second test employs sand as the filtering medium. The filter cavity is filled in order with a six-layer coarse supporting screen, 20 ml of 60 to 80 mesh sand, 45 ml of 25/50 mesh sand and a single-layer coarse covering screen. The two tests give the results shown on the following page.

	Fe/35% Cr/3% Si	Sand
Spinning breaks/1,000 lb	1.3	2.1
Pack life (days)	12	1.3

In the case of the Fe/35% Cr/3% Si test, life of the pack is determined as the time at which pack pressure rises to a preset pressure (5,000 psi). In the case of the sand test, pack life is determined as the time when the level of broken filaments during subsequent drawing and winding exceeds acceptable levels. Careful analysis shows the breaks to result from gel particles in the filaments.

Filter Bed Element for Use in Melt-Spinning

H. Toda; U.S. Patent 3,630,384; December 28, 1971; assigned to Teijin Limited, Japan describes a filter bed element for use in melt-spinning, which consists of a filter sand which has previously been molded using as a binder a synthetic high polymer having a melting point equal to or lower than that of the synthetic high polymer to be filtered.

Example: Polycaprolactam having a limiting viscosity of 1.45 as measured in m-cresol at 25°C, and molten at 260°C, was passed through a filter bed element in the process of extrusion, which filter bed element had been molded of a filter sand using polyethylene as a binder. The results are shown in the table below.

	Performance			
Filter Bed	Ratio of Polymer Leakage (percent)	Ratio of Unsatisfactory Discharge (percent)	Acceptance Ratio of Products (percent)	Service Life (hours)
Conventional filter bed	4	3	93	200-250
Process filter bed	0	1	98	500-800

The filter bed is provided by a filter bed element which has previously been produced by molding a filter sand in an optional thickness, using as a binder a synthetic high polymer having a melting point equal to or lower than that of a synthetic high polymer to be filtered.

Therefore, a filter bed can be formed in a spinning apparatus readily and uniformly without the possibility of displacement of the filter bed, which would otherwise be caused during transportation, preheating and mounting in the spinning apparatus of the filter bed subsequent to the assembly of a filtering device.

Accordingly, the filter bed is prevented from becoming nonuniform. Furthermore, since the binder is dissolved in the process of melt-spinning, a filtering capacity the same as that of the conventional filter bed can be obtained.

The method for preparing a filter bed comprises simply forming the filter sand to be used to the shape of the filter bed, then binding the same with the binding synthetic high polymer, then placing the molded filter bed in the apparatus for melting, and finally filtering the synthetic high polymer to be filtered through the molded bed. In the final step, binder is dissolved or flushed out of the filter bed, leaving only the filter sand.

OTHER APPLICATIONS

High Temperature Filtration—Power Plants

G. Hirs; U.S. Patent 3,669,882; June 13, 1972; assigned to Hydromation Engineering Company describes the use of granular, finely divided alumina (Al_2O_3) as an insoluble, chemically inert, high temperature filtration medium in systems requiring ultrafine filtration, such as in steam turbine power plants, chemical processing and pharmaceutical manufacturing.

Unfortunately, alumina or aluminum oxide is a fairly heavy material and is extremely abrasive. Consequently, special handling apparatus and methods were developed and are described which make practical the utilization of alumina as a filter medium. The specific characteristics of the filtration apparatus and the specific manipulative techniques utilized in the formation of an acceptable aluminum oxide medium, in rejuvenating the medium for reuse, and in reconstituting the filter medium are described below.

In Figure 9.4a, reference numeral **10** refers generally to a tank or pressure vessel of welded steel construction including a cylindrical side wall **11**, an upper spherical end cap **12** and a lower hemispherical end cap **13**.

Disposed within the tank are a plurality of outlet tubes **15** and inlet tubes **16** radially spaced from one another. These tubes **15** and **16** are substantially identical in structure and, in a preferred case, are formed standard pipe sections of stainless steel or other similar material provided with a plurality of radial perforations **17**, these perforations preferably being on the order of less than about 0.003 inch in diameter. Enveloping each of the tubes **15** and **16** is a stainless steel mesh wire or cloth **18** forming a tight mesh upon which the filter medium will precoat or accrete.

The outlet tubes **15** communicate with an upper, horizontal, manifold outlet pipe **20** projecting radially through the upper cap **12** and generally overlying the tubes **15** to be connected thereto through arcuate manifolds **21**. Suitable adapters **22** and connecting piping **23** join the tubes **15** to the manifolds **21** and the outlet line **20**. Similarly manifolds **24**, adapters **25** and connecting piping **26** interconnect the inlet tubes **16** with outlet line **27** projecting through the cap **12**. The bottom ends of the outlet pipes only are similarly connected through manifold **28** in appropriate piping **29** with a return pipe or line **30** projecting through the lower cap **13**. A drain line **32** is provided at the medial, lowermost portion of the bottom cap **13** and a fill line **33** communicates with the apex of the upper cap **12**.

As best shown in Figure 9.4b, a plurality of pipes **15** and **16** are provided, the pipes being retained in peripheral, radially and circumferentially spaced array so that the minimum distance between any inlet pipe **16** and any outlet pipe **15** is not less than 1 inch, and preferably from 3 to 6 inches.

Also as shown in Figure 9.4a, the tank or casing **10** is filled with a body of particulate filter medium indicated generally at **35**, the preferred filter medium of the process comprising alumina or aluminum oxide (Al_2O_3). As a specific example, the alumina which is preferred is calcined and typically has one of the chemical analyses as shown on the following page.

	Sample 1	Sample 2
Na_2O	0.60	0.50
SiO_2	0.02	0.02
Fe_2O_3	0.02	0.02
TiO_2	0.002	0.002
Loss on ignition	0.60	0.20
Al_2O_3 (by difference)	98.70	99.20
Sorbable water at 50% RH	0.40	0.20

Thus it will be seen that the calcined aluminum which is preferred is of a high degree of purity (on the order of 99% Al_2O_3).

FIGURE 9.4: FILTERING APPARATUS

a.

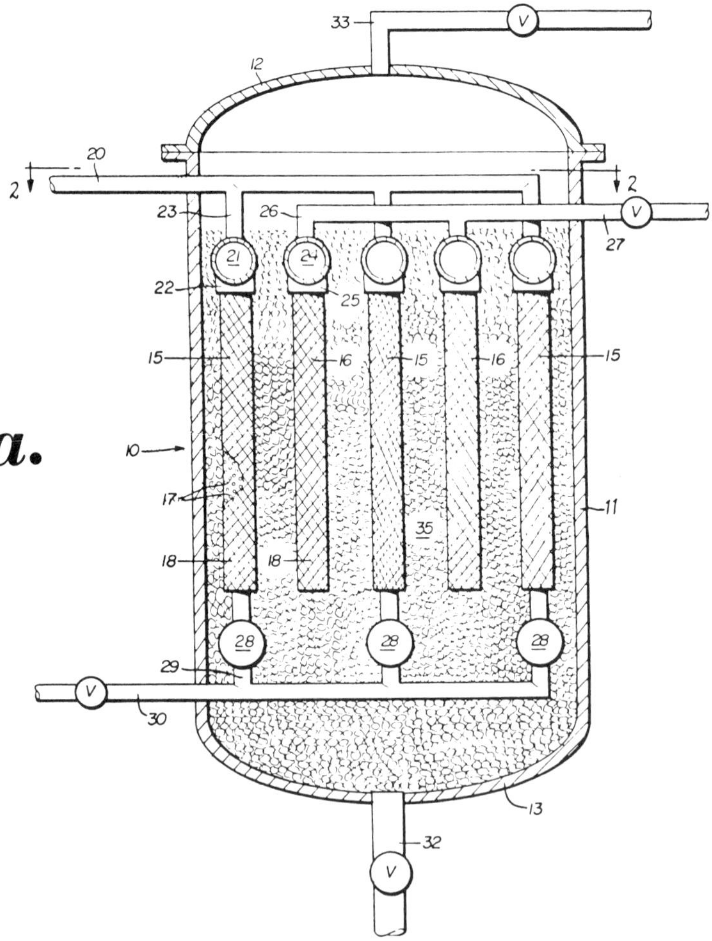

(a) Vertical Sectional View of Deep Bed Filter with Parts Shown in Elevation

(continued)

FIGURE 9.4: (continued)

b.

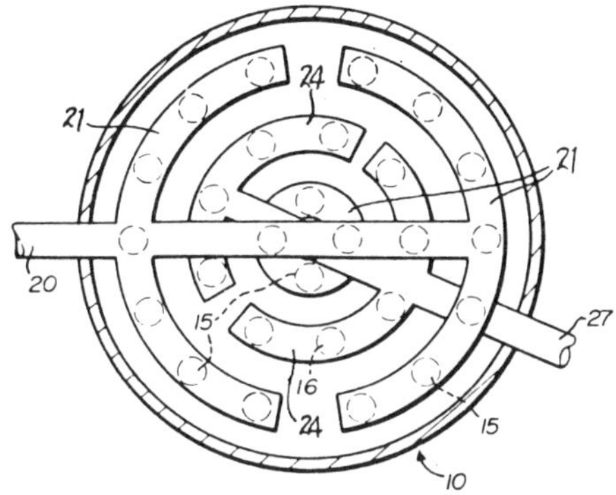

c.

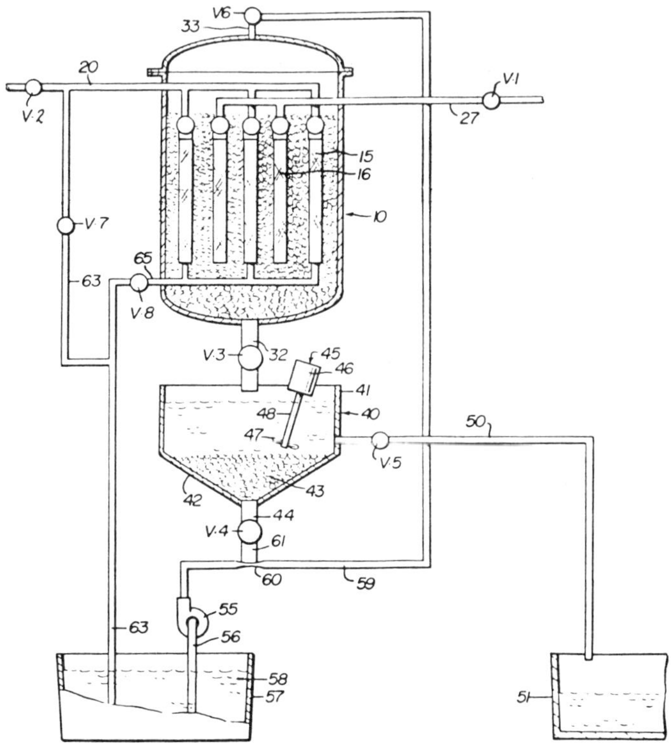

(b) Sectional View Taken Along Plane **2–2** of Figure 9.4a
(c) Schematic Representation of an Overall Filtration Apparatus Utilizing the Deep Bed of Figure 9.4a

Source: U.S. Patent 3,669,882

The calcined aluminum which is preferred is also finely divided and has the following typical screen analysis (Tyler Standard Screen):

	On 100 Mesh	On 200 Mesh	On 325 Mesh	Through 325 Mesh
Sample 1	5 - 15	70 - 90	97 - 99	1 - 3
Sample 2	5 - 15	70 - 90	97 - 99	1 - 3

Calcined aluminas of the above-defined typical chemical analysis and typical screen analysis or the equivalent are commercially available from various sources. By reference to a U.S. sieve chart under the standard ASTM E-11-61, it will readily be determined that the alumina of the preferred embodiment is of a predominate or average particle size of less than 149 microns and greater than 14 microns.

Turning to the filtration system illustrated in Figure 9.4c, it will be noted that the tank **10** is filled with granular, finely divided alumina, as above-defined, to a level above the level of the manifolds **21, 24** so that the tubes **15, 16** are completely embedded in the filter material and are separated from one another through a depth of alumina of at least 1 inch.

The line **27** is provided with an off-on valve **V1** through which the line **27** is connected to a source of dirty liquid, e.g., a source of condensate liquid in a steam generation plant. This dirty liquid will flow through the line **27** and the connecting lines **16** into the manifold **24** and the connectors **25** into the individual, radially spaced inlet tubes **16**. The liquid will flow along the length of the inlet tubes **16** and will escape through the perforations **17** therein for flow through the granular filter medium **35** into the outlet tubes **15**.

The liquid exits from the tubes **15** by flowing upwardly through the outlet tubes **15**, the connectors **24**, the manifolds **21** and the connecting piping **23** for exit through the outlet line **20** in which an off-on valve **V2** is interposed. This is the normal operating flow of liquid through the filter apparatus **10**, and any solid contaminants in the incoming dirty liquid in the pipe **27** will be entrapped within the interstices of the finely divided filter medium **35** during such flow.

It has been found through actual experimentation that, with a filter influent or dirty liquid having a dirt content of 10 parts per million, the effluent or clean liquid flow from the filter consistently contains less than 5 parts per billion. When the influent was on the order of 100 parts per billion, the effluent was consistently below 3 parts per billion. These results confirm the filtration properties of aluminum oxide and well meet the operating requirements for power plant condensate filtration and the like.

Normal flow of influent into the filter medium and effluent from the filter medium will continue until such time as sufficient foreign material has accreted in the body **35** of filter medium to substantially inhibit the flow of liquid or until such time as the effluent in the line **20** is of an intolerable degree of turbidity. Either the increase in resistance to the flow of liquid through the filter medium or such increased turbidity may be used to determine the necessity for rejuvenation of the filter medium. In the event that it is desired to rejuvenate the filter medium in the event of a rise of pressure drop across the medium, a simple pressure gauge may be installed in the inlet line **27**, which gauge would sound an alarm, give a visual indication or even automatically initiate the filter rejuvenation cycle. Alternatively, a continuous turbidity monitoring system may be installed in the clean outlet or effluent line **20**, such a monitoring system would again give audible or visual alarm or initiate automatically the filter rejuvenation cycle.

To rejuvenate the filter medium, the first step required is to close the valves **V1** and **V2**, shutting off the flow of liquid through the tank **10**. Next, the valve **V3** interposed in the drain line **32** is opened and the filter medium drains through the drain line **32** into a lower medium-dirt separation tank **40**.

This tank **40** comprises generally an open-topped, cylindrical upper section **41** and a lower drainage section having sloping or conical side walls **42** defining a lower settling area **43** provided with a drain line **44** controlled by a valve **V4**. The drainage of the filter medium through the line **32** into the tank is accomplished with the valve **V4** closed. A mixer indicated generally at **45** is provided to stir the contents of the tank **40**, this mixer having a drive motor **46** and a lower agitator or propeller **47** at the lower extremity of an actuating shaft **48**.

Upon actuation of the mixer, the propeller **47** will stir up the mixture of media, liquid and admixed dirt or contaminants within the separation tank **40**. The alumina is far more dense than the finely divided particulate or dirt contaminants admixed with the water and a gravity separation of the alumina filter medium to the bottom region **43** of the tank very quickly occurs. A decanting line **50** leads from the upper portion of the tank **40** to a waste tank **51** and a valve **V5** is interposed in the line **50** to accommodate opening of the line and decanting of the liquid and mixed dirt through the line.

Next, a refill pump **55** is started, this pump having its inlet line **56** immersed within a body of clean water **58** contained within a clean water tank **57** and has its outlet line **59** connected through a valve **V6** with the refill line heretofore described and communicating with the extreme free end of the tank **10** through the cap **12**. A venturi-type eductor, schematically illustrated at **60**, is interposed in the outlet line **59** of pump **55** and this eductor has its throat connected, as by line **61**, to the valve **V4** controlling the flow of medium through the drain line **44** to the tank **40**.

Upon opening of the valve **V4**, the flow of clean water through the pump outlet line **49** will educt or aspirate the media from the tank **40** through the line **44**, the open valve **V4** and the line **61** into the line **59** for travel to the tank **10** when the valve **V6** is open. At the same time the valve **V7** positioned in a bypass line **63** is opened to provide a return path through which liquid flows through the interior of the outlet tubes **15** back to the clean water tank **58**.

In this manner, the media **35**, which settled to the bottom of the medium and dirt separation tank **40**, is pumped by flow from the pump **55** through the line **59** into the interior of the tank **10**. The liquid is vented from the tank through the outlet tubes back to the clean water tank **57**, and the media accretes or accumulates on the exterior surfaces of the outlet tubes **15**. More specifically, the media **35** is precoated upon the cloth covering **18** of the outlet tubes **15**.

Inasmuch as this is a conventional precoating-type operation, some of the media may well travel through the cloth covering **18** on the outlet tubes **15** and into the tubes **15** for flow back through the lines **20** and **63** to the clean water tank **57**. In the event that this happens, the much denser medium soon settles to the bottom of the tank and does not interfere with the intake of clean water through the pump intake **56**. More likely, the flow of liquid upwardly through the outlet tubes **15** will be at a velocity which is insufficient to flush upwardly any media entering the tubes **15** and the medium particles will either remain suspended in the liquid flowing upwardly through the tubes **15** or will settle to the bottom of the tubes.

The cycling of liquid from the tank **57** through the line **59** and into the tank **15** and thence through the lines **20** and **63** back to the tank **58** continues for a period of time sufficient (1) to refill the tank **10** with all of the medium which is collected in the lower portion **43** of the separating tank **40**, (2) until the level of the medium is substantially as shown in Figure 9.4a and (3) until the outlet tubes **15** are completely precoated.

At this time valve **V7** is closed and valve **V8** interposed in drain line **65** is opened. This line interconnects the lower manifolds **28** communicating with the lower open ends of the outlet tubes **15** with the drain line **63** to the clean water tank **58**, and any media within the tubes **15** will be flushed therefrom by the flow of liquid into the tank from the pump **55**, through the line **59** and valve **V6**. Once again, this media will readily gravitationally separate within the tank **58** for subsequent reuse, if desired.

After sufficient time has elapsed for substantially all of any such particles to have escaped to the clean water tank **58** from the bottom of the outlet tube **15**, the valve **V8** is closed. The valve **V2** is opened, the valve **V4** is closed, the pump **55** is shut off and the valve **V6** is closed, while the valve **V1** is opened. The filtration apparatus **10** is now back on its filtering cycle with dirty liquid entering the tank **10** through the line **12** and clean liquid exiting from the tank **10** through the line **20**.

While the precoating of the outlet tubes **15** has been described in connection with the complete rejuvenation cycle, it will be appreciated that substantially this same precoating operation will be carried out following initial filling of the tank with media to initiate filtration operation. In other words, the upper cap **12** is initially removed, the granular, finely divided body of filter media **25** is placed within the tank to completely cover the tubes and the manifolds **21** and **24**, valve **V2** and **V7** are opened, the pump **55** is started and the valve **V6** is opened. Thus, the initial precoating is carried out.

At the conclusion of the initial precoating, the valves **V2** and **V7** are closed and valve **V8** is opened to drain any medium from the interior of the tubes **15**. Finally, valves **V8** and **V6** are closed, the pump **55** is stopped and the valves **V1** and **V2** are opened to initiate the actual filtration cycle.

The process provides apparatus and medium manipulation techniques effective to rejuvenate the heavy, abrasive, difficultly handled filter medium for reuse. The rejuvenation of the medium is important, not only for economy, but also to avoid the necessity of disposing of additional radioactive waste from atomic power plants or the like. Of course, the medium can be thrown away when clogged with contaminants or a simple reverse flow cleaning can be used.

Aluminum oxide or alumina also has one other outstanding characteristic. Due to the extreme surface areas which are present in the extremely finely divided filter medium, wet aluminum oxide acts almost like a solid and will not drain through an aperture such as provided by drain line **32** in the tank **10**. However, if an excess of water is present in the form of a layer of water over the aluminum oxide, it has been found that the aluminum oxide will readily flow through even a restricted aperture. By maintaining a body of water over the filter medium inside the container **10**, the filter medium can be readily drained through the restricted drain line **32** into the lower separation tank **40**. Similarly, by retaining a liquid level over the medium as it is present in the lower section **43** of the tank **40**, it is possible to drain the media through the line **44** of the valve **V4** and the line **61** into the throat of the eductor **60**.

In the drainage of the tank **10**, complete drainage can be assured by simply actuating the pump **55** and opening the valve **V6** during drainage or even after drainage to flush down the last remnants of the filter medium.

By utilizing the suspended tubes **15** and **16** embedded within the filter medium, it is possible to provide a large amount of filtration area within the tank **10** of relatively small size. Since a typical power plant might use 20,000 gallons per minute of steam condensate and since the flow rate through aluminum oxide ranges from about 5 gallons per minute per square foot to about 10 gallons per minute per square foot, it is necessary to create filtration area cheaply and effectively. The concentric, multiple tube design of the process creates this area in a relatively small tank **10**.

The density of aluminum oxide prevents its suspension in a slurry of liquid for effective movement by a pump or other pressure generating means. Thus, the natural flow characteristics of the material are utilized by gravitationally dumping the tank **10** into the separation tank **40** and by gravitationally flowing from the separation tank to the throat of the eductor **60**. The transfer of the dense media back to the pressure vessel **10** is accomplished by creating a large flow of liquid, by means of the pump **55** at a high velocity and feeding the dense medium into this line through the eductor at a slower and more sluggish pace. By utilizing 4 to 5 times as much water flowing through the line **59** as there is media flowing into the line **59**, it is possible to transport the dense media for deposition within the filter tank or at the bed location.

Filter and Drier Block for Refrigerants

A.A. Sakhnovsky, W.O. Walker and R.B. Tilney; U.S. Patent 3,545,622; December 8, 1970; assigned to Alco Controls Corporation describe a permeable filter and drier block comprised of silica gel and other inorganic desiccants, such as sodium aluminum silicate, bound together into a self-supporting block, not subject to decrepitation, by an organic plastic binder, such as an epoxy resin, a silicone or polyvinyl alcohol resin. The block does not char at temperatures of 300° to 450°F and can be reactivated at high temperatures and is temperature stable. The block can be compacted at relatively low pressures, such as 200 pounds. In the preferred example, larger sodium aluminum silicate particles, which are difficult to bind, are mechanically trapped in the organic resin-bonded silica gel mass. The composition finds an important use in the drying of various fluids and the removal of water therefrom and in filtering out solid contaminants. It is especially useful as a drier in refrigeration systems.

The preferred form of the process comprising a block made of silica gel, sodium aluminum silicate and an epoxy resin binder, can be as follows wherein each component is set forth, followed by a size range where appropriate, a preferred specific quantity and finally a range in percentages. The first two percentages are related to each other, by weight, and the third percentage is related to the silica gel by weight.

Ingredient	Size Range	Specific Amount, grams	Range, %
Silica gel	6 - 60 mesh	232	95 - 75
Sodium aluminum silicate	⅛" - 30 mesh	11	5 - 25
Epoxy resin binder	-	7.7	1 - 18

A suitable sodium aluminum silicate is Molecular Sieve MS4A sold by Linde Company. A suitable binder is Epon 828 sold by Shell Chemical Company. It is a liquid at room temperature, and has a viscosity of 100 to 160 p at 25°C and an epoxide equivalent of 175 to 210. Epon 815 and 820 may also be used.

Epoxy resins are reaction products of epichlorohydrin and 2,2-bis(4-hydroxyphenyl)propane. The epoxide equivalent is the number of grams of resin to completely esterify 1 gram mol of monobasic acid. The preferred curing agent for use with the epoxy resin is 4,4'-methylenedianiline (MDA). This curing agent needs to be one that will not be adsorbed by the desiccants before it can react with the epoxide. If it does adsorb to a high degree, an undesirably large amount of binder must be used. As noted herein, the named curing agent allows the use of a low percentage of binder.

Where ease in handling and pot-life are important considerations, it is preferable to dissolve the 4,4'-methylenedianiline in acetone to give a solution containing 40 to 60% of the former. The MDA preferably should be from 26.8 to 31.1% by weight of the resin. An amount of epoxide resin should be added to this MDA solution to give 28.5 grams of 4,4'-methylenedianiline per 100 grams of resin. The amount of acetone should be kept at a minimum for economic as well as safety reasons. The maximum of 60% 4,4'-methylenedianiline in acetone approaches the practicable solubility limit. As an example of the process of manufacturing the block, there is shown in Example 1 a method of manufacture.

Example 1: Silica gel in the amount of 232 grams, 20 to 40 mesh, is mixed with 7.7 grams of binder made as described above. To 187 grams of this mixture is added 23 grams of sodium aluminum silicate spheres, 4 to 8 mesh. The mixture is tamped to 200 psi. The block is cured at about 420°F, then activated for 3 hours at about 375°F, heated air being passed through as a purge gas. The final block has about 2.5% binder.

It is desirable to surround the foregoing block with a layer of the originally mixed silica gel and binder, which can be done by tamping a first layer of the same into the mold, followed by a layer of the entire mix, and finally by a layer of the silica gel and binder, prior to curing. This laminating is especially desirable where the silicate particles are large, and might become dislodged in handling or use.

It appears that the binder has greater adhesive effect on the silica gel particles, and that the silicate particles, which are larger and smoother, are less subject to the binder, and are primarily held by being mechanically trapped in the bonded silica gel mass. By this combination a strong filter-desiccant block, of both silica gel and sodium aluminum silicate, is attained. In general, if larger particles of silica gel are used, more binder is required.

Example 2: In this example an epoxy binder made as previously described is employed with molecular sieves and silica gel. The molecular sieve is used in the amount of 20% of the total desiccant while silica gel is used in the amount of 80%. The epoxy binder is employed in a sufficient amount to provide 3% total solids after curing. In the process 161 grams of silica gel are thoroughly mixed with the epoxy resin binder previously described.

Subsequently the molecular sieve is added. It is to be noted that if it is added earlier, it tends to adsorb the epoxy resin and this should be avoided. After this last addition the resultant composition to surround the molecular sieve particles by coated silica gel, which locks the former particles in a lattice type of structure when the composition is finally cured, is mixed.

The composition is then molded and impacted in a jolter at 90 psi to provide a uniform density. While still in the mold the composition is baked in an oven at 325°F for a period of 2½ to 3 hours. The resultant block is then transferred to a cylinder and painted and baked for a final period of about 4 hours at 390°F to cure the paint and fully activate the filter and drier block. In the baking stages the block is purged with heated air and a vacuum is applied at 29 inches of mercury as an example. No moisture is required for activation.

Example 3: In this example an epoxy resin was employed that was a Shell Epon 828 as before. It was employed with molecular sieves and silica gel. The molecular sieve was used in the amount of 22 grams in the form of 4 to 12 mesh beads, preferably in the 8 to 12 mesh range. 80 grams of silica gel was employed of a 20 to 40 mesh size. The epoxy binder consisted of 2.33 grams of the Epon 828 resin mixed with two-thirds gram MDA which had been previously diluted with 1 gram acetone.

The drying ability of the blocks made with the various inorganic desiccants and an organic binder combination, as set forth above, has been evaluated over different temperature ranges with different types of organic refrigerants. Tests have demonstrated the satisfactory operation and pickup of water in such refrigerants and have also demonstrated the desirable filtration aspects of these blocks.

Shaped Polyurethane Filter and Drying Element

A process described by *K. Kuhn and M. Boedecker; U.S. Patent 3,687,297; August 29, 1972; assigned to Kali-Chemie AG, Germany* involves forming a shaped porous body which comprises a solid adsorptive drying agent or adsorbent and between 10 and 0.3% of a binding agent relative to the weight of the drying agent or adsorbent, the binding agent consisting essentially of the reaction product between a di- or polyisocyanate and a polyester or polyetherpolyol which reaction product is in the form of a hardened polyurethane.

In making this product, a solid adsorptive drying agent or adsorbent is mixed with (a) a di- or polyisocyanate and (b) a polyester or polyetherpolyol or mixtures of the latter two compounds, the amount of the components (a) and

(b) being 0.3 to 10 grams for each 100 grams of drying agent or adsorbent, whereupon this mixture is poured in a mold and subjected to curing to a hard polyurethane at a temperature between 20° and 120°C.

The formed bodies may contain, for instance, bentonite, activated coal and/or quartz sand in addition to the drying agent. These mixtures of different adsorbent materials can be used as such or the different materials can be incorporated in separate layers. In the first case the mixtures of all the materials and, in the latter case, the individual components of each layer are mixed with the mixture which is then hardened to a polyurethane.

For the total amount of solid adsorbents not more than 10 grams binding agent for each 100 grams of solid material is necessary and, preferably, the amount is not higher than 5 grams binding agent. If the individual components are arranged in layers, the amount of binding agent must be adjusted to the material present in the layer. If a layer consists only of activated coal, the amount of binding agent must be increased in excess of 10%. Depending on the granulation of the activated coal, between 10 and 20 weight percent of binding agent should be used. However, the amount of binding agent can be reduced if sand is added to the activated coal. The following examples illustrate the process.

Example 1: 100 weight parts of dried commercially available pearl-shaped silicic acid gel with a diameter between 2 and 6 mm were mixed with 2 weight parts of a mixture of stoichiometric amounts of a polyether on trimethylolpropane basis having a hydroxyl number of about 520 and diphenylmethane-4,4'-diisocyanate. The mixture was effected above the solidification point of the isocyanate. The mixture after pouring into a mold was then hardened at 45°C. The hardening was complete after 30 minutes. The obtained formed body had a high stability.

Example 2: A drying agent-containing formed body was made as described in Example 1, except that the mixture of polyether and diphenylmethane-4,4'-diisocyanate was employed in an amount of 3 weight parts instead of 2 weight parts. A formed body having a high form stability was obtained.

The drying agent bodies of Examples 1 and 2 possessed a high permeability for liquids. They were resistant to fluorinated hydrocarbons. In tests of their stability by placing them in monofluorotrichloromethane at 20°C for 5 weeks no swelling occurred.

The water adsorption capacity of the bodies of Examples 1 and 2 was tested by passing air of varying relative humidity at 25°C through the body and comparing the capacity with that of the loose drying agent. The water vapor adsorption at equilibrium at 25°C was as follows:

Relative Humidity, %	Example 1, wt %	Example 2, wt %	Loose Drying Agent, wt %
10	6.4	6.3	6.6
20	9.7	9.7	10.6
40	21.1	21.0	21.9
60	32.1	32.2	33.6
80	35.5	35.3	37.1

Example 3: 100 weight parts of a dry commercially available pearl-shaped silicic acid gel were mixed with 2 and 3 weight parts, respectively, of a mixture of stoichiometric amounts of a trimethylolpropane-based polyether with a hydroxyl number of about 520 and an isomer mixture at a ratio of 80:20 of 2,4- and 2,6-toluylenediisocyanate and were hardened at 45°C after pouring into a mold. In both cases stable bodies with high permeability for liquids and gases were obtained.

The water adsorption capacity of the formed bodies was only slightly different from that of the loose drying agent as shown by the following table. This table shows the water vapor adsorption at 25°C at equilibrium:

Relative Humidity, %	2% Mixture, wt %	3% Mixture, wt %
10	5.9	6.5
20	9.4	9.9
40	20.8	21.2
60	29.8	31.0
80	32.7	33.8

Drycleaning Purifier

J.R. Conlisk and C.A. Filliben; U.S. Patent 3,619,120; November 9, 1971 describe a drycleaning purifier composition which comprises in a major fraction an activated carbon having a particle size intermediate that of powdered activated carbon, which is ordinarily sufficiently fine to pass through a mesh screen No. 325 (U.S. Bureau of Standards, Standard Screen Series, 1919) and that of conventional granular activated carbon ordinarily having a particle size ranging from about No. 8 mesh to about No. 40 mesh, and in a minor fraction an activated carbon or other particulate material having a particle size which is substantially greater than the particle size of the activated carbon of the major fraction.

The major fraction of the purifier composition comprises particles of activated carbon having a wet density equal to or greater than the density of the drycleaning solvent in which the activated carbon is used and the minor fraction of the purifier composition comprises, at least in part, particles of particulate material having a wet density less than the density of the drycleaning solvent in which the particulate material is used.

The major fraction of the purifier composition comprising activated carbon particles of a size intermediate between the size of the particles of powdered carbon and the size of the particles of conventional granular carbon constitutes not less than about 85% by weight of the composition nor more than about 95% by weight of the composition while the minor fraction of the purifier composition of this process comprising particles of particulate material of a size substantially greater than the size of the particles of the major fraction constitutes not more than about 15% by weight of the composition nor less than about 5% by weight of the composition.

The drycleaning solvents for the purification of which the purifier composition is suitable are those synthetic and petroleum drycleaning solvents in ordinary commercial use. Of these, the only synthetic drycleaning solvent of significance is perchlorethylene, which has a density of about 1.62 g/ml. The petroleum drycleaning solvent in commonest use is Stoddard solvent, having a density of about 0.78 g/ml. Other petroleum solvents which may be used have densities not far from that of Stoddard solvent and may range from about 0.75 g/ml to about 0.8 g/ml. Accordingly, the drycleaning solvents, the purification of which is contemplated by the purifier composition of this process will have densities within the range of about 0.75 g/ml to about 1.65 g/ml.

Ordinarily, the purifier composition is used in the solvent systems of drycleaning apparatus supplied with a filter system of the conventionally tubular type. Typical examples of this type of drycleaning filter are Per-Mite and Per-Flex. The following is a specific example of the purifier composition of the process.

Example 1:

Ingredient	Percent by Weight
Activated carbon particles of U.S.S. No. 40 to 140 mesh size and 1.9 g/ml average wet density	95
Activated carbon particles of U.S.S. No. 4 to 12 mesh size and 1.55 g/ml average wet density	5

A filter tube (flexible type) of a glass filter was made defective by deforming the tube to provide several breaks near the top thereof approximately 1/16 inch by 1/8 inch in dimension. Thereafter the defective tube was placed in the filter apparatus and perchlorethylene drycleaning solvent was arranged to circulate through in the manner normal to the solvent systems of conventional drycleaning apparatus. Thereafter, the purifier composition was added to the perchlorethylene solvent along with filter aid in the conventional manner. The particles of the minor fraction of the purifier composition were observed to float in the perchlorethylene solvent while the particles of the major fraction were observed to be suspended therethrough.

When the filter apparatus was started, the large, light particles of the minor fraction of the purifier composition were carried promptly to the holes of the defective filter tube and effectively blocked the passage of the smaller, denser particles of the major fraction of the purifier composition, as evidenced by the fact that the filtered solvent became clear after about 7 minutes of the operation of the filter whereas normally, using only particles of the same size and density as those contained in the major fraction of the composition, the drycleaning solvent did not clear for more than 40 minutes. The following are further examples of the purifier composition as provided in accordance with this process.

Example 2:

Ingredient	Percent by Weight
Activated carbon particles of U.S.S. No. 40 to 140 mesh size and 1.9 g/ml average wet density	90
Activated carbon particles of U.S.S. No. 4 to 12 mesh size and 1.55 g/ml average wet density	5
Activated carbon particles of U.S.S. No. 4 to 12 mesh size and 2.0 g/ml average wet density	5

Example 3:

Ingredient	Percent by Weight
Activated carbon particles of U.S.S. No. 40 to 140 mesh size and 1.9 g/ml average wet density	85
Activated carbon particles of U.S.S. No. 4 to 12 mesh size and 1.55 g/ml average wet density	10
Cinder particles of U.S.S. No. 4 to 12 mesh size and average wet density in excess of 1.62 g/ml	5

Sand Filter for Oil Wells

W.G. Bearden and C.A. Powers; U.S. Patent 3,729,337; April 24, 1973; assigned to Amoco Production Company describe a specially designed screen for use in wells producing oil from unconsolidated sands. The apparatus consists essentially of a sleeve fabricated from a ground hydraulic cement clinker, coated with a heat-resistant film from material such as sodium silicate to impart compressive strength to the set clinker when the latter is exposed to oil. The clinker material which may be calcium silicate, calcium aluminate or high alumina cement may have a particle size ranging from –4 to +40 mesh.

Briefly stated, the process involves the discovery that the cement clinker in filter screens such as the type described in U.S. Patent 3,244,229 when exposed to oil tends to undergo a substantial reduction in compressive strength, particularly after the screen has been exposed to elevated temperatures. This difficulty can be overcome by coating the set or cured clinker mass with a material that is oil- and water-insoluble.

The treatment of cement clinker forms in accordance with this process not only renders such forms stable on extended periods of exposure to oil, but also unexpectedly increases the compressive strength of these forms.

In carrying out the process, a ground hydraulic cement clinker is silicate prepared from calcium silicate, high alumina cement or calcium aluminate. The ground material may be placed in position in essentially the same manner as a gravel pack, i.e., introduced in dry form. However, the pack resulting from the use of cement clinker differs markedly from that made of gravel since the particles of clinker in the presence of water, which is ordinarily standing in the well, hydrate and bond to one another at points of contact.

After the cement clinker has set, it is treated with a material capable of leaving or forming a high temperature (e.g., 1500° to 2000°F) resistant film on the surface of the cement particles. Typical examples of such materials are potassium and sodium silicates, saueresin (a liquid porcelain) and the like. This material is placed down the well either in the form of a solution or a slurry to contact the set cement clinker.

Thereafter a hardened film is formed on the cement by subjecting the resulting mass to a drying operation by the use of an inert gas such as air. The gas employed may be either substantially at formation temperature or, if desired, can be heated to higher temperatures if a shorter treating period is felt necessary.

In this way a strong, rigid filter is formed which does not lose its compressive strength on exposure to oil over long periods of time. This filter is capable of preventing sand from entering the well while at the same time allowing fluids to be produced in a normal manner. The increased strength of the resulting filter pack is beneficial for the reason that it lends support to the incompetent formation and thereby prevents movement of the sand grains therein.

As a source of suitable ground material it has been found that high alumina cement clinker is particularly effective. The clinker may be converted for use in this process in a number of different ways. For example, it can be passed through a coarse grinder, then screened to provide the desired particle size distribution. In this regard, tests have shown that high alumina cement (Lumnite—a calcium aluminate cement) clinker ground to a particle size ranging from about -4 to about +40 mesh, when allowed to set properly, can have a tensile strength in excess of 200 psi, a compressive strength of 1,500 psi and a permeability as high as 60 darcys.

Generally speaking, and particularly for thick producing sections, it is preferred to prepare the filter screens from Lumnite cement clinker or its equivalent having a particle size range of from about -4 to about +12 mesh. Such screens function more satisfactorily and retain more of their original permeability after installation than do screens made from particles of a smaller size range. For example, in using screens made with the smaller particle size range, i.e., -10 to +20 mesh, in wells with a thick, e.g., 200 feet, producing section such as found in the tar sand deposits, these screens lose approximately 75% of their initial permeability as a result of the installation procedure.

However, for wells having a relatively thin producing section, e.g., 20 feet, screens made from the -10 to +20 mesh particles operate satisfactorily because the installation procedure is less severe and requires substantially less fluids in the "washing in" step than are necessary where the screen is longer and the producing zone is thicker.

In installing these liners, they may be run into the well on tubing and placed opposite the incompetent oil producing zone. Methods for running screens of this kind into a well have been practiced for many years. Typical of such procedures are those described in U.S. Patents 2,154,461 and 2,167,190 as well as 2,205,422.

The advantages afforded by this process are further illustrated in columns 2 and 3 of the following table which reports the results of a series of tests made to establish strength and permeability characteristics of Lumnite cement clinker, for example, -12 to +20 mesh, for use in connection with the manufacture of well liner sand screens. The clinker was tested in the form of 2-inch cubes and placed in water usually for 48 hours to cure at room temperature, i.e., about 75°F. After setting, these cubes were then treated under the various conditions referred to below. The principal objectives of these tests were to determine the effect of oil on the strength and permeability of the clinker screen after firing, i.e., heating in an oven to about 800°F for 48 to 72 hours. The results are shown in column 2 and demonstrate that the strength of the clinker is reduced substantially on contact with oil after the heating step.

Comparative compressive strength data obtained on the clinker sample cured first in water and then soaked in oil before firing showed that the oil causes a strength reduction of about 43%. After firing, the compressive strength of the clinker first cured in water and then soaked in oil was reduced by approximately 68% compared to that of the initially cured material. The strength reduction is reflected in the higher permeabilities obtained after firing.

In another series of tests, shown in column 3 of the table, two-inch cubes of the cement clinker after being cured in water for 48 hours at room temperature were saturated with 41° Baumé sodium silicate solution. The saturation step was effected by flowing 1 cube pore volume of the sodium silicate solution through the clinker and thereafter blowing this material dry with air (100°F). Next, the permeability and compressive strength of the treated clinker were measured after which it was exposed to oil, and permeability and compressive strength observed. The specimens were then heated at 800°F for a period of 48 hours and exposed to oil, following which permeability and compressive strength measurements were again taken.

These data indicate that appreciable permeability reduction can be expected by treatment with a relatively concentrated sodium silicate solution but that greater strength is imparted to the clinker by such treatment.

	Plain Clinker	Clinker Treated with Silicate
After initial curing:		
Permeability to water, darcys	50	28
Compressive strength, psi	1,200	1,780
After exposure to oil—before firing:		
Permeability to oil, darcys	56	28
Compressive strength, psi	785	1,290
After firing at 800°F and exposure to oil:		
Permeability to oil, darcys	71	20
Compressive strength, psi	390	1,155

An effort was made to determine whether or not the water curing step could be avoided by attempting to cure the cement clinker in solutions of sodium silicate. However, when the samples were dried and heated to temperatures of 300°F in some cases and 800°F in others, it was found that the cubes disintegrated.

To demonstrate the ability of these alkali metal silicates to increase the compressive strength of a high alumina cement form, still another series of tests were carried out. In these tests, 2-inch cubes of –4 to +12 mesh of Lumnite cement clinker were used and prior to treatment with the various materials indicated below were cured in water at room (75°F) temperature for 96 hours and then air dried at room (75°F) temperature for 48 hours.

Thereafter, some of these cubes were saturated with clear furniture lacquer, some with clear spar varnish, some with 37% sodium silicate solution, some with a 10% sodium silicate solution and some were untreated. In the treating process the cubes were saturated with the film-forming material, then immediately blown dry with air. Thereafter, they were placed in water for 120 hours and then dried for 168 hours. The results obtained are shown in the table on the following page.

	Compressive Strength, psi
Plain clinker	1,066
Clinker treated with:	
Clear lacquer	766
Clear varnish	756
37% sodium silicate solution	1,226
10% sodium silicate solution	1,262

From the above data it is evident that the compressive strength of the clinker cement is not improved merely by coating it with any film-forming material but that such strength can be substantially increased by the use of an alkali metal silicate solution in accordance with the process.

As a result of the work described above, acceptable permeabilities and high compressive strengths unaffected by oil can be secured by treating the clinker material with a suitable liquid porcelain or alkali metal silicate before the material is subjected to high temperature.

Filter Plugs for Cigarettes

A process described by *H.-J. Block; U.S. Patent 3,971,695; July 27, 1976; assigned to Hauni-Werke Korber & Co., KG, Germany* relates to a method and apparatus for the production of filter plugs which can be used in the manufacture of filter cigarettes, filter cigars and like smokers' products.

One feature of the process resides in the provision of a method of producing filter plugs which comprises conveying a continuous tow of filamentary filter material lengthwise, subjecting the tow to a variable stretching action to thereby change the mass of filamentary material per unit length of the tow, applying a web of wrapping material around the thus stretched tow to form a wrapped filter rod, subdividing the rod into filter plugs, conveying a gaseous testing fluid axially through the filter plugs, measuring the resistance which the filter plugs offer to the flow of such fluid, and regulating the stretching action as a function of deviations of measured resistance from a predetermined value which is indicative of satisfactory resistance.

The filter plugs are preferably transported sideways during testing and the measuring step preferably comprises determining the average resistance of several filter plugs; the regulating step then comprises varying the stretching action as a function of deviations of the average resistance from the predetermined value.

Referring to Figure 9.5a, there is shown an apparatus for the production of filter plugs **14**. The apparatus comprises a plug making machine **1**, a filter tow processing machine **2** and a pneumatic testing unit **3** for filter plugs **14**. The filter plug machine **1** comprises the following components.

A bonded layer or tow **6** of fibers is introduced into a gathering funnel **4** which converts the layer into a cylindrical filler and feeds the filler into a wrapping mechanism **7** including an endless garniture conveyor belt **9**. The mechanism **7** applies around the cylindrical filler a web **8** of cigarette paper or like wrapping material to produce a continuous wrapped filter rod **12** which passes along a heater or sealer **11** serving to complete the seam between the overlapping marginal portions of the wrapper.

FIGURE 9.5: APPARATUS FOR THE PRODUCTION OF FILTER PLUGS

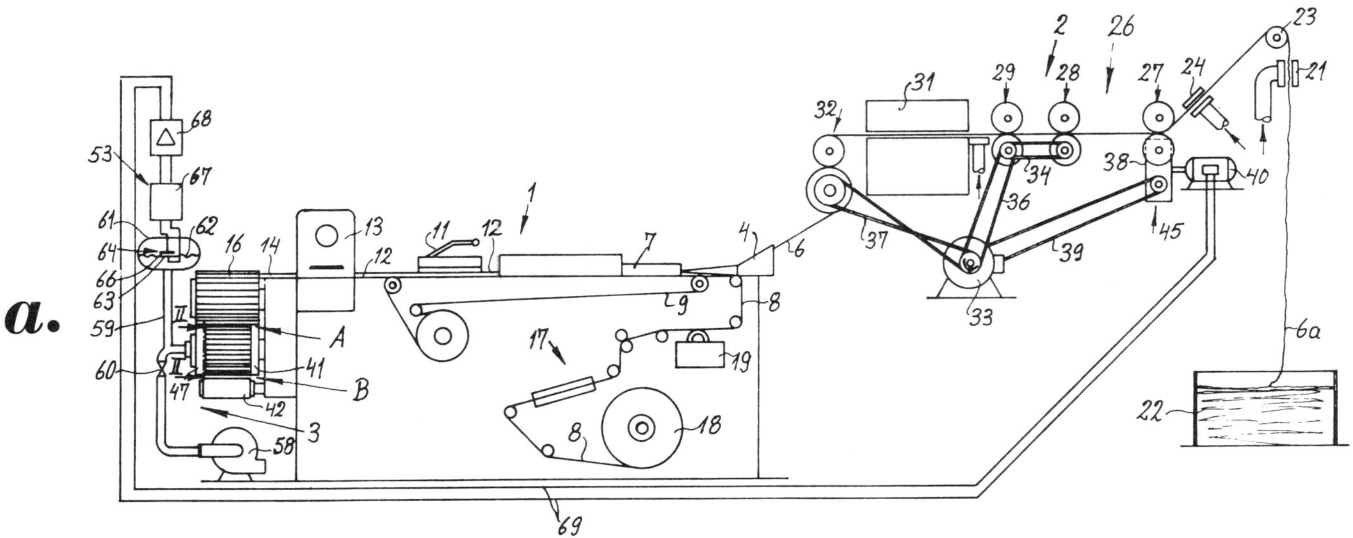

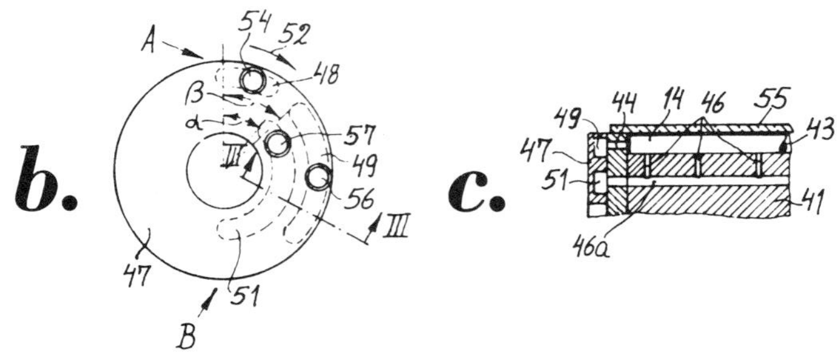

(a) Diagrammatic Side Elevational View of Process Apparatus
(b) Enlarged Fragmentary Sectional View as Seen in the Direction of Arrows from the Line II-II of Figure 9.5a
(c) Fragmentary Sectional View as Seen in the Direction of Arrows from the Line III-III of Figure 9.5b

Source: U.S. Patent 3,971,695

The rod **12** passes along a cutoff **13** which severs it at regular intervals to subdivide it into filter plugs **14** which are admitted into the flutes of a braking drum **16**. The latter serves to change the direction of travel of filter plugs, i.e., it causes the filter plugs to move sideways and delivers them into successive flutes of a travelling testing conveyor **41** here shown as a drum which forms part of the aforementioned testing unit **3**. The machine **1** further comprises an advancing mechanism **17** which draws the web **8** of wrapping material from a bobbin **18** and moves it along the applicator roller of a paster **19** which coats one side (or at least one marginal zone) of the web **8** with a suitable adhesive.

The filter tow processing machine **2** includes a first banding device **21** which preferably includes one or more nozzles for directing jets of air against the raw tow **6a** of fibrous material to thereby insure that the individual filaments are separated from each other. The thus treated raw tow is caused to pass over a deflecting roller **23** and past a second banding device **24**. The raw tow is drawn from a bale **22**.

The machine **2** further comprises a feeding and stretching device **26** including a first pair of stretching rolls **27** and a second pair of stretching rolls **28**. The stretching rolls **28** are followed by a pair of spreading rolls **29** which convert the raw tow **6a** into a layer upstream of an impregnating device **31** which applies to the fibers a suitable plasticizer of known composition. The layer which is formed by the spreading rolls **29** is located in a plane making a right angle with the plane of Figure 9.5a. The bonded layer **6** is thereupon caused to pass around a pair of deflecting rollers **32** and into the gathering funnel **4** of the machine **1**.

The rolls **27, 28** and **29** and the rollers **32** are driven by an electric motor **33** in the following way: the output shaft of the motor **33** drives endless chains **34** and **36** which rotate the rolls **28** and **29** at a constant speed. The rollers **32** are driven by an endless belt **37**. A further endless chain **39** drives the input member of a variable-speed transmission **38** for the stretching rolls **27**. The ratio of the transmission **38** is adjustable by a variable-speed electric motor **40**; the transmission drives the stretching rolls **27** at a speed which is less than the speed of the rolls **28** and **29**.

The pneumatic testing unit **3** comprises the fluted testing drum **41** which receives filter plugs **14** from the braking drum **16** of the machine **1**. Certain details of the testing drum **41** are shown in Figures 9.5b and 9.5c. This drum has axially parallel peripheral flutes **43** which receive plugs **14** from the flutes of the drum **16** and deliver them onto the upper stretch of an endless conveyor belt **42**. The drum **41** is further provided with axially parallel bores or channels **44** each of which communicates with one axial end of a flute **43**, and with radially extending groups of bores or channels **46**, each such group being in communication with one of the flutes **43** as shown in Figure 9.5c. A stationary valve plate **47** is adjacent to one axial end of the testing drum **41** and is provided with arcuate grooves **48** (shown only in Figure 9.5b) and **49, 51** (shown in Figures 9.5b and 9.5c).

The grooves **48** and **49** can communicate with the axially parallel bores **44** and the groove **51** can communicate with groups of radial bores **46** by way of axially parallel passages **46a** in the drum **41**. The relative positions of grooves **48, 49** and **51** are illustrated in Figure 9.5b, and the direction in which the drum **41** is driven is indicated by the arrow **52**. The groove **48** begins at a first transfer station A where the flutes **43** of the drum **41** receive filter plugs **14** from the braking drum **16**; such plugs remain in the respective flutes **43** while the drum completes about one-half of a revolution to leave the flutes at a second transfer station B where the plugs descend onto the belt **42**. The purpose of the groove **48** is to effect proper positioning of filter plugs **14** in the axial direction of the respective flutes **43**.

The groove **49** serves to connect several filter plugs **14** at a time with an air flow resistance measuring transducer **53** and to insure that, during testing of a group of filter plugs, the air pressure at one axial end of each tested filter plug is different from air pressure at the other axial end. The purpose of the groove **51** is to connect a suction generating device **58** with those radial bores **46** which communicate with flutes **43** for filter plugs which are being tested. A stationary shroud **55** overlies those flutes **43** of the testing drum **41** which are located between the transfer stations A and B.

Conduits **54, 56** and **57** respectively serve to connect the grooves **48, 49** and **51** of the valve plate **47** with the suction generating device **58**. The latter is preferably a conventional suction fan. The conduit **56** between the fan **58** and groove **49** has a branch **59** which connects it with the measuring transducer **53**. A flow restrictor **60** in the conduit **56** throttles the outflow of air from the groove **49**.

The transducer comprises a suction chamber **61** which accommodates a diaphragm **62** provided with a preferably plate-like electrode **63** which forms part of a testing capacitor **64**. The fixed electrode of the capacitor **64** is shown at **66**. This capacitor is connected in an oscillator circuit **67** whose oscillation frequency varies as a function of changes in the distance between the electrodes **63** and **66**, i.e., as a function of the position of electrode **63** in the suction chamber **61**. Such distance, in turn, is a function of the resistance which the filter plugs **14** being tested offer to the flow of air from one axial end to the other axial end.

A measuring transducer which can be used in the testing unit **3** of the apparatus shown in Figure 9.5a is described, for example, in British Patent 1,083,111 (see particularly Figures 2 and 3 of the patent). The patented transducer is actually designed for testing of wrapped tobacco rod sections and operates with compressed air; however, it can be readily adapted for operation with suction in a manner as described above.

The oscillator circuit **67** transmits signals to an amplifier **68** whose output is connected with the electric motor **40** by means of electrical connections **69**. The motor and the variable-speed transmission **38** constitute a control means **45** for regulating the elongation of filamentary filter material in the stretching device **26**.

The operation is as follows. The raw tow **6a** is continuously withdrawn from the bale **22** and its filaments are separated from each other by the banding devices **21** and **24** upstream of the stretching device **26**. The tow **6a** is extended between the stretching rolls **27** and **28** because the rotational speed of the rolls **27** is less than that of the rolls **28**. This reduces variations in distribution of the mass in the tow **6**. The spreading rolls **29** convert the raw tow **6a** into a layer which advances in a plane making a right angle with the plane of Figure 9.5a and through the impregnating device **31** wherein the filaments are impregnated with a plasticizer. Such plasticizer is a softening agent which is preferably sprayed onto the filaments. Since the filaments are separated from each other (banding devices **21** and **24**) and are thereupon spread apart by the rolls **29**, the softening agent is capable of reaching all or nearly all filaments of the layer **6**. The layer **6** is then fed into the gathering funnel **4** which delivers successive increments of the resulting filler to the wrapping mechanism **7**.

This mechanism applies the web **8** (which was coated with adhesive during travel along the paster **19**) around the cylindrical filler to form the wrapped filter rod **12** whose seam is heated by the sealer **11**. The cutoff **13** severs the rod **12** to produce a series of filter plugs **14** which are caused to enter successive flutes of the braking drum **16** and travel sideways toward the transfer station A where they enter successive flutes **43** of the testing drum **41**. Suction in the groove **48** of the valve plate **47** causes the filter plugs **14** to move axially and to assume predetermined axial positions before they reach the front end of the groove **51** (angle alpha in Figure 9.5b).

The groove **51** draws air from the corresponding passages **46a** and bores **46** so that the filter plugs **14** are held against axial movement during testing. Each filter plug **14** which has covered an angle beta (shown in Figure 9.5b) with the associated flute **43** is thereupon subjected to the action of suction in the groove **49** and in the corresponding axially parallel bore **44**. Depending on the resistance which the plug **14** offers to flow of air therethrough (as considered in axial direction of its filler), the pressure in the suction chamber **61** of the measuring transducer **53** rises or drops to cause a commensurate deformation of the diaphragm **62**. The oscillator circuit **67** sends a corresponding signal to the amplifier **68** which causes the motor **40** to adjust the transmission **38** and to thus influence the action of the stretching device **26** whenever the average resistance of simultaneously tested plugs **14** deviates from a predetermined value.

If the resistance of filter plugs **14** is less than such predetermined value, the transmission **38** begins to drive the stretching rolls **27** at a higher speed so that the mass of filamentary material in the filler of the rod **12** increases, i.e., the stretching action of the device **26** is reduced. This brings about an increase in the resistance to flow of air in the axial direction of resulting filter plugs. If the resistance to flow of air through the filter plugs **14** which are tested between the transfer stations A and B is higher than normal, the transmission **38** is caused to reduce the speed of the stretching rolls **27** so that the device **26** produces a more pronounced stretching action and the mass of filamentary filter material in the filter plugs **14** decreases. This results in the production of filter plugs whose resistance to the flow of testing air in axial direction (during travel past the groove **49** of the valve plate **47**) is reduced and closely approximates or equals the desired resistance.

Since the apparatus of Figure 9.5a regulates the stretching action as a result of measurements of resistance which the filter plugs **14** offer to axial flow of the testing fluid, such measurements are more reliable than the measurements of fluid flow across the tow. Thus, the measured resistance is identical with that which a smoker encounters in drawing the smoke through the filter plug of a cigarette or cigar.

Another advantage of the apparatus is that the transducer **53** measures the average resistance of several filter plugs **14** on the drum **41**. This is desirable to insure that a single filter plug which exhibits a pronounced defect (e.g., an open seam in its wrapper) cannot unduly affect the action of the stretching device **26**.

Algicidal Filter for Aquarium

E.A. Abbott, A.J. Isquith and P.A. Walters; U.S. Patent 3,865,728; February 11, 1975; assigned to Dow Corning Corporation have found a method of inhibiting the growth of algae which comprises contacting the organism with a solid material having bonded to the surface thereof an algicidal amount of a coating consisting essentially of an organosilicon compound of the unit formula

$$(R'SiO_{\frac{3-n}{2}}R_n)_a\overset{\displaystyle Me_{3-a}}{\overset{|}{N^+}}R''\ Y^-$$

in which Y is chloride or bromine, R'' is a methyl radical or an aliphatic hydrocarbon radical of from 11 to 22 carbon atoms, R' is a divalent hydrocarbon radical of 1 to 10 carbon atoms, R is a monovalent hydrocarbon radical of 1 to 6 carbon atoms or the $CF_3CH_2CH_2$— radical, n is an integer from 0 to 2, and a is an integer from 1 to 2.

This process also relates to a filter comprising a fibrous material, the fibers of which are coated with an algicidal amount of the above defined organopolysiloxane.

The quaternary ammonium organosiloxanes are known materials and they are prepared by hydrolysis of the corresponding hydrolyzable silanes which are also known. The surfaces employed in this process can be coated in any desired manner. One can apply a solvent solution of the siloxane and thereafter allow the coated material to air dry or one can follow the coating step with a heating step. Preferably, however, the substrates are coated with an aqueous solution of the organosilicon compound.

These solutions are prepared by dissolving the corresponding hydrolyzable silanes in water at any desirable concentration. It has been found that concentrations of from 0.4 to 10% by weight siloxane in the treating bath give satisfactory results. The treated surface is then allowed to dry which causes a film of the siloxane to be deposited on the solid surface. If desired the surface can thereafter be heated at temperatures from 65° to 100°C or above in order to more firmly fix the siloxane to the surface. Also if desired, the known siloxane condensation catalysts can be used to enhance the condensation of the siloxane and to fix it more firmly on the surface. The following examples illustrate the process. Me is an abbreviation of the methyl radical and Ph of the phenyl radical.

Example 1: This example shows the scope of the type of compounds operative in this process. The test compounds shown below were applied in the form of a 1% by weight aqueous solution to the lower half of cleaned glass slides and allowed to air dry. The coated slides were then washed under running tap water for 15 to 60 minutes and redried. Using this method the upper portion of each slide served as an untreated control. The slides were embedded in a vertical position in paraffin floats and suspended in algal solutions. Each solution was placed in a covered aerated beaker under controlled constant illumination. The slides were removed from the solution in from 7 to 14 days and examined for growth and attachment of algae on both the treated and untreated portion of the slide. In each case the untreated portion of the slide was covered with algal growth whereas the treated portion was either clear of algal cells (as shown by microscopic examination) or it was demonstrated that the cells attached to the glass were nonviable on subsequent subculture. The killing of algae in solution was not observed, indicating that the compounds were permanently attached to the slide surface.

For comparison glass slides were treated in like manner with a 1% solution of a mixture of benzyl dimethyl alkyl ammonium chlorides in which the alkyl groups contained from 8 to 18 carbon atoms. No algicidal activity was observed on the glass slide. The organosilicon compounds tested and found active are shown below.

$$\mathrm{MeO(Me_2)SiCH_2N^+Me_2(C_{12}H_{25})Cl^-}$$
$$\mathrm{(MeO)_2MeSiCH_2N^+(Me_2)C_{12}H_{25}Cl^-}$$
$$\mathrm{(MeO)_2(CF_3CH_2CH_2)Si(CH_2)_3N^+(Me_2)C_{12}H_{25}Cl^-}$$
$$\mathrm{(MeO)_2PhSi(CH_2)_3N^+(Me_2)C_{12}H_{25}Cl^-}$$
$$\mathrm{(MeO)_3Si(CH_2)_3N^+Me_3Cl^-}$$
$$\mathrm{[(MeO)_3Si(CH_2)_3]_2N^+(Me)C_{11}H_{23}Br^-}$$
$$\mathrm{(MeO)_3Si(CH_2)_3N^+(Me_2)C_{12}H_{25}Cl^-}$$
$$\mathrm{(MeO)_3Si(CH_2)_3N^+(Me_2)C_{12}H_{25}Br^-}$$
$$\mathrm{(MeO)_3SiCH_2\overset{\overset{CH_2}{\|}}{C}CH_2N^+(Me)_2C_{12}H_{25}Cl^-}$$
$$\mathrm{(MeO)_3SiC_6H_4CH_2N^+(Me_2)C_{12}H_{25}Cl^-}$$
$$\mathrm{(MeO)_3Si(CH_2)_3N^+(Me_2)C_{14}H_{29}Cl^-}$$
$$\mathrm{(MeO)(H_{21}C_{10}O)_2Si(CH_2)_3N^+(Me_2)C_{18}H_{37}Cl^-}$$
$$\mathrm{(MeO)_3Si(CH_2)_3N^+(Me_2)C_{22}H_{45}{}^*Cl^-}$$
$$\mathrm{(MeO)_3Si(CH_2)_3N^+(Me_2)C_9H_{18}CH{=}CH_2Br^-}$$
$$\mathrm{(MeO)_3Si(CH_2)_3N^+(Me_2)C_{11}H_{23}Cl^-}$$
$$\mathrm{(MeO)_3SiC_6H_4CH_2N^+(Me_2)C_{16}H_{33}Cl^-}$$
$$\mathrm{(MeO)_3Si(CH_2)_3N^+(Me_2)C_{18}H_{37}Cl^-}$$

*Actually a mixture of silanes in which the alkyl groups vary from 20 to 22 carbon atoms.

Example 2: This example shows the efficacy of the method with various substrates. Each of the substrates shown below was impregnated with a 1% aqueous solution of the silanes shown below and the fibrous material squeezed free of excess solution and then allowed to air dry. Each of the treated fibrous materials was introduced into a circulating side mounted aquarium filter. The aquarium tanks were filled with aged tap water and aerated for 24 hours previous to attachment of the control and treated filters. The experiments were conducted in 5-gallon landscaped aquariums and each tank contained 15 liters of water and 3 species of fish. Each species was represented by three specimens. The fish were angel fish, neon tetra and guppy. Each aquarium was inoculated with 1 cc of an algal solution containing 300,000 cells per cc. All tanks were covered and continuously illuminated with a Gro-Lux lamp. Normal tank maintenance included aeration, daily fish feeding, maintaining water level, etc. The following compounds were employed:

$$(MeO)_3Si(CH_2)_3N^+(Me_2)C_{12}H_{25}Cl^-$$
$$(MeO)_3Si(CH_2)_3N^+(Me_2)C_{14}H_{29}Cl^-$$
$$(MeO)_3Si(CH_2)_3N^+(Me_2)C_{16}H_{33}Cl^-$$
$$(MeO)_3Si(CH_2)_3N^+(Me_2)C_{18}H_{37}Cl^-$$
$$(MeO)_3Si(CH_2)_3N^+(Me_2)(C_{20}H_{41} \text{ to } C_{22}H_{45})Cl^-$$

Each of the following substrates was employed and in each case there was excellent algal control against a broad spectrum of the algae kingdom including Cyanophyta (blue-green algae), Chlorophyta (green algae) and Crysophyta (yellow-green algae) and mixed populations thereof. The substrates were glass-wool, cotton, wool, linen, felt, acrylic, modacrylic, polyester, rayon, cellulose acetate, polyurethane and vinyl alcohol-vinyl acetal copolymer.

Example 3: This example shows the effectiveness of sand treated with the compositions of this process to kill algae. Two columns were filled with sand having a particle size range of 4 to 5 mm diameter. Into one column was passed a 1% aqueous solution of $(MeO)_3Si(CH_2)_3N^+(Me_2)C_{18}H_{37}Cl^-$ and into the other column was passed distilled water. Each column was allowed to air dry. Four liters of an algal solution containing 3,500,000 cells per cc was circulated through each column for 24 hours. The effluent from each was then examined microscopically. The effluent from the siloxane coated sand showed 0 algae count while that from the distilled water treated sand showed 2,100,000 cells per cc. The drop in count in the blank was due to the filtering action of the sand.

Example 4: This example gives a comparison between the use of the treated filters of this process, the use of no treatment on the filter and the use of an untreated filter together with a commercial algicide in which the active ingredient is 3-(p-chlorophenyl)dimethyl urea. The algicide was used in tablet form and in the amount specified by the manufacturer. The filter employed in this experiment was polyester fluff and it was treated by immersing it in a 0.8% solution of $(MeO)_3Si(CH_2)_3N^+(Me_2)C_{18}H_{37}Cl^-$. The pH of the treating solution was varied in order to ascertain if pH has any appreciable effect on the durability or effectiveness of the coating. In each case the polyester was allowed to soak for 30 minutes and the fluff was then blotted and air dried. Each of the treated filters plus the controls was employed in 5-gallon aquariums and a triplicate run was made in each case.

The aquariums employed were first aerated and then water was circulated through an aquarium filter for 48 hours to remove the iron. Seven species of fish were introduced into these aquariums prior to introduction of the treated filters. The fish employed were marble hatchet, angel fish, guppy, coolie loach, cardinal tetra, bottom catfish and cherry barb.

Each aquarium was inoculated with 1 cc of an algal solution containing 300,000 cells per cc. Daily aquarium maintenance, feeding, pH adjustment and water height adjustment were carried out and daily observations were recorded on the algal count. In no case was any fish toxicity observed. The results are shown in the table on the following page.

	---- Number of Algal Cells per cc Expressed as 1×10^4 ----				
Days of Treatment	Untreated Filter	Untreated Filter plus Algicide	pH 3.8	pH 7.1	pH 9.0
0	0	0	0	0	0
1	1	0	0	0	0
2	2	0	0	0	0
3	2	0	1	0	0
4	3	1	2	0	0
5	44	17	4	1	4
6	62	21	9	1	1
7	66	27	22	3	5
8	73	36	38	2	11
9	142	69	38	7	10
10	146	83	41	20	16
11	108	81	35	17	20
12	113	80	67	46	31
13	108	104	73	54	40
14	134	102	74	81	79

An algae count of 0 means that there are less than 10,000 cells per cc. An algae count of above 100,000 cells per cc becomes objectionable in the aquarium. From the above data it can be seen that the treated filters prevented algae buildup to an objectionable point twice as long as either the untreated filter or the untreated filter plus the algicide. It is also to be noted that there are only minor differences in algae control as between a pH of 3.8, 7.1 and 9.0. Slightly better results were obtained at a neutral pH or a slightly basic pH.

Artificial Mineral Water

A process described by *A. Nishino, H. Kumano and K. Sonetaka; U.S. Patent 3,932,578; January 13, 1976; assigned to Matsushita Electric Industrial Co., Ltd., Japan* provides a reaction bed material for use in the production of artificial mineral water.

The reaction bed material comprises $40<CaCO_3<100$ mol percent and which may further comprise $0<MgCO_3<60$ mol percent. Furthermore, a carbonate or carbonates of manganese, iron and/or zinc, or a hypogene rock which contains water-insoluble but acid-soluble compounds of Mn, Fe, Zn, K and/or Na may be added in an amount from more than 0 to less than 10 mol percent. The reaction bed material can be obtained by mixing powdered metal carbonate or carbonates with water, granulating and molding the mixture in a suitable form and drying the formed material to obtain the ultimate water-insoluble reaction bed material.

In order to improve bonding properties of the carbonate powder to be granulated, a suitable water-soluble bonding material may be added to the mixture by preliminarily dissolving it in water, including starch, carboxymethylcellulose, polyvinyl alcohol or gelatin. The molded material may be formed into granules or particles in one step, or may be first formed into cakes or blocks and then, before or after drying, broken into particles. The following examples illustrate the process.

Example 1: Preparation of a Reaction Bed Having a Granular Size of 3 mm Diameter – 5,000 grams of $CaCO_3$ was placed in a kneader for agitation in a dry state for 5 minutes. 2,200 grams of water was added to the material, which was then transferred to another kneader for wet-mixing for 10 minutes. The mixture was then molded and dried to obtain pellets having a diameter of 3 mm at a yield of 95% by weight of the starting material.

Example 2: 5,000 grams of $CaCO_3$ was placed in a kneader for dry-mixing for 5 minutes. 2,100 grams of a binder mixture consisting of 98% by weight of water and the balance of carboxymethylcellulose was added to the dry-mixed starting material which had been transferred into another kneader. The mixture was thoroughly agitated in the kneader for 10 minutes. Then, the mixture was molded and dried to obtain pellets having a diameter of 3 mm, at a yield of 97% by weight of the starting material.

The pellets of Examples 1 and 2 did not break when immersed in water for a long period of time. Moreover, when the pellets were used as a reaction bed material supported by a filter cloth, artificial mineral water was readily obtained simply by passing carbonated water therethrough, without causing clogging of the filter cloth.

COMPANY INDEX

INVENTOR INDEX

U.S. PATENT NUMBER INDEX

NOTICE

Nothing contained in this Review shall be construed to constitute a permission or recommendation to practice any invention covered by any patent without a license from the patent owners. Further, neither the author nor the publisher assumes any liability with respect to the use of, or for damages resulting from the use of, any information, apparatus, method or process described in this Review.

HOW TO REMOVE POLLUTANTS AND TOXIC MATERIALS FROM AIR AND WATER A PRACTICAL GUIDE 1977

by Marshall Sittig

Pollution Technology Review No. 32

This book is designed to provide a one volume ready reference for the handling of noxious materials emerging into the air and water as a result of industrial processes or from running machinery of any kind.

The descriptions are based almost entirely on U.S. patents dealing with practical environmental control methods and systems. The book surveys some 500 patents in the 1973 to 1976 period with exhaustive coverage up to November 1st, 1976. Since environmental patents are given priority handling by the U.S. Patent Office, some applications for these patents were filed only in late 1975 and even early 1976. This book therefore contains substantial technical information on very late developments in their field.

This book is addressed to industrialists who must keep abreast of the latest pollution removal techniques, to legislators and public health officials intent upon sensible rules and regulations, to interested conservationists and to those eager students who can foresee permanent and brilliant careers in the fields of pollution abatement and engineering.

Subject entries are in alphabetical sequence. Because of the encyclopedic nature of the book, only those entries which are at the beginning of the alphabet can be shown here without bias.

The book contains a total of 293 subject entries. The subject name refers to the polluting substance and the text underneath each entry tells how to combat pollution by said substance.

ISBN 0-8155-0654-6

621 pages

MEMBRANE TECHNOLOGY AND INDUSTRIAL SEPARATION TECHNIQUES 1976

by Peter R. Keller

Chemical Technology Review No. 70

Membrane technology is assuming an increasingly greater role in industrial separations. Electrodialysis and reverse osmosis (molecular ultrafiltration) units using semipermeable membranes are being installed the world over to recover drinking water from brackish water.

So-called "composite" membranes are now in use, where semipermeability results from an ultrathin polymeric film of one material which is formed separately and is supported by an underlying porous substrate made from a different polymer.

The applications are legion: aside from the desalination of sea water, there is demineralization of fresh water and of whey, general purification of industrial effluents and rivers, and the recovery of dissolved or colloidally dispersed substances, or the concentration of serum proteins, red blood cells, and other medical applications.

This book presents over 280 processes described in the U.S. patent literature. Fabrication techniques as well as end-use applications are given in detail. A partial and very condensed table of contents follows here. Numbers in parentheses indicate the number of processes per topic. Chapter headings and some important subtitles are given here.

ISBN 0-8155-0621-X **412 pages**